LONDON MATHEMATICAL SOCIETY LECTURE NOTE SERIES

Managing Editor: Professor N.J. Hitchin, Mathematical Institute, University of Oxford, 24-29 St Giles, Oxford OXI 3LB, United Kingdom

The titles below are available from booksellers, or from Cambridge University Press at www.cambridge.org/mathematics

London Mathematical Society Lecture Note Series: 341

Ranks of Elliptic Curves and Random Matrix Theory

Edited by

J. B. Conrey
American Institute of Mathematics

D. W. Farmer
American Institute of Mathematics

F. Mezzadri
University of Bristol

N. C. Snaith
University of Bristol

CAMBRIDGE
UNIVERSITY PRESS

University Printing House, Cambridge CB2 8BS, United Kingdom

Published in the United States of America by Cambridge University Press, New York

Cambridge University Press is part of the University of Cambridge.

It furthers the University's mission by disseminating knowledge in the pursuit of education, learning and research at the highest international levels of excellence.

www.cambridge.org
Information on this title: www.cambridge.org/9780521699648

© Cambridge University Press 2007

First published 2007, 2011
Second Edition 2012
Reprinted 2013

A catalogue record for this publication is available from the British Library

ISBN 978-0-521-69964-8 Hardback

Contents

Introduction

J. B. Conrey, D. W. Farmer, F. Mezzadri and N. C. Snaith

The group of rational points on an elliptic curve is a fascinating number theoretic object. The description of this group, as enunciated by Birch and Swinnerton-Dyer in terms of the special value of the associated L-function, or a derivative of some order, at the center of the critical strip, is surely one of the most beautiful relationships in all of mathematics; and it's understanding also carries a \$1 million dollar reward!

Random Matrix Theory (RMT) has recently been revealed to be an exceptionally powerful tool for expressing the finer structure of the value-distribution of L-functions. Initially developed in great detail by physicists interested in the statistical properties of energy levels of atomic nuclei, RMT has proven to be capable of describing many complex phenomena, including average behavior of L-functions.

The purpose of this volume is to expose how RMT can be used to describe the statistics of some exotic phenomena such as the frequency of rank two elliptic curves. Many, but not all, of the papers here have origins in a workshop that took place at the Isaac Newton Institute in February of 2004 entitled "Clay Mathematics Institute Special week on Ranks of Elliptic Curves and Random Matrix Theory." The workshop began with the Spittalsfield day of expository lectures, highlighted by reminiscences by Bryan Birch and Sir Peter Swinnerton-Dyer on the development of their conjecture. The week continued with a somewhat free-form workshop featuring discussion sessions, groups working on various problems, and spontaneous lectures. The idea for this volume arose at this workshop. The intention is to gather together a number of articles to assist someone wishing to begin work in this area.

One of the hightlights of this volume is the collection of beautiful expository papers and surveys: Kowalski's introduction to elliptic curves, Silverberg on ranks of elliptic curves, Ulmer's discussion of zeta-functions over function fields, Gamburd's explanation of symmetric function theory, Rodriguez-Villegas on the theta series associated with special values, Delaunay on probabilistic group theory, Farmer on families, and Young on exotic families of elliptic curves. There are an amazingly rich variety of topics arising from this one focus.

The most important invariant of an elliptic curve is the rank of its (Mordell-Weil) group of rational points; it is a non-negative integer, believed to be 0 or 1 for almost all elliptic curves. The catalyst for the Newton Institute workshop was a conjecture (see [CKRS]) about how often the rank is 2 for the family of quadratic twists of a given elliptic curve. Each elliptic curve has an L-function associated with it; this is an entire function which satisfies a functional equation. The Birch and Swinnerton-Dyer conjecture asserts, among other things, that the order of vanishing at the central point of the L-function associated with an elliptic curve is equal to the rank. It is generally conjectured that almost all elliptic curves have rank zero or one according to whether the sign of the functional equation of the related L-function is $+1$ or -1. Rank two curves should occur with L-functions that have a $+1$ sign of their functional equation but vanish nevertheless at the central point. These are expected to be rare; the question of how rare is the subject here.

If the elliptic curve is given by $E : y^2 = x^3 + Ax + B$, and if d is a fundamental discriminant, then the quadratic twist of E by d is the elliptic curve $E_d := dy^2 = x^3 + Ax + B$. The conjecture, derived from RMT and number theory, is that E_d will have rank 2 for asymptotically $c_E x^{3/4} (\log x)^{b_E}$ values of d with $|d| \leq x$. Here b_E is one of four values described in the article by Delaunay and Watkins, whereas c_E is yet to be determined but depends on a mix of RMT, number theory, and probabilistic group theory (see the article of Delaunay on class groups and Tate-Shafarevich groups).

This conjecture, while interesting, is not as compelling as it might be because of our ignorance of c_E. However, an absolutely convincing case for RMT can be given by considering rank 2 curves as above but divided into *arithmetic progressions* of d modulo some prime p.

Using RMT arguments combined with a number theoretic discretization of the problem, one is led to predict that if a is a quadratic residue mod p and b is a quadratic non-residue then the ratio of rank 2 twists among $d \equiv a \bmod p$ to $d \equiv b \bmod p$ is, in the limit,

$$\sqrt{\frac{p + 1 - a_p}{p + 1 + a_p}},$$

where $L(s) = \sum_{n=1}^{\infty} a_n n^{-s}$ is the L-function associated with E. Those familiar with the conjecture of Birch and Swinnerton-Dyer might not be surprised to see the ratio

$$\frac{p + 1 - a_p}{p + 1 + a_p}$$

show up; however, it is the square-root, contributed by RMT, that is the surprise.

The basic calculation to obtain this result involves a ratio of conjectures for

$$\sum_{\substack{d \equiv a \bmod p \\ d \leq x}} L_{E_d}(1/2)^{-1/2};$$

the reason that one takes the -1/2 power here is due to the rightmost pole at $s = -1/2$ of the s'th moment of characteristic polynomials of matrices chosen randomly from $SO(2N)$ with respect to Haar measure. The description of this calculation and the compelling numerical evidence is in the paper [CKRS]. In this volume, the calculation is taken a step further in the paper of Conrey, Rubinstein, and Watkins where lower order terms for the moments are incorporated and lead to an even more precise evaluation of these ratios.

The conjectures about quadratic twists can be generalized to cubic twists in two different ways. One involves the frequency of rank 2 elliptic curves within the classical family $E_m := x^3 + y^3 = m$. See the interesting paper of Watkins to understand why it is precisely twice as likely that a number which is 2 mod 7 is a sum of two rational cubes compared with a number which is 3 mod 7.

The other way to do a cubic twist is to take a fixed elliptic curve E and a Dirichlet character χ of order 3 and consider the twisted L-function, $L_E(s, \chi) = \sum_{n=1}^{\infty} a_n \chi(n) n^{-s}$. David, Fearnley, and Kisilevsky [DFK] have shown, very surprisingly, that such twists vanish for about $x^{1/2}$ cubic twists of modulus $\leq x$, and have given precise conjectures, based on RMT, for the asymptotic frequency of this event. They also consider quintic twists (see their paper in this volume) and conclude that there are (barely!) infinitely many order five characters for which the twisted L-function vanishes at the central point. These predictions are based on calculations with random unitary matrices, whereas the previously mentioned conjectures arise from considering groups of orthogonal matrices.

It is interesting to begin with a weight 4 modular newform f, with integer Fourier coefficients, and similarly ask about vanishing of, say, quadratic twists of the associated L-function. In this case it is expected that there will be asymptotically $c_f x^{1/4} (\log x)^{b_f}$ vanishings at the central point of the quadratically twisted L-functions . The possible values of b_f have not been worked out here; however, if one restricts to prime discriminants, then the power on the log is expected to be $-5/8$ in both this case and the case of twists of elliptic curve L-functions. If one considers weight 6 or higher, it is expected that there will only be finitely many vanishings of quadratic twists of the associated L-functions. It is not clear whether one accumulates infinitely many vanishings if one considers all such weight 6 forms and all of their twists. There is an arithmetic significance to the vanishings of the twists of the weight 4 modular forms: it is related to the rank of an associated Chow group, about which we hope to say more at a later time.

In the twists mentioned in the cases above we only consider the twists for which there is a plus sign in the functional equation.

The numerical evidence for many of the above conjectures has been accumulated by a combination of people: Tornaria, Rodriguez-Villegas, Rosson, Mao, and Rubinstein. Much of it is based on an algorithm of Gross for finding the half-integral weight form, as a theta series involving ternary quadratic forms, whose Fourier coefficients yield the values of the twisted L-series at the central point. Prior to the February workshop, only a handful of such theta series were known. During that workshop, the first four people above worked out the obstacles to further progress and produced literally thousands of examples for Rubinstein who computed hundreds of millions of values for each; this provides a nice data bank for testing conjectures.

Matt Young has considered the situation of the "family of all elliptic curves." Basically he parametrizes this family as $E_{A,B} : y^2 = x^3 + Ax + B$ and allows (A, B) to run over a rectangle. He is concerned not only with the distribution of ranks in this family, but also with statistics such as the 'one-level density' of the zeros. He considers other more exotic families as well, such as E_{A,B^2} which is forced to have rank at last one. Such families play a role in Iwaniec' approach to the Riemann Hypothesis.

All of the above discussion has been focused on rank two. The question of modelling rank 3 members of a family is much more difficult; in fact it is not at all satisfactorily addressed. In the case of quadratic twists, to conjecture the number of rank 2 curves the application of random matrix theory relies on a discretization arising from the beautiful formula, due in this form to Kohnen and Zagier:

$$L_{E_d}(1/2) = \kappa_E \frac{c_E(|d|)^2}{\sqrt{|d|}},$$

where $c_E(|d|)$ is an integer and $\kappa_E > 0$. In the case of rank 3, we consider the conjectural formula of Birch and Swinnerton-Dyer for the value of the derivative of an odd $L_{E_d}(s)$:

$$L'_{E_d}(1/2) = \frac{h_{E_d}|Sha_{E_d}|}{\sqrt{d}},$$

where h_{E_d} is the height of a generating point. (Change this to the formula of Gross-Zagier.) The problem is that we don't know what kind of discretization to give h_P. It could conceivably be as small as $\log|d|$ but statistically this does not seem to be the correct model. By the work of Snaith (in this volume), the right-most pole of the derivative of the sth moment of characteristic polynomials of odd orthogonal matrices occurs at $s = -3/2$. This might suggest, if one uses the discretization $(\log|d|)/\sqrt{|d|}$, that there are only about $x^{1/4}$ rank 3 curves among the family of twists with conductor smaller than x. However, Rubin and Silverberg give examples of E which have many more rank 3 quadratic twists, suggesting that this discretization is not correct. In examining the limited data we have for rank 3 twists, an interesting phenomenon seems to appear: it looks as though $L'_{E_d}(1/2)$ cannot be as small as $(\log|d|)/\sqrt{|d|}$. Is it possible that when Sha is small then the height of a generating point is big and vice-versa? This linkage does not seem unnatural if one compares for example to the situation of the class number of a real quadratic field. There one finds that the product of the regulator times the size of the class group is always about the size of the square root of the discriminant. However, this analogy may not be correct, since this involves L-functions at the edge of the critical strip whereas we are discussing values at the center. The paper of Conrey, Rubinstein, Snaith, and Watkins discusses the so-called 'Saturday night conjecture' about the possible sizes of this product. Much more data is needed to make an informed conclusion.

All of the above and more is contained in this volume. Other directions yet to be considered are odd weight modular forms, Siegel modular forms, and

Chow groups and we hope this collection of papers will attract new researchers to this field and inspire those well acquainted with it to explore further.

References

[1] J.B. Conrey, J.P. Keating, M. Rubinstein, N.C. Snaith, On the frequency of vanishing of quadratic twists of modular L-functions, Number theory for the Milennium I, 301–315, Editors: M. A. Bennett, B. C. Berndt, N. Boston NH. G. , Diamond, A. J. Hildebrand, W. Philipp, A. K. Peters Ltd, Natick, 2002.

[2] C. David, J. Fearnley and H. Kisilevsky, On the Vanishing of Twisted L-Functions of Elliptic Curves, *Experimental Mathematics*, 13 (2004), 185–198.

Elliptic curves, rank in families and random matrices

E. Kowalski

This survey paper contains two parts. The first one is a written version of a lecture given at the "Random Matrix Theory and L-functions" workshop organized at the Newton Institute in July 2004. This was meant as a very concrete and down to earth introduction to elliptic curves with some description of how random matrices become a tool for the (conjectural) understanding of the rank of Mordell-Weil groups by means of the Birch and Swinnerton-Dyer Conjecture; the reader already acquainted with the basics of the theory of elliptic curves can certainly skip it. The second part was originally the write-up of a lecture given for a workshop on the Birch and Swinnerton-Dyer Conjecture itself, in November 2003 at Princeton University, dealing with what is known and expected about the variation of the rank in families of elliptic curves. Thus it is also a natural continuation of the first part. In comparison with the original text and in accordance with the focus of the first part, more details about the input and confirmations of Random Matrix Theory have been added.

Acknowledgments. I would like to thank the organizers of both workshops for inviting me to gives these lectures, and H. Helfgott, C. Hall, C. Delaunay, S. Miller, M. Young and M. Rubinstein for helpful remarks, in particular for informing me of work in process of publication or in progress that I was unaware at the time of the talks. In fact, since this paper was written, a number of other relevant preprints have appeared; among these we mention [Sn], [Mil2], with no claim to exhaustivity!

Notation. We use synonymously the two notations $f(x) = O(g(x))$ and $f(x) \ll g(x)$ for $x \in X$, where X is some set on which both f and $g \geqslant 0$ are defined; it means that for some "implied" constant $C \geqslant 0$ (which may depend on further parameters), we have $|f(x)| \leqslant Cg(x)$ for all $x \in X$. On the other hand, we use $f = o(g)$ as $x \to x_0$, for some limit point x_0, to mean that the limit of f/g exists and is 0 as $x \to x_0$, and similarly $f \sim g$ for $x \to x_0$ means $f/g \to 1$ as $x \to x_0$.

1   A concrete introduction to elliptic curves

Before embarking on our journey, we refer in general to Silverman's book [AEC] for a very good and readable discussion of the topics covered here, with complete proofs for all but the most advanced. Each subsection will include references to the parts of this book that corresponds, and other references if necessary.

1.1 Elliptic curves as algebraic curves, complex tori and the link between the two

Elliptic curves can be seen in a number of different ways. We will present the two most geometric. First, an *affine plane cubic curve* over the field $\mathbf{C}$ of complex numbers is simply the set of complex solutions $(x, y) \in \mathbf{C} \times \mathbf{C}$ of an equation

$$y^2 + a_1 xy + a_3 y = x^3 + a_2 x^2 + a_4 x + a_6 \qquad (1.1)$$

(called a *general Weierstrass equation*), where a_1, a_2, a_3, a_4 and a_6 are arbitrary complex numbers. If all the a_i are rational numbers, the curve is said to be *defined over* $\mathbf{Q}$. It is those curves which are most relevant for number theory, and especially one is concerned with the basic diophantine question which is to find all rational solutions $(x, y) \in \mathbf{Q} \times \mathbf{Q}$ to the equation (1.1).

For many reasons, it is usually more convenient to present the equation (1.1) in homogeneous form

$$Y^2 Z + a_1 XYZ + a_3 Y Z^2 = X^3 + a_2 X^2 Z + a_4 X Z^2 + a_6 Z^3 \qquad (1.2)$$

(which defines a *projective cubic curve*) and look for triplets of solutions (X, Y, Z) in the projective plane $\mathbf{P}_2(\mathbf{C})$ instead of the place $\mathbf{C}^2$, which means looking for non-zero solutions $(X, Y, Z) \neq (0, 0, 0)$ and identifying two solutions (X, Y, Z) and $(\alpha X, \alpha Y, \alpha Z)$ for any non-zero $\alpha \in \mathbf{C}^\times$.

If in a triplet (X, Y, Z) satisfying (1.2) we have $Z \neq 0$, then we can replace (X, Y, Z) by the equivalent solution $(X/Z, Y/Z, 1)$ and this satisfies (1.2) if and only if the pair $(x, y) = (X/Z, Y/Z)$ satisfies the original equation (1.1). So the homogeneous solutions with $Z \neq 0$ are in one-to-one correspondence with the points on the affine cubic curve. However, if $Z = 0$, the equation (1.2) gives $X = 0$, so the solutions are $(0, Y, 0)$ with $Y \neq 0$ arbitrary. All those are in fact equivalent to a single solution $(0, 1, 0)$, which is called the *point at infinity*, often denote ∞. Note in particular that this point always has rational coordinates.

Plane cubic curves provide the first "picture" of elliptic curves, that as algebraic curves. However, there is a necessary condition imposed on an equation (1.1) before it is said to be the equation of an elliptic curve, namely it must define a smooth curve in $\mathbf{C} \times \mathbf{C}$. This means that the partial derivatives

$$2y + a_1 x + a_3 \quad \text{and} \quad a_1 y - 3x^2 - 2a_2 x - a_4$$

must not have a common zero (x, y) which is also a point on the cubic curve. There is an explicit "numeric" criterion for this to hold (see [AEC, p. 46]); in the slightly simpler case where $a_1 = a_3 = 0$ (we will see that one can reduce to this case in most situations), the smoothness expresses simply that the cubic polynomial $x^3 + a_2 x^2 + a_4 x + a_6$ has three distinct roots in $\mathbf{C}$, equivalently that the *discriminant* $\Delta = -16(4a_4^3 + 27a_6^2)$ is non-zero. Thus, this will be true for a "random" equation (1.1).

To summarize this definition: an elliptic curve, as an algebraic curve, is the set of projective solutions (X, Y, Z) to an equation (1.2) which defines a smooth curve.

Example 1.1. • The plane cubic curve with equation

$$y^2 = x^3$$

is not an elliptic curve: the point $(0, 0)$ is a singular point (the curve looks like a "cusp" in the neighborhood of $(0, 0)$).

• Similarly, the curve with equation

$$y^2 = x^3 + x^2$$

is not an elliptic curve; again $(0, 0)$ is singular, and the curve looks like a node in the neighborhood of $(0, 0)$.

• The curve with equation

$$y^2 = x^3 - x = x(x - 1)(x + 1)$$

is an elliptic curve, since the right-hand side has three distinct roots in $\mathbf{C}$. This curve is defined over $\mathbf{Q}$. It is often called the *congruent number* curve, for reasons we will explain below; it is also a so-called *CM curve*, and this terminology will also be explained.

• Let $\ell > 2$ be a prime number. *If* (a, b, c) were non-zero rationals such that $a^\ell + b^\ell = c^\ell$, then the cubic curve

$$y^2 = x(x - a^\ell)(x + b^\ell)$$

would be a very remarkable elliptic curve (defined over $\mathbf{Q}$), in fact so remarkable that it cannot possibly exist: this is the "highest level" summary of how Wiles proved Fermat's Great Theorem.

The other view of elliptic curves is more analytic in flavor, and identifies them with *complex tori*. Namely, let ω_1, ω_2 be non-zero complex numbers, with $\omega_1/\omega_2 \notin \mathbf{R}$. Let $\Lambda = \omega_1 \mathbf{Z} \oplus \omega_2 \mathbf{Z}$; this is an abelian subgroup of $\mathbf{C}$, and it generates $\mathbf{C}$ as an $\mathbf{R}$-vector space. Those two properties characterize the *lattices* in $\mathbf{C}$, and all of them are given as described.

Now consider the quotient group $X = \mathbf{C}/\Lambda$ which one views as a compact Riemann surface (it is compact because, for instance the compact set $\{a\omega_1 \mid b\omega_2 \mid (a, b) \in [0, 1] \times [0, 1]\}$ projects surjectively to X). Topologically, this is a torus, and as a group, this is $(\mathbf{R}/\mathbf{Z})^2$. Now the analytic definition of an elliptic curve is simply that it is one such quotient $\mathbf{C}/\Lambda$ for some lattice $\Lambda \subset \mathbf{C}$. We will now discuss how this definition and that as smooth plane cubic curve are compatible. A small warning: although it is tempting to think so at first, taking ω_i with rational coordinates does not give the analogue of cubic curves defined over $\mathbf{Q}$! In fact, for a curve defined over $\mathbf{Q}$, the ratio ω_2/ω_1 is almost always transcendental, see e.g. [Ba, Ch. 6].

It is always natural to look for meromorphic functions defined on a Riemann surface (for instance, think that on a cubic curve we have two natural rational functions, $(x, y) \mapsto x$ and $(x, y) \mapsto y$ which are used to give the equation of the curve). Very concretely, this means we wish to consider meromorphic functions

$$f : \mathbf{C} \to \mathbf{C}$$

which are ω_1 and ω_2-periodic:

$$f(z + \omega_1) = f(z) \text{ and } f(z + \omega_2) = f(z).$$

Those f are called *elliptic functions*; this is where the history began in fact, since it was found, over a long period, that the arc-length on an ellipse can be expressed in terms of (inverses of) such functions (see [AEC, 168–170] for a sequence of exercises explaining this).

Now for a given Λ, one can construct an elliptic function $\wp$ which has a pole of order 2 at points of Λ and no other singularities, and satisfies the algebraic differential equation

$$\wp'^2 = 4\wp^3 - g_2\wp - g_3$$

for some $g_2, g_3 \in \mathbf{C}$. In fact, this is the Weierstrass $\wp$-function of Λ which is given explicitly by the series

$$\wp(z) = \frac{1}{z^2} + \sum_{\substack{\omega \in \Lambda \\ \omega \neq 0}} \left(\frac{1}{(z - \omega)^2} - \frac{1}{\omega^2} \right),$$

and g_2 and g_3 are the absolutely convergent series

$$g_2 = 60 \sum_{\substack{\omega \in \Lambda \\ \omega \neq 0}} \frac{1}{\omega^4}, \quad g_3 = 140 \sum_{\substack{\omega \in \Lambda \\ \omega \neq 0}} \frac{1}{\omega^6}.$$

Sending $z \mapsto (2\wp(z), \sqrt{2}\wp'(z))$ gives points on the plane cubic

$$y^2 = x^3 - g_2 x - 2g_3 \tag{1.3}$$

with $0 \mapsto \infty$ since $\wp$ has a pole at $z = 0$. One shows that this map is bijective, and that this cubic curve is smooth, hence is an elliptic curve "as plane curve". Moreover, one shows that all elliptic curves with $a_1 = a_3 = a_2 = 0$ arise in this manner, and also that simple changes of variables can bring any Weierstrass equation (1.2) to the form (1.3).

References: [AEC, III.1,VI]

1.2 The group law on elliptic curves and maps between elliptic curves

The quotient $\mathbf{C}/\Lambda$ has a natural abelian group structure. So there must be a corresponding group structure for the points in the incarnation of the elliptic curve as a smooth plane cubic curve. This "group law" turns out to have a very nice geometric description, which is that if $P = (x_1, y_1)$, $Q = (x_2, y_2)$, $R = (x_3, y_3)$ are distinct points on the curve (i.e. distinct solutions to (1.1)), then we have $P + Q + R = 0$ for this group law if and only if P, Q and R are collinear.

Here is a quick sketch that this does indeed correspond, via the link presented in the previous section, to the addition on $\mathbf{C}/\Lambda$: the equation $F(x, y) = 0$ of the line joining P, Q and R gives an elliptic function $f(z) = F(\wp(z), \wp'(z))$ such that the *divisor* of f is $(p) + (q) + (r) - 3(0)$ where p, q, $r \in \mathbf{C}/\Lambda$ correspond respectively to P, Q, R by the analytic parameterization. (This means that f has three zeros p, q and r modulo Λ, and a triple pole at 0). By integrating zf'/f along the boundary of a fundamental parallelogram, one gets $p + q + r \in \Lambda$.

It is an essential fact that this group law can be expressed by algebraic formulae: for instance, one finds for $y^2 = x^3 + a_4 x + a_6$ that $-(x, y) = (x, -y)$, and $(x_1, y_1) + (x_2, y_2) = (x_3, y_3)$ with

$$x_3 = \left(\frac{y_2 - y_1}{x_2 - x_1}\right)^2 - x_1 - x_2, \tag{1.4}$$

$$y_3 = \frac{y_2 - y_1}{x_2 - x_1}(x_3 - x_1) - y_1 \tag{1.5}$$

if $x_1 \neq x_2$. The case $x_1 = x_2$ is treated by a limit process (in other words, replace the line joining the two points by the tangent). It is in fact essential here to use the projective model (1.2) because the origin for the group law is the point at infinity.

This algebraic description shows that if the curve is defined over $\mathbf{Q}$, then the points with rational coordinates on an elliptic curve (those that we wish to determine as the basic diophantine question) form a subgroup of the group of complex-valued points.

In addition to considering a single elliptic curve, it is also important to study maps between elliptic curves (also called *morphisms* of elliptic curves). They are most easily described in the analytic description: given two lattices Λ_1 and Λ_2 in $\mathbf{C}$, we are looking for holomorphic maps $\mathbf{C}/\Lambda_1 \to \mathbf{C}/\Lambda_2$. It is easy to see that there exists complex numbers α and β such that $\alpha\Lambda_1 \subset \Lambda_2$ and $f(z) = \alpha z + \beta$ for $z \in \mathbf{C}/\Lambda_1$.

On the algebraic side, those maps become expressed by polynomials, or more often rational functions; thus it may be necessary to use two or more formulas to describe $f(x, y)$, depending on whether a certain expression is well-defined at (x, y), as we saw already for the case of the group law itself (the formula (1.4) is valid only for $x_1 \neq x_2$). So an algebraic map between two elliptic curves, seen as algebraic curves, is really a collection of applications defined by rational functions, one of which at least is valid at any given point (including at infinity), and which coincide in case there is more than one possibility. If all those rational functions can be chosen with coefficients in $\mathbf{Q}$, the map is said to be defined over $\mathbf{Q}$.

Example 1.2. Let E be an elliptic curve (1.2).

• For a fixed point $P_0 \in E$, defining $f(P) = P + P_0$, where $+$ is the group law defined above, gives a map $E \to E$.

• For any integer $n \in \mathbf{Z}$, the application

$$P \mapsto \underbrace{P + \cdots + P}_{n}$$

(again with $+$ the group law on E) is a map $[n] : E \to E$. It is defined over $\mathbf{Q}$ if E itself is defined over $\mathbf{Q}$.

• Let a, b be complex numbers with $a^2 \neq 4b$. The map

$$\{y^2 = x^3 + ax^2 + bx\} \to \{w^2 = v^3 - 2av^2 + (a^2 - 4b)v\}$$
$$(x, y) \mapsto \left(\tfrac{y^2}{x^2}, \tfrac{y(b-x)^2}{x^2}\right)$$

"is" a map between those two elliptic curves, with $(0, 0) \mapsto \infty$.

• Let $E : y^2 = x^3 - x$. Then $[i] : (x, y) \mapsto (-x, iy)$ is a map (not defined over $\mathbf{Q}$, but over $\mathbf{Q}(i)$, in an obvious sense).

As in the case of complex tori (and as it should be!), any map $E \to F$ between elliptic curves is of the form $f(x) = g(x) + x_0$ where g is a map that preserves the group law, i.e. $g(P + Q) = g(P) + g(Q)$. Such a map is called an *isogeny*.

The isogenies from a given curve to itself form a ring $\mathrm{End}(E)$, where the product is composition of maps and the addition is performed pointwise using the group law: $(f + g)(P) = f(P) + g(P)$. Similarly, if E is defined over $\mathbf{Q}$, the isogenies defined over $\mathbf{Q}$ form a subring $\mathrm{End}_{\mathbf{Q}}(E) \subset \mathrm{End}(E)$ which can be smaller than $\mathrm{End}(E)$, as the fourth example above illustrates.

Usually one has $\mathrm{End}(E) = \mathbf{Z}$, where $n \in \mathbf{Z}$ corresponds to the "multiplication by n" map. This is most easily seen using the analytic description: if a lattice Λ and $\alpha \in \mathbf{C}$ satisfy $\alpha\Lambda \subset \Lambda$, using a basis (ω_1, ω_2) of Λ one has for some integers n_i

$$\begin{cases} (n_1 - \alpha)\omega_1 + n_2\omega_2 = 0 \\ n_3\omega_1 + (n_4 - \alpha)\omega_2 = 0 \end{cases}$$

hence $(n_1 - \alpha)(n_4 - \alpha) - n_2 n_3 = 0$, which shows that α is either an integer or the root of a quadratic polynomial; and by solving the system, if $\alpha \notin \mathbf{Z}$, one sees that ω_1/ω_2 is also a quadratic number, so the lattice is very special.

If $\mathrm{End}(E) \neq \mathbf{Z}$, one says that E has *complex multiplication*, abbreviated CM. For instance, the curve $y^2 = x^3 - x$ above is a CM curve since the map $[i]$ described in the example is not multiplication by n for any n.

One shows (again, it is obvious in the analytic description) that a non-zero isogeny $E_1 \to E_2$ is necessarily surjective. Its kernel $\ker f = \{x \in E_1 \mid f(x) = 0\}$ is a finite abelian group. For instance, for $f = [n]$ with $n \neq 0$, the kernel of f is called the group of n-torsion points on E, denoted $E[n]$. Using the group structure for a complex torus $\mathbf{C}/\Lambda \simeq (\mathbf{R}/\mathbf{Z})^2$, it is clear that $E[n] \simeq (\mathbf{Z}/n\mathbf{Z})^2$. All these facts can in fact be proved algebraically.

If the elliptic curve E is defined over $\mathbf{Q}$, the n-torsion points on E have the important property that their coordinates are algebraic numbers. One can think of those points as analogues of the classical roots of unity, since they are solutions to an equation $nx = 0$, similar to the equation $z^n = 1$ in the multiplicative group $\mathbf{C}^\times$. There are indeed numerous analogies from the arithmetic point of view.

Example 1.3. Let E have equation $y^2 = x^3 + a_4 x + a_6$. Then

$$E[2] = \{\infty, (e_1, 0), (e_2, 0), (e_3, 0)\}$$

where e_i, $1 \leqslant i \leqslant 3$, are the distinct complex roots of $x_3 + a_4 x + a_6 = 0$. The group structure on $E[2]$ is described by $(e_i, 0) + (e_j, 0) = (e_k, 0)$ if $i \neq j$, with k the element in $\{1, 2, 3\} - \{i, j\}$.

An isomorphism of elliptic curves is an isogeny f which is one-to-one, or equivalently with $\ker(f) = 0$. It is natural to try to classify all elliptic curves up to isomorphism.

Over $\mathbf{C}$, by simple changes of variable, any Weierstrass equation (1.1) can be brought to the form $y^2 = x^3 + c_4 x + c_6$ for some c_4, c_6. As already mentioned, such equations define an elliptic curve if $\Delta = -16(4c_4^3 + 27c_4^2) \neq 0$. Two such equations can define isomorphic curves only if (with obvious notation) $c_4' = u^4 c_4$ and $c_6' = u^6 c_6$ for some $u \in \mathbf{C}^\times$. This shows easily that the so-called j-invariant $j = 1728(4c_4)^3/\Delta$ completely describes the isomorphism class of the elliptic curve. Moreover, the curves

$$y^2 + xy = x^3 - \frac{36}{j - 1728}x - \frac{1}{j - 1728} \text{ for } j \notin \{0, 1728\}$$
$$y^2 = x^3 - 1 \text{ for } j = 0$$
$$y^2 = x^3 - x \text{ for } j = 1728$$

show that every complex number is the j-invariant for some elliptic curve.

In arithmetic, it is also important to notice that elliptic curves E_1 and E_2 defined over $\mathbf{Q}$ might be isomorphic over $\mathbf{C}$ (i.e., have the same necessarily rational j-invariant) without being isomorphic over $\mathbf{Q}$, in which case E_1 and E_2 are called *twists* of each other. For instance, if $\alpha^2 \in \mathbf{Q}$, $(x, y) \mapsto (\alpha x, \alpha^{3/2} y)$ gives an isomorphism over $\mathbf{C}$ between $y^2 = x^3 - x$ and $y^2 = x^3 - \alpha^2 x$, and those two curves are not usually isomorphic over $\mathbf{Q}$, for instance because the elements of $E[2]$ are rational points on $y^2 = x^3 - x$, whereas they are not on $y^2 = x^3 - \alpha^2 x$ if α is not itself rational.

On the analytic side, where isomorphic tori correspond to homothetic lattices, one shows quite easily that any lattice $\omega_1 \mathbf{Z} \oplus \omega_2 \mathbf{Z} \subset \mathbf{C}$ can be brought by homothety to the form $\mathbf{Z} \oplus \tau \mathbf{Z}$ for some τ which can be chosen in the upper half-plane $\mathbf{H} = \{z \in \mathbf{C} \mid \mathrm{Im}(z) > 0\}$. Then two such lattices $\mathbf{Z} \oplus \tau \mathbf{Z}$ and $\mathbf{Z} \oplus \tau' \mathbf{Z}$ define isomorphic complex tori if and only if there exists some $\gamma = \begin{pmatrix} a & b \\ c & d \end{pmatrix} \in SL(2, \mathbf{Z})$ such that

$$\tau' = \gamma \tau = \frac{a\tau + b}{c\tau + d}.$$

The j-invariant can then be described as a holomorphic map $\mathbf{H} \to \mathbf{C}$ which is $SL(2, \mathbf{Z})$-invariant; it is the prototypical example of a modular function.

References: [AEC, III.2,3,9,VI.4,C.12]

1.3 The arithmetic of elliptic curves: the Mordell-Weil group

We now come to our main concern, which is the arithmetic properties of elliptic curves, and in particular the structure of the set of rational points. Let $E/\mathbf{Q}$ be an elliptic curve defined over $\mathbf{Q}$. As already mentioned, the fact that $a_i \in \mathbf{Q}$ implies immediately that the set of rational points on E, denoted $E(\mathbf{Q})$, is in fact a subgroup of E. It is called the *Mordell-Weil group* of E. The fundamental structure theorem is due to Mordell in this case.

Theorem 1.4. *For any elliptic curve $E/\mathbf{Q}$, the group $E(\mathbf{Q})$ is a finitely generated abelian group.*

This means that one has an isomorphism

$$E(\mathbf{Q}) \simeq \mathbf{Z}^r \oplus F$$

for some integer $r \geqslant 0$, called the *rank* of E (over $\mathbf{Q}$), and some finite group F, which is simply the torsion subgroup of $E(\mathbf{Q})$, i.e., the subgroup of elements of finite order.

The current proof of the theorem is still much the same as Mordell's. It proceeds in two steps: in the first step, one shows that $E(\mathbf{Q})/mE(\mathbf{Q})$ is finite (for some integer $m \geqslant 2$, $m = 2$ gives quite elementary proofs). Then, given representatives for the finite group $E(\mathbf{Q})/mE(\mathbf{Q})$, one shows one to construct a finite set of generators of $E(\mathbf{Q})$.

From a diophantist's point of view, the problem with the proof is that the first part is *ineffective*: it does not provide (and no other argument is proved to yield) the representatives for $E(\mathbf{Q})/mE(\mathbf{Q})$ which are required for the second step (on the other hand, given the representatives, the second step is completely effective). More precisely, one does get an upper bound on r, but no bound for the "height" (i.e., the size) of elements filling up $E(\mathbf{Q})/mE(\mathbf{Q})$. (See below in Section 1.5 for the rigorous definition of the height; here you can think simply of the largest of the number of digits of the numerator and denominator of the x-coordinate of a point on $E(\mathbf{Q})$).

On the other hand, the finite torsion group F can be computed efficiently, and in fact it has been possible to find a complete list of all finite abelian groups which arise in this way (this is due to Mazur). Here is an example of each torsion group (one can show that each arises for infinitely many elliptic curves over $\mathbf{Q}$):

- $y^2 = x^3 - 2$, torsion $= \{0\}$.

- $y^2 = x^3 + 8$, torsion $= \mathbf{Z}/2\mathbf{Z}$.

- $y^2 = x^3 + 4$, torsion $\simeq \mathbf{Z}/3\mathbf{Z}$.

- $y^2 = x^3 + 4x$, torsion $\simeq \mathbf{Z}/4\mathbf{Z}$.

- $y^2 - y = x^3 - x$, torsion $\simeq \mathbf{Z}/5\mathbf{Z}$.

- $y^2 = x^3 + 1$, torsion $\simeq \mathbf{Z}/6\mathbf{Z}$.

- $y^2 - xy - 4y = x^3 - x^2$, torsion $\simeq \mathbf{Z}/7\mathbf{Z}$.

- $y^2 + 7xy = x^3 + 16x$, torsion $\simeq \mathbf{Z}/8\mathbf{Z}$.

- $y^2 + xy + y = x^3 - x^2 - 14x + 29$, torsion $\simeq \mathbf{Z}/9\mathbf{Z}$.

- $y^2 + xy = x^3 - 45x + 81$, torsion $\simeq \mathbf{Z}/10\mathbf{Z}$.

- $y^2 + 43xy - 210y = x^3 - 210x^2$, torsion $\simeq \mathbf{Z}/12\mathbf{Z}$.

- $y^2 = x^3 - 4x$, torsion $\simeq \mathbf{Z}/2\mathbf{Z} \times \mathbf{Z}/2\mathbf{Z}$.

- $y^2 + xy - 5y = x^3 - 5x^2$, torsion $\simeq \mathbf{Z}/4\mathbf{Z} \times \mathbf{Z}/2\mathbf{Z}$.

- $y^2 + 5xy - 6y = x^3 - 3x^2$, torsion $\simeq \mathbf{Z}/6\mathbf{Z} \times \mathbf{Z}/2\mathbf{Z}$.

- $y^2 + 17xy - 120y = x^3 - 60x^2$, torsion $\simeq \mathbf{Z}/8\mathbf{Z} \times \mathbf{Z}/2\mathbf{Z}$.

Before continuing, here is a beautiful instance of the intrusion of elliptic curves in a very classical problem: what are the rationals (so-called *congruent numbers*) r such that there is a right-triangle with rational lengths a, b, c and area r.

Proposition 1.5. *A squarefree integer $n \geqslant 1$ is a congruent number if and only if the elliptic curve*

$$E_n : y^2 = x^3 - n^2 x$$

has rank $r_n \geqslant 1$.

The j-invariant of E_n is $j(E_n) = 1728$, which shows that all the curves E_n are isomorphic over $\mathbf{C}$ (not over $\mathbf{Q}$!), i.e., they are all twists of each other.

The arithmetic of elliptic curves has led J. Tunnell to a very simple algorithm for checking whether a given squarefree integer n is a congruent number; it is however still conditional on the Birch and Swinnerton-Dyer Conjecture described below.

Theorem 1.6 (Tunnell). *If the Birch and Swinnerton-Dyer Conjecture holds, then (for odd squarefree n), n is a congruent number if and only if the number of triples of integers (x, y, z) such that $2x^2 + y^2 + 8z^2 = n$ is twice the number of triples such that $2x^2 + y^2 + 32z^2 = n$.*

Example 1.7. Let's check that $n = 41$ is congruent:

$$41 = \overbrace{2(\pm4)^2 + (\pm3)^2}^{4} = \overbrace{(\pm3)^2 + 8(\pm2)^2}^{4} = \overbrace{2(\pm4)^2 + (\pm1)^2 + 8(\pm1)^2}^{8}$$

$$= \overbrace{2(\pm2)^2 + (\pm5)^2 + 8(\pm1)^2}^{8} = \overbrace{2(\pm2)^2 + (\pm1)^2 + 8(\pm2)^2}^{8}$$

$$41 = \overbrace{2(\pm4)^2 + (\pm3)^2}^{4} = \overbrace{(\pm3)^2 + 32(\pm1)^2}^{4} = c\,\overbrace{2(\pm2)^2 + (\pm1)^2 + 32(\pm1)^2}^{8}$$

Note that Tunnell's theorem does not provide the lengths a, b, c of the right triangle with area 41, but they can be derived easily from the proof of the proposition, provided one knows a point with infinite order on E_n.

References: [AEC, VIII, X], [K].

1.4   Reduction modulo primes and the Hasse-Weil L-function of an elliptic curve

Let $E/\mathbf{Q}$ be an elliptic curve. By change of variable one can assume (i.e., E is isomorphic to a curve such) that the Weierstrass equation (1.1) has integral coefficients. For any prime p, one can reduce modulo p and look at solutions (x, y) in the finite field $\mathbf{Z}/p\mathbf{Z}$ of

$$y^2 + a_1 xy + a_3 y = x^3 + a_2 x^2 + a_4 x + a_6 \pmod{p}. \tag{1.6}$$

For any prime p that does not divide the discriminant Δ_E, this equation, when one looks for solutions in an algebraic closure of $\mathbf{Z}/p\mathbf{Z}$, "defines" an elliptic curve over $\mathbf{Z}/p\mathbf{Z}$ (but we haven't really said what this means and this requires some care).

It is a simpler diophantine question to find the solutions to (1.6). In fact, there are certainly only a finite number of them in $(\mathbf{Z}/p\mathbf{Z})^2$, or in homogeneous coordinates in the projective plane over $\mathbf{Z}/p\mathbf{Z}$. The main fact is then the following result which says quite precisely how many solutions there can be:

Theorem 1.8 (Hasse). *Let $p \nmid \Delta_E$. The number N_p of projective solutions modulo p to the equation defining E can be written $N_p = p + 1 - a_p$ with*

$$|a_p| \leqslant 2\sqrt{p}. \tag{1.7}$$

This is also called the Riemann Hypothesis for the curve E reduced modulo p.

Remark 1.9. If $a_1 = a_3 = 0$ then

$$a_p = - \sum_{x \,(\mathrm{mod}\, p)} \left(\frac{x^3 + a_2 x^2 + a_4 x + a_6}{p} \right) \tag{1.8}$$

with $\left(\frac{y}{p}\right)$ the Legendre symbol, i.e., $\left(\frac{y}{p}\right)$ is equal to 0 for $y = 0$, and otherwise is equal to 1 if y is a square modulo p and -1 is y is not a square modulo p; note that $1 + \left(\frac{y}{p}\right)$ is the number of solutions to the equation $X^2 = y$ in $\mathbf{Z}/p\mathbf{Z}$, which gives quickly the formula stated from the definition $N_p = p + 1 - a_p$ and the fact that there is a single point at infinity.

It is reasonable to expect on probabilistic grounds that the size of this sum should be about $\sqrt{p}$, because there is about the same chance that the value of $x^3 + a_2 x^2 + a_4 x + a_6$ be a square as a non-square modulo p (for p odd, there are as many squares as non-squares among non-zero integers modulo p, namely $(p-1)/2$).

Example 1.10. In general, there is no simpler explicit formula for a_p. However, there is an elementary description if the curve has complex multiplication. For instance, let E be the congruent number curve with equation $y^2 = x^3 - x$, which has complex multiplication by i. We have $\Delta_E = 64$. Then a_p is given as follows: if $p \equiv 3 \,(\mathrm{mod}\, 4)$, then $a_p = 0$; if $p \equiv 1 \,(\mathrm{mod}\, 4)$, then (Fermat) one can write $p = a^2 + b^2$ with a odd, b even, and $a + b \equiv 1 \,(\mathrm{mod}\, 4)$; then $a_p = 2a$.

For a few non-CM elliptic curves, one can give an "implicit" description. For instance, consider the curve

$$X_1(11) : y^2 + y = x^3 - x^2$$

with discriminant -11, then define $a(n)$ for $n \geqslant 1$ by the formal power series identity

$$\sum_{n \geqslant 1} a(n) q^n = q \prod_{n \geqslant 1} (1 - q^n)^2 (1 - q^{11n})^2 = q - 2q^2 - q^3 + 2q^4 + q^5 + \cdots$$

Then the a_p for the curve $X_1(11)$ is the coefficient $a(p)$.

Finding the points on the curve modulo primes is fairly easy, and can provide information on the rational points (i.e., the Mordell-Weil group). However, using only one prime is clearly not sufficient (for instance, because there are usually many points modulo p which are not obtained by reduction of rational points). An important idea in number theory is to construct a "global" invariant that encompasses information obtained modulo all primes. In the case of elliptic curves, this takes the form of the so-called Hasse-Weil zeta function (or L-function) of an elliptic curve $E/\mathbf{Q}$.

We define first a naive version, namely

$$\ell(E, s) = \prod_{p \nmid \Delta_E} (1 - a_p p^{-s} + p^{1-2s})^{-1},$$

where p runs over the primes not dividing the discriminant of E. This product converges absolutely for $\mathrm{Re}(s) > 3/2$ by Hasse's Theorem (the precise shape of the product may seem strange, but it is very well explained by looking at the points on the elliptic curve, not only after reduction modulo p, but also in finite extension fields of $\mathbf{Z}/p\mathbf{Z}$).

As a first statement indicating some kind of nice behavior of the various reductions modulo primes, having to do with the fact that they have a single "global" origin over $\mathbf{Q}$, Hasse conjectured that $\ell(E, s)$ has an analytic continuation to $\mathbf{C}$. This is now seen as an imprecise form of the *modularity* of elliptic curves over $\mathbf{Q}$, which was proved by Wiles, Taylor-Wiles, Breuil-Conrad-Diamond-Taylor.

To explain the precise form, one must first refine the definition to obtain the "right" L-function. This requires the insertion in the product of correct factors at the primes $p \mid \Delta_E$.

First one may remark that because Δ_E is not an isomorphism-invariant of E, one can have $p \mid \Delta_E$ for some Weierstrass equation but not for another. So one defines the *conductor* of E, an integer $\mathfrak{f}(E) \geqslant 1$ such that $p \nmid \mathfrak{f}(E)$ if and only if E has a smooth reduction modulo p, possibly after some change of variable (isomorphism over $\mathbf{Q}$). For $p \mid \mathfrak{f}(E)$, the exponent f_p of p in $\mathfrak{f}(E)$ is dictated by the geometry of the singular reduction, in ways that can be quite complicated. But here are the simplest cases which are often sufficient:

• If the reduction of E modulo p has a node, then $f_p = 1$ ("multiplicative reduction").

• If $p > 3$ and the reduction of E modulo p has a cusp, then $f_p = 2$ ("additive reduction").

• If $p = 2$ or $p = 3$ and the reduction of E modulo p has a cusp, the definition of f_p is much more intricate. In all cases, one shows that $2 \leqslant f_p \leqslant 11$.

If $p \mid \mathfrak{f}(E)$, define

$$a_p = \begin{cases} 0 & \text{if } f_p \geqslant 2, \\ -1 & \text{if } f_p = 1, \text{ and the slopes of the node are in } \mathbf{Z}/p\mathbf{Z}, \\ 1 & \text{otherwise.} \end{cases}$$

(for the second case, the meaning is that a node can be such that the two "tangent directions" are either in $\mathbf{Z}/p\mathbf{Z}$ or generate a quadratic extension of $\mathbf{Z}/p\mathbf{Z}$; one speaks of *split* or *non-split* multiplicative reduction).

Example 1.11. • For the curve $X_1(11)$ or for the curve $y^2 + y = x^3 - x$, one has $\mathfrak{f}(E) = 11$. This is the smallest possible conductor for an elliptic curve $E/\mathbf{Q}$.

• If E is given by $y^2 = x^3 + ax + b$ and E_d is its *quadratic twist* given by $dy^2 = x^3 + ax + b$, where d is a squarefree integer, then $\mathfrak{f}(E_d)$ divides $d^2 \mathfrak{f}(E)$, with equality if d is coprime with $\mathfrak{f}(E)$.

Using $\mathfrak{f}(E)$ and those a_p, the Hasse-Weil zeta function is defined by

$$L(E, s) = \prod_{p \mid \mathfrak{f}(E)} (1 - a_p p^{-s})^{-1} \prod_{p \nmid \mathfrak{f}(E)} (1 - a_p p^{-s} + p^{1-2s})^{-1}.$$

The meaning of the modularity of E can now be stated precisely. Denote by $a_E(n)$ the coefficients in the expansion of the Euler product $L(E, s)$ in Dirichlet series

$$L(E, s) = \sum_{n \geqslant 1} a_E(n) n^{-s}$$

and define

$$f(z) = \sum_{n \geqslant 1} a_E(n) e^{2\pi i n z} \text{ for } z \in \mathbf{H}, \text{i.e}, \operatorname{Im}(z) > 0.$$

The series converges absolutely and uniformly on compacts to define a holomorphic function $f : \mathbf{H} \to \mathbf{C}$. Then modularity of E means that we have

$$f\left(\frac{az + b}{cz + d}\right) = (cz + d)^2 f(z)$$

for all $a, b, c, d \in \mathbf{Z}$, $ad - bc = 1$, with $\mathfrak{f}(E) \mid c$, and moreover that $\operatorname{Im}(z)|f(z)|$ is bounded on $\mathbf{H}$ (those conditions express that f is a cusp form of weight 2 for the Hecke congruence subgroup $\Gamma_0(\mathfrak{f}(E))$).

From this one deduces that $L(E, s)$ has analytic continuation to an entire function by means of the formula

$$(2\pi)^{-s} \Gamma(s) L(E, s) = \int_0^\infty f(iy) y^{s-1} dy,$$

which is due to Hecke (and applies to all cusp forms of weight 2). Thus Hasse's Conjecture follows in this indirect manner.

The theory of Hecke gives more information, which is also very important: by means of the so-called Fricke involution and multiplicity one for Hecke operators, one shows that $L(E, s)$ also satisfies a functional equation

$$\Lambda(E, s) = w_E \, \mathfrak{f}(E)^{1-s} \Lambda(E, 2 - s) \tag{1.9}$$

for some $w_E \in \{\pm 1\}$ (called the sign of the functional equation), where

$$\Lambda(E, s) = (2\pi)^{-s} \Gamma(s) L(E, s).$$

In addition, one can prove that the sign w_E factorizes as a product over $p \mid \mathfrak{f}(E)$ of "local" signs $w_{E,p} \in \{\pm 1\}$. It is also important to know that w_E is effectively computable. For instance, if $\mathfrak{f}(E)$ is squarefree, one can show that

$$w_E = \mu(\mathfrak{f}(E))a_E(\mathfrak{f}(E))$$

where $\mu(\mathfrak{f}(E))$ is the Möbius function, which is simply here $(-1)^k$, k being the number of distinct prime factors of $\mathfrak{f}(E)$.

Remark 1.12. When the curve $E/\mathbf{Q}$ happens to be a CM curve (for instance, the congruent number curve $y^2 = x^3 - x$), then one can give a much more elementary proof of the modularity of $L(E, s)$ than the general one provided by Wiles et al.

References: [AEC, V,C.16], [I2, 8], [K, II].

1.5   The Birch and Swinnerton-Dyer conjecture

Let E be an elliptic curve defined over $\mathbf{Q}$. Recall that the hope is still to solve, as much as possible, the diophantine question of finding the Mordell-Weil group of E. There is an intuition that the L-function of E, which has been built from "local" information about the various reductions of E modulo primes, should provide some help. It is not at all clear how to make this precise. However, there is a beautiful conjecture that provides a very clean link.

By modularity, we know that $L(E, s)$ is holomorphic, in particular defined at $s = 1$. Then the simplest form of the Birch and Swinnerton-Dyer Conjecture is

Conjecture 1.13. *We have*

$$\operatorname{rank} E(\mathbf{Q}) = \operatorname{ord}_{s=1} L(E, s),$$

i.e, $L(E, s)$ has a zero at $s = 1$ with order equal to the rank of the Mordell-Weil group of E.

Remark 1.14. To indicate the amazing consequences of such a statement, notice that if the sign w_E of the functional equation happens to be -1 (one often speaks of "odd" functional equation, or "odd" curve), then by (1.9) we find that $L(E, 1) = 0$, hence, under the Birch and Swinnerton-Dyer Conjecture, we have $\operatorname{rank} E(\mathbf{Q}) \geqslant 1$ in that case. However, the condition $w_E = -1$, as we have remarked, is a *local* condition, which in fact only involves the behavior of the curve at primes dividing $\mathfrak{f}(E)$ (primes of bad reduction). So this very simple-looking local condition should imply the global consequence that there is a non-trivial point of infinite order in $E(\mathbf{Q})$. The challenge is then to find a way to obtain concretely such a point; no algorithm is known to solve that problem.

There is also a more refined form of the conjecture, which takes the following form:

Conjecture 1.15. *We have*

$$L(E, s) \sim \alpha (s - 1)^r \text{ as } s \to 1,$$

where $r = \operatorname{rank} E(\mathbf{Q})$ *and*

$$\alpha = \frac{\Omega |\text{III}(E)| R(E) c}{|E(\mathbf{Q})_{tors}|^2} > 0, \tag{1.10}$$

the various terms Ω, $\text{III}(E)$, $R(E)$, c *being all strictly positive real numbers which are described below.*

Here are short descriptions of the unexplained quantities in this conjecture.

• $|E(\mathbf{Q})_{tors}|$ is the cardinality of the set of rational torsion points on E. As we have already mentioned, it is easy to compute, and in fact it is well-understood theoretically. (In particular, it takes only finitely many values).

• c (the *Tamagawa number*) is given by the product over primes of the local Tamagawa numbers $c_p = |E(\mathbf{Q}_p)/E_0(\mathbf{Q}_p)|$, where $E_0(\mathbf{Q}_p)$ is the set of points which have non-singular reduction modulo p. If E has good reduction at p, we have $c_p = 1$, so that the product really has only finitely many terms. There is an efficient algorithm to compute c_p. This algorithm is described by K. Rubin in his paper in this volume.

• Ω is the *real period* of E, an elliptic integral of the type

$$\int_{E^0(\mathbf{R})} \frac{dx}{2y}$$

where $E^0(\mathbf{R})$ is the infinite component of the real points of the curve. It is also easily computable.

• $R(E)$ is the *elliptic regulator*: let $x_1, \ldots, x_r$ be a basis for the free part of $E(\mathbf{Q})$. Then

$$R(E) = \det(\langle x_i, x_j \rangle)$$

where $\langle \cdot, \cdot \rangle$ is the *canonical height* on $E(\mathbf{Q})$, the bilinear form coming from the following quadratic form:

$$\|p\| = \lim_{n \to +\infty} 4^{-n} h([2^n]p)$$

where, for a point $p = (x, y) \in E(\mathbf{Q})$, the "naïve" height $h(p)$ is defined by

$$h(p) = h((x, y)) = \frac{1}{2} \log H(x)$$

with

$$H(x) = \max(|u|, |v|), \text{ if } x = \frac{u}{v} \text{ with } u, v \text{ integers and } (u, v) = 1.$$

The regulator R is hard to compute, since it involves finding a basis of the Mordell-Weil group, but because there are explicit and efficiently computable formulas for the height function, one can indeed compute it very quickly given the generators x_i. Note in particular that if $r = 0$, we have $R = 1$ by definition.

References: [AEC, VIII.7,8,9C.16]

1.6   The Tate-Shafarevitch group

There only remains to explain the term III in the refined Birch and Swinnerton-Dyer Conjecture. This is the so-called Tate-Shafarevitch group of E, which is in many ways the most mysterious component of the formula. For instance, although it is implicit in the statement that this must be a *finite* group, this is not known in general![1]

We will spend a few paragraphs trying to explain a bit more where this group comes from and why it is so elusive. We do this partly because it is quite a beautiful object in its own right, and partly from the belief that some progress could be made on its study (e.g., the finiteness conjecture) if more people, especially with an analytic frame of mind, looked at it more carefully...

First, here is a sketch explaining how the elements of $\mathrm{III}(E)$ arise. Let $E/\mathbf{Q}$ be an elliptic curve, and assume its equation is of the type

$$y^2 = (x - e_1)(x - e_2)(x - e_3)$$

with $e_i \in \mathbf{Q}$ (which means that the 2-torsion points $(0, e_i)$ are in $E(\mathbf{Q})$). We are looking for a way to find all rational solutions.

Let $(x, y) \in E(\mathbf{Q})$. Note that for $p \nmid \Delta_E$, the smoothness modulo p implies that at most one of the $x - e_i$ can be divisible by p. Since their product is the square y^2, we see that p occurs with an even exponent in the factorization of each $x - e_i$. This holds for all $p \nmid \Delta_E$, hence putting everything together, we can write

$$x - e_i = c_i w_i^2$$

for some $z_i \in \mathbf{Q}$, where the numerator and denominator of c_i are divisible only by primes dividing Δ_E. Now if a $p \mid \Delta_E$ has even exponent, we can change w_i and get a similar relation with c_i coprime with this p. If $p \mid \Delta_E$ has odd exponent (say $2k + 1$), we can still "remove" similarly the p^{2k} part. Hence we have a relation

$$x - e_i = b_i z_i^2$$

with $z_i \in \mathbf{Q}$ and b_i is an integer which is product of some primes dividing Δ_E, each with exponent at most 1.

Obviously the set of choices for b_i is finite. We do not know which b_i actually occur, but we can make a list of all those which can conceivably arise from a rational point on E. Let us call T this finite, effectively computable, set of triples $b = (b_1, b_2, b_3)$ of non-zero squarefree integers (i.e., those where each prime divisor divides Δ_E). We have shown that given $(x, y) \in E(\mathbf{Q})$, there is

---

[1] In fact, for elliptic curves over function fields over finite fields, the full Birch and Swinnerton-Dyer Conjecture is now a theorem of Kato and Trihan, *if* the Tate-Shafarevitch group is always finite.

some $b \in T$, and rationals z_1, z_2, z_3, such that

$$\begin{cases} y^2 = (x - e_1)(x - e_2)(x - e_3) \\ x - e_1 = b_1 z_1^2 \\ x - e_2 = b_2 z_2^2 \\ x - e_3 = b_3 z_3^2. \end{cases} \qquad (1.11)$$

We now consider the set C_b of solutions to these equations for a fixed b; thus there are 5 variables (x, y, z_1, z_2, z_3). The set C_b is a curve in affine 5-space (since there are 4 relations). It is easy to see, by eliminating some unknowns, that C_b is isomorphic (over $\mathbf{Q}$) to the curve in 3-space with coordinates (z_1, z_2, z_3) given by the two equations

$$\begin{cases} b_1 z_1^2 - b_2 z_2^2 = (e_2 - e_1) \\ b_1 z_1^2 - b_1 b_2 z_3^2 = (e_3 - e_1). \end{cases} \qquad (1.12)$$

The crucial point is that for given $b \in T$ (recall that T is not simply the set of those (b_1, b_2, b_3) that actually arise from a rational point, but may be larger), the curve C_b may have a rational point or not. Certainly, if it does not, this particular b could not in fact arise from a rational point on E in the way described above. But conversely, if a rational point $(z_1, z_2, z_3) \in C_b(\mathbf{Q})$ exists, then using (1.11) one clearly gets at least one point in $E(\mathbf{Q})$. Now one shows (by further elementary algebraic manipulations for instance) that finding a rational point on each of the curves C_b (for which one exists) is tantamount to finding representatives of the quotient group $E(\mathbf{Q})/2E(\mathbf{Q})$. As the discussion of Mordell's Theorem recalled, it is then known how to find generators for $E(\mathbf{Q})$.

The above method of computing $E(\mathbf{Q})$ can indeed be implemented in many situations. However, in general it is confronted with the problem that there is no algorithm known to check whether the curves C_b have a rational point or not.

The common method of dealing with this has been to remark that one can, on the other hand, compute quite easily the subset $S \subset T$ of those b for which C_b has "locally" a point at all p (essentially, a point modulo p for all p), and a real-valued point. This is useful because, obviously, the set of b for which C_b has a rational point is a subset of S. (The S is for Selmer; this set can also be equipped with a group structure and is then called the 2-Selmer group of E).

However, there may be elements of S which still do not have a rational point (one says that the "Hasse principle" fails for C_b). Those elements "are" exactly the non-zero elements of order 2 in $\mathrm{III}(E)$. Note that, as a subset of S, it is a finite set, but the point is that this is simply a subset of the full Tate-Shafarevitch group.

The full group is not as easy to define in concrete terms. Here is a more abstract definition, as a set (the group structure is not obvious): one says that a curve $C/\mathbf{Q}$ is a principal homogeneous space for $E/\mathbf{Q}$ if one can define an action of E on C, i.e. an algebraic map (denoted $+$ here)

$$E \times C \to C$$
$$(P, p) \mapsto p + P$$

such that $p + (P + Q) = (p + P) + Q$ and $p + P = q$ has a unique solution (denoted $q - p$) for all (p, q). (Note the similarity with the notion of an affine space with its associated vector space in elementary geometry). There is always a "trivial" homogeneous space, namely E itself with the action being given by the addition on E.

It is not quite obvious, but in fact the curves C_b with equations (1.12) are examples of homogeneous spaces for E. A definition of $\text{III}(E)$ is then as the set of all homogeneous spaces $C/\mathbf{Q}$ for which $C(\mathbf{R})$ and $C(\mathbf{Q}_p)$, for all p, are non-empty, modulo the relation of isomorphism as homogeneous spaces, which means that $C \sim C'$ if there exists an isomorphism $f : C \to C'$ defined over $\mathbf{Q}$ with $f(p + P) = p + f(P)$.

Once the group structure on $\text{III}(E)$ is defined, one sees that E is the identity element in $\text{III}(E)$. Then, to make the link with the previous curves C_b, notice that $C \in \text{III}(E)$ is trivial if and only if $C(\mathbf{Q}) \neq \emptyset$: first, if C is trivial, it is isomorphic to E, so has a rational point corresponding to the origin 0 of the group law of E. Conversely, if $p_0 \in C(\mathbf{Q})$ is a rational point, the map $p \mapsto p - p_0$ gives the required isomorphism $C \simeq E$.

There is no reason (and it often happens that this is not the case) that $\text{III}(E)$ should contain only the elements of order 2 which have already been described. One can show that $\text{III}(E)$ is a torsion group (i.e., every element is of finite order), and also that for any integer $n \geqslant 1$, the subgroup of n-torsion elements in $\text{III}(E)$ is finite (in ways at least similar in spirit to the case of $n = 2$). However, we have no a priori bound on the order of an element of $\text{III}(E)$; to have such a (finite) bound would be equivalent to proving that $\text{III}(E)$ is finite. This we state formally as a conjecture due to Tate and Shafarevich:

Conjecture 1.16. *For all $E/\mathbf{Q}$, the Tate-Shafarevich group $\text{III}(E)$ is a finite group.*

The refined form of the Birch and Swinnerton-Dyer Conjecture does not make sense without assuming this statement. For quite a long time, it was the case that not a *single* elliptic curve $E/\mathbf{Q}$ with $\text{III}(E)$ finite was known, but the work of Rubin, Kolyvagin and others have provided many examples in cases where the order of vanishing of the L-function of E is $\leqslant 1$.

A further useful known fact is that there is a (highly non-obvious!) symplectic pairing (due to Cassels)

$$\text{III}(E) \times \text{III}(E) \to \mathbf{Q}/\mathbf{Z} \tag{1.13}$$

which is perfect if $\text{III}(E)$ is finite; this gives information on the group structure.

Example 1.17. Let E be the elliptic curve $y^2 = x^3 - 24300$, $j = 0$. The rank is 0, the regulator is 1, the torsion group is trivial, the Tamagawa number is 1, we have

$$L(E, 1) = 4.061375813927\ldots$$
$$\Omega = 0.451263979325\ldots$$

and so $|\text{III}(E)| = 9$, which means (by the existence of the Cassels pairing) that $\text{III}(E) \simeq (\mathbf{Z}/3\mathbf{Z})^2$. In fact, the following are equations for all locally trivial homogeneous spaces under E:

$$\begin{aligned}
C \simeq E \quad & x^3 + y^3 + 60x^3 = 0 \\
C_1 \quad & 3x^3 + 4y^3 + 5z^3 = 0 \\
C_2 \quad & 12x^3 + y^3 + 5z^3 = 0 \\
C_3 \quad & 15x^3 + 4y^3 + z^3 = 0 \\
C_4 \quad & 3x^3 + 20y^3 + z^3 = 0
\end{aligned}$$

(each of the four equations C_i above corresponds to two opposite elements of $\text{III}(E)$, equivalently to a line in $(\mathbf{Z}/3\mathbf{Z})^2$). See [Ma] for more details.

Remark 1.18. (1) Here is the cohomological definition of $\text{III}(E)$, which makes the group structure apparent, but gives little information related to finiteness:

$$\text{III}(E) = \ker\left\{ H^1(G_\mathbf{Q}, E) \to \prod_v H^1(G_{\mathbf{Q}_v}, E) \right\}.$$

Similarly the set called S above can be introduced more generally as the *Selmer group* for any prime ℓ

$$\text{Sel}_\ell(E) = \ker\left\{ H^1(G_\mathbf{Q}, E[\ell]) \to \prod_v H^1(G_{\mathbf{Q}_v}, E) \right\}$$

and then the elementary computations that have been sketched correspond to the case $\ell = 2$ of the following short exact sequence:

$$0 \to E(\mathbf{Q})/\ell E(\mathbf{Q}) \to \text{Sel}_\ell(E) \to \text{III}(E)[\ell] \to 0. \tag{1.14}$$

(2) The author believes that analytic number theory should be brought to bear on the finiteness conjecture for $\text{III}(E)$. Here is one reason for this: there is a well-known analogy between $\text{III}(E)$ and the class group of number fields, and in the latter case, the finiteness is known, but all proofs, in one way or another, depend on *inequalities*, whether from geometry of numbers or from the use of L-functions. Here is a wildly off-hand suggestion[2]: can one associate to a given $C \in \text{III}(E)$ some kind of holomorphic function f_C, in such a way that the f_C are linearly independent, but belong to a finite dimensional space? (Think of theta functions associated to ideal classes in an imaginary quadratic field, which are modular forms of a fixed type, hence live in a space that can be proved to be of finite dimension in a completely independent way...)

---

[2] Of course, the author *has* tried to make something out of it without success...

References: [AEC, Chapter X], see also the article by Swinnerton-Dyer [Sw] in this volume.

1.7   Enter random matrices...

The Birch and Swinnerton-Dyer Conjecture is still unproved, but much evidence exists in its favor (certainly for the simple form), so that it is very reasonable to take it as an assumption if one wishes to study the "general" behavior of the rank of elliptic curves over $\mathbf{Q}$. One may expect it to help understand, for example, whether there are elliptic curves $E/\mathbf{Q}$ with arbitrarily large rank.[3]

The conjecture certainly helps explain why such curves, if they exist, are very hard to find: it is easy to show that if the Birch and Swinnerton-Dyer Conjecture holds, there exists an absolute constant $c > 0$ such that for any $E/\mathbf{Q}$ with rank r we have

$$\mathfrak{f}(E) \geqslant e^{cr}; \tag{1.15}$$

this is obtained by bounding the order of vanishing of $L(E,s)$ at $s = 1$ by the number $N(E,1)$ of zeros ρ of $L(E,s)$ with $|\operatorname{Im}(\rho)| \leqslant 1$ (counted with multiplicity), which is well-known to satisfy

$$\operatorname{ord}_{s=1} L(E,s) \leqslant N(E,1) \ll \log \mathfrak{f}(E)$$

with an absolute implied constant. If one assumes the Generalized Riemann Hypothesis, there even exists $c > 0$ such that

$$\mathfrak{f}(E) \geqslant e^{cr \log r} \tag{1.16}$$

(for both facts, see e.g. [IK, 5.8]; the analogue of this inequality is known to be sharp over function fields, see [U1], and it may also be over $\mathbf{Q}$).

Unfortunately, at the present moment at least, our knowledge about L-functions and the distribution of their zeros is still quite limited, and we do not have very many unconditional results. It is worth mentioning one striking application of the Birch and Swinnerton-Dyer Conjecture, but one that goes the other way: Goldfeld showed how one could use an L-function with a zero at $s = 1$ of order $\geqslant 3$ to solve effectively the class number problem for imaginary quadratic fields, and thus, instead of using L-functions to study elliptic curves, it was elliptic curves which were used by Gross-Zagier to produce such an L-function in confirmation with Goldfeld's expectation.

However, for many questions, the recent development of Random Matrix Models for families of L-functions offers a new, unexpected, way of probing the diophantine mystery that is the Mordell-Weil group. As we will see in the second part, new phenomena and conjectures are appearing and it can be hoped that besides new insight, they will yield new ideas by comparison with the viewpoints of algebraic geometry.

---

[3]In early May, 2006, N. Elkies announced having found a curve with rank $\geqslant 28$, improving the previous record of 24.

2 Variation of the rank in families of elliptic curves

The purpose of this second part is to describe some of the known results concerning the variation of the rank of elliptic curves, mostly over $\mathbf{Q}$, when a large number of curves are taken together and considered as a whole, not as individuals. We discuss what analytic methods, especially based on L-functions (and hence on assuming the Birch and Swinnerton-Dyer conjecture), have been able to produce. Thus some other techniques (such as the use of sieve methods to produce twists with "large" rank) will not be considered, although they are certainly interesting.

It will be seen that even with quite deep assumptions, the outcome remains in some ways disappointing. It may be that currently the most remarkable achievements are the conjectures that arise out of the random matrix models concerning the order of vanishing of L-functions, hence conjecturally concerning the rank; this will be our second main topic.

For readers who skipped the first part, we recall some relevant notation: $\mathfrak{f}(E)$ is the conductor of an elliptic curve $E/\mathbf{Q}$, and $a_E(n)$ denotes the coefficients of its Hasse-Weil zeta function $L(E, s)$.

2.1 Families and invariants

Although the term "family" has a number of well-defined and deep meanings in algebraic geometry and arithmetic, we will only need a very weak notion here, amounting to hardly more than walking through a (multi)set of elliptic curves with some indexing. (This is not the same definition discussed by D. W. Farmer in his paper [Fa] in this volume; it may be said that we exploit here some concrete features of the specific L-functions of elliptic curves, or of modular forms, and benefit from the fact that for some problems, it is only necessary to have means of "comparing" two curves taken in the family, which is done efficiently using the Rankin-Selberg convolution; also, a useful survey of the case of function fields, to which we will make passing references, is contained in D. Ulmer's paper [U2]).

Precisely, a *family* $\mathcal{E}$ of elliptic curves over $\mathbf{Q}$ is the data, for any $T \geqslant 1$, of a finite (multi)set[4] $\mathcal{E}(T)$ of elliptic curves $E/\mathbf{Q}$, which we subject to the following simple conditions:

(a) There exist constants c_1, $c_2 \geqslant 0$ and $\alpha, \beta > 0$ such that

$$c_1 T^\alpha \leqslant |\mathcal{E}(T)| \leqslant c_2 T^\beta \tag{2.1}$$

for $T \geqslant 1$. (The curves are counted with their multiplicity in $|\mathcal{E}(T)|$, if one $E/\mathbf{Q}$ appears more than once in $\mathcal{E}(T)$.)

---

[4] We permit some of the curves to come "with multiplicity"; see the examples below, especially algebraic families.

(b) There exist constants $c_3 \geqslant 0$ and $A \geqslant 0$ such that for any $T \geqslant 1$ and $E \in \mathcal{E}(T)$, we have

$$\mathfrak{f}(E) \leqslant c_3 T^A, \tag{2.2}$$

where $\mathfrak{f}(E)$ is the conductor of E.

We will now give several examples to indicate more precisely what we have in mind. For simplicity we often write $E \in \mathcal{E}$ to indicate that $E \in \mathcal{E}(T)$ for some T; similarly a map $f : \mathcal{E} \to X$ for any set X is a family of maps $f_T : \mathcal{E}(T) \to X$ (so if a curve E belongs to both $\mathcal{E}(T_1)$ and $\mathcal{E}(T_2)$, one may have $f_{T_1}(E) \neq f_{T_2}(E)$, e.g. if $f_T(E) = T$.)

(1) <u>Algebraic families</u>: Consider polynomials a_1, a_2, a_3, a_4, $a_6 \in \mathbf{Z}[t]$ and for any $t \in \mathbf{Z}$ let E_t be the curve given by the equation

$$y^2 + a_1(t)xy + a_3(t)y = x^3 + a_2(t)x^2 + a_4(t)x + a_6(t). \tag{2.3}$$

If $\Delta(t)$, the discriminant of this curve, is not identically 0, there will be a finite set S of $t \in \mathbf{Z}$ such that E_t is an elliptic curve for $t \notin S$. If the j invariant $j(t)$ is also non-constant, then letting

$$\mathcal{E}(T) = \{E_t \mid |t| \leqslant T \text{ and } t \notin S\}$$

we get a family of elliptic curves; here there will be multiplicity if $E_t \simeq E_s$ for some $t \neq s$.

(2) <u>Quadratic twists</u>: This is partly a special case of the previous one. Fix an elliptic curve $E/\mathbf{Q}$ and for all quadratic fundamental discriminants d let E_d be the corresponding quadratic twist of E: if E is given by $y^2 = x^3 + a_4 x + a_6$, then E_d is the curve with equation

$$dy^2 = x^3 + a_4 x + a_6.$$

For any d, this is an elliptic curve and putting

$$\mathcal{E}_E(T) = \{E_d \mid |d| \leqslant T \text{ and } d \text{ is a fundamental quadratic discriminant}\}$$

gives an example of a family (recall $\mathfrak{f}(E_d) \mid d^2 \mathfrak{f}(E)$, with equality if d is coprime to $\mathfrak{f}(E)$).

(3) <u>All curves indexed by height</u>: This was considered by Brumer [B]: for any integers a_4 and a_6 such that $4a_4^3 + 27a_6^2 \neq 0$ and such that $p^4 \mid a_4$ implies that $p^6 \nmid a_6$, let E_{a_4,a_6} be the curve

$$E_{a_4,a_6} : y^2 = x^3 + a_4 x + a_6,$$

given by the corresponding Weierstrass equation. Then let

$$\mathcal{E}_H(T) = \{E_{a_4,a_6} \mid |a_4|^3, \ |a_6|^2 \leqslant T\} \tag{2.4}$$

This is a family. It is known that every elliptic curve over $\mathbf{Q}$ occurs exactly once among the E_{a_4,a_6}.

(4) <u>All curves indexed by conductor</u>: In this case we simply take

$$\mathcal{E}_c(T) = \{E/\mathbf{Q} \mid \mathfrak{f}(E) \leqslant T\}$$

A variant consists in taking only one representative in each isogeny class; in the corresponding family $\mathcal{E}'_c$, one can identify

$$\mathcal{E}'_c(T) = \{f \in S_2^*(q, \mathbf{Z}) \mid q \leqslant T\}$$

the set of primitive forms of weight 2 for $\Gamma_0(q)$, $q \leqslant T$, which have integral coefficients (this by modularity of elliptic curves over $\mathbf{Q}$ and the isogeny theorem).

(5) <u>A counter-example</u>: Here is a set of elliptic curves that (conjecturally) fails to define a family in our sense. For any $n \geqslant 0$, let $E_n/\mathbf{Q}$ be a curve (if it exists) with smallest conductor such that rank $E_n(\mathbf{Q}) = n$ and $\mathcal{E}(T) = \{E_n \mid 0 \leqslant n \leqslant T\}$. Two things prevent this from being a family: either E_n does not exist for n large enough (although this is not currently expected to be the case); or even if it exists, then on B-SD we have (1.15) so the conductor grows exponentially, contradicting (2.2).

Obviously one can generalize the definitions above. Particularly, one could consider elliptic curves over an arbitrary global field, or abelian varieties over a global field. In the case of $\mathbf{Q}$ a narrower but very natural generalization is to consider arbitrary primitive modular forms (of weight 2) instead of those associated with Hasse-Weil L-functions of elliptic curves, or more geometrically (as described by Shimura) the isomorphism (or isogeny) classes of abelian varieties which are quotients of the jacobians $J_0(q)$ of the modular curves $X_0(q)$. We will only give the briefest remarks below about the "general" generalizations, but we will sometimes mention in more detail the case of the "family" $\mathcal{E}_0$ with $\mathcal{E}_0(q)$ the set of primitive weight 2 cusp-forms of level q (often restricted to primes for simplicity). Indeed much stronger analytic results have been obtained in this case, which can usefully serve as reference points in investigations of families elliptic curves.

For example it is worth mentioning (correcting slightly my remark quoted at the end of [U1]) that for an abelian variety $A/\mathbf{Q}$ of dimension $g \geqslant 1$, the analogue of (1.16), expressed in terms of the order of vanishing instead of the conductor, is that

$$\operatorname{ord}_{s=1} L(A, s) \ll \frac{\log \mathfrak{f}(A)}{\log \frac{1}{g} \log \mathfrak{f}(A)} \tag{2.5}$$

on GRH, the implied constant being absolute, if the conjectured analytic continuation and functional equation of $L(A, s)$ hold (note that $\mathfrak{f}(A) \geqslant 3^g$ also follows from the latter); see e.g. [IK, 5.14], [M]. The bound (2.5) is sharp because one can take $A = E^g$, $g \to +\infty$, for some elliptic curve $E/\mathbf{Q}$ of rank $\geqslant 1$. One may suspect that it is possible to improve (2.5) if A is simple. This is indeed the case for $J_0(q)$, q prime, for which $\log \mathfrak{f}(J_0(q)) \leqslant q \log q$,

$$\operatorname{ord}_{s=1} L(J_0(q), s) \geqslant \tfrac{q}{24} + o(q) \tag{2.6}$$

(by very easy sign considerations) and in fact

$$\text{rank } J_0(q) \geqslant \tfrac{7q}{192} + o(q) \tag{2.7}$$

using Heegner points and fairly difficult non-vanishing results for L-functions [KM1].

Given a family $\mathcal{E}$, we are interested in the average behavior of various invariants related to the rank of the Mordell-Weil group of the curves $E \in \mathcal{E}$. For this we introduce the notation

$$A_{\mathcal{E}}(T, f) = \sum_{E \in \mathcal{E}(T)} f(E)$$

for any function $f : \mathcal{E} \to \mathbf{C}$. If $f = 1$ we denote simply $A_{\mathcal{E}}(T, 1) = A_{\mathcal{E}}(T)$, the number of elements in the family of "index" T (by (2.1), it grows polynomially). This is the natural comparison function with respect to which one can speak of "the average value" of f on $\mathcal{E}$: for instance, if f is real valued, this average value will be said to be $\leqslant M$ for some $M \in \mathbf{R}$ if

$$A_{\mathcal{E}}(T, f) \leqslant M A_{\mathcal{E}}(T) + o(A_{\mathcal{E}}(T))$$

for $T \to +\infty$, and similarly with $\geqslant M$, $= M$...

Among many interesting functions, we will mention the following:

(1) The rank, $\mathrm{rk}(E) = \mathrm{rank}\, E(\mathbf{Q})$.

(2) The "analytic rank", $\mathrm{ord}(E) = \mathrm{ord}_{s=1} L(E, s)$, well-defined since $E/\mathbf{Q}$ is always modular. The B-SD conjecture implies $\mathrm{rk}(E) = \mathrm{ord}(E)$.

(3) The special value, $L(E) = L(E, 1)$, and more generally the moments and derivatives: $L^{(k)}(E)^m = L^{(k)}(E, 1)^m$, or the characteristic functions $v^{(k)}$ of k-th order vanishing and $V^{(k)}$ of order of vanishing $\geqslant k$: $v^{(k)}(E) = 1$ if $\mathrm{ord}(E) = k$, and 0 otherwise, $V^{(k)}(E) = 1$ if $\mathrm{ord}(E) \geqslant k$ and 0 otherwise.

(4) The root number $w(E) = \pm 1$, i.e. the sign of the functional equation (1.9).

(5) The parity of the rank, $p(E) = (-1)^{\mathrm{rk}(E)}$; conjecturally $p(E) = w(E)$, and this is now known if the Tate-Shafarevitch group $\mathrm{III}(E)$ is finite [N].

(6) For a prime ℓ, the order $m_\ell(E)$ of $E(\mathbf{Q})/\ell E(\mathbf{Q})$ or the order $s_\ell(E)$ of the ℓ-Selmer group $\mathrm{Sel}_\ell(E)$. One has $\ell^{\mathrm{rk}(E)} \leqslant m_\ell(E) \leqslant s_\ell(E)$, and if $E[\ell](\mathbf{Q}) = 0$, then $m_\ell(E) = \ell^{\mathrm{rk}(E)}$ (see (1.14)).

In addition to the relations already indicated, an important observation (going back to Shimura) is that $\mathrm{ord}(E) \geqslant \tfrac{1}{2}(1 - w(E))$: the functional equation (1.9) imposes $\mathrm{ord}(E) \geqslant 1$ if $w(E) = -1$, and otherwise $\mathrm{ord}(E) \geqslant 0$ (the latter is *not* a trivial fact: it needs modularity to be mentioned, and the fact that Hecke L-functions are entire). In particular we derive for a family $\mathcal{E}$

$$A_{\mathcal{E}}(T, \mathrm{ord}) \geqslant A_{\mathcal{E}}(T, \tfrac{1}{2}(1 - w)),$$

and on B-SD

$$A_{\mathcal{E}}(T, \mathrm{rk}) \geqslant A_{\mathcal{E}}(T, \tfrac{1}{2}(1 - w)),$$

which is one of the ways to get lower-bounds for the average rank. Without B-SD, recall that the Gross-Zagier formula and Kolyvagin's results give

$$A_{\mathcal{E}}(T, \mathrm{rk}) \geqslant A_{\mathcal{E}}(T, \tfrac{1}{2}(1 - w)v')$$

(this is the starting point towards (2.7) for instance.)

2.2 Conjectures and heuristics

Most of the current heuristics and conjectures on the variation of the rank involve assuming B-SD and then using some of the well-known (or emerging) conjectures about L-functions and the distribution of their zeros, a very extensively studied subject. One can also try to argue from the arguments leading to the proof of the Mordell-Weil theorem, but it is hard to make precise predictions because of the subtlety of issues involved.[5] (A third method is to make a guess, as in the conjecture that the rank of elliptic curves over $\mathbf{Q}$ is unbounded; neither from the point of view of L-functions and B-SD, nor from the Mordell-Weil theorem, does there appear convincing evidence at this time).

The basic heuristic about the order of vanishing of $L(E, s)$ at $s = 1$ has already been mentioned: it is that if $w(E) = -1$, then $\mathrm{ord}(E) \geqslant 1$, and that this should in general be the only way to produce an L-function vanishing at the central point, the order of vanishing being then the minimal compatible with $w(E)$, i.e. $L(E)$ should be non-zero if $w(E) = 1$ and $L'(E)$ should be non-zero if $w(E) = -1$. Note that this heuristic can only be true in a suitable average sense since there certainly exist families of curves with $\mathrm{rk}(E) > 1$ for $E \in \mathcal{E}$.

To transform this principle into more precise predictions for the average rank, one needs another ingredient: namely, we are led to expect that in an "unbiased" family $\mathcal{E}$ one has

$$A_{\mathcal{E}}(T, \mathrm{ord}) \sim A_{\mathcal{E}}(T, \tfrac{1}{2}(1 + w)) = \frac{A_{\mathcal{E}}(T)}{2} + \frac{A_{\mathcal{E}}(T, w)}{2},$$

and one has to treat the average of the root number. A second heuristic principle is that, again "in general", one should have about equal chance to have $w(E) = 1$ as $w(E) = -1$ in a family $\mathcal{E}$, leading to the vague conjecture:

Conjecture 2.1. *Let $\mathcal{E}$ be an "unbiased" family of elliptic curves over $\mathbf{Q}$. Then we have*

$$A_{\mathcal{E}}(T, w) = o(A_{\mathcal{E}}(T))$$
$$A_{\mathcal{E}}(T, \mathrm{rk}) = A_{\mathcal{E}}(T, \mathrm{ord}) \sim \tfrac{1}{2}A_{\mathcal{E}}(T)$$

as $T \to +\infty$.

---

[5] E.g. problems with class groups or ramification in the torsion fields of the curve.

It remains to find a better idea of what should be an unbiased family. For an algebraic family (Example 1 above), it is known that it is possible to find $\mathcal{E}$ such that $w(E)$, $E \in \mathcal{E}$, is constant. However the results of Helfgott [He] give a much clearer picture of the situation: in case the elliptic surface which "is the family" has at least one place v of multiplicative reduction (a generic situation), it shows that the even distribution of root numbers holds under some deep conjectures on the squarefree numbers represented by polynomials (and in fact are more or less equivalent with those), which are widely expected to hold – in particular, are consistent with general principles of cancellation in sums involving the Möbius function.

For the family $\mathcal{E}_E$ of quadratic twists of a given curve E, one has

$$w(E_d) = \chi_d(-\mathfrak{f}(E))w(E) \tag{2.8}$$

for $(d, \mathfrak{f}(E)) = 1$, where $\chi_d = (\frac{d}{\cdot})$ is the Kronecker symbol associated with d. It follows that $w(E_d)$ takes the values ± 1 depending on $d \,(\mathrm{mod}\, 4\,\mathfrak{f}(E))$, and does so asymptotically equally often. In this case, Conjecture 2.1 was formulated by Goldfeld.

Certainly the families $\mathcal{E}_H$ and $\mathcal{E}_c$ are expected to be unbiased. It is known that $w(E)$ is evenly distributed for $\mathcal{E}_H$, but it is not yet known for $\mathcal{E}_c$.

Remark 2.2. It is not necessarily the case that the family $J_0(q)$ is always simpler to deal with than families of elliptic curves. For instance consider the problem of averaging over q (prime) the root number $w(q)$ of $J_0(q)$. One finds using the Selberg trace formula and the formula for the dimension of the space of weight 2 cusp forms of level q that proving even distribution of ± 1 is equivalent to proving the even distribution of the value $h_{odd}(p) \,(\mathrm{mod}\, 4)$ of the odd part of $h(\mathbf{Q}(\sqrt{-q}))$ modulo 4! This is a special case of Cohen-Lenstra predictions but is completely open. One also has

$$h_{odd}(p) \,(\mathrm{mod}\, 4) = \Gamma_p(\tfrac{1}{2}) \,(\mathrm{mod}\, 4)$$

where Γ_p is the p-adic Gamma function and this makes sense because $\Gamma_p(\frac{1}{2})^2 = 1$. (This was remarked by H. Cohen who happened to be computing the right-hand side using GP exactly as I came in his office asking about the distribution of the left-hand side...)

2.3   Random matrix models

Quite recently, a deeper understanding of zeros of various L-functions has been obtained from Montgomery's first study of pair-correlation of zeros of the Riemann zeta function and the observed relation with the density of spacings of eigenvalues of random hermitian matrices of large rank, studied by Wigner for entirely different reasons. This has been set in a general framework by Katz and Sarnak [KS1], [KS2], who also gave very strong evidence by proving an analogue over function fields. Although the fundamental principle (that the distribution of zeros of "families"[6] of L-functions should be governed in some sense by the corresponding distributions of eigenvalues of large random matrices of some type, the latter being dictated by an hypothetical "symmetry group" of the family) remains entirely conjectural over number fields, a body of evidence exists, both numerical (see e.g. [O], [R1]), and theoretical: with certain restrictions, one has verified predictions based on this principle and a heuristic determination of the symmetry group, see e.g. [RoS] and [ILS] (the latter is significant because it treats situations where the type of symmetry group matters, and gets answers in perfect agreement). We will mention in Section 2.5 some similar results for families of elliptic curves.

There are more detailed surveys in the papers and books already mentioned. For us, it will be enough to state that by making various identifications of symmetry group (supported by the analogue cases over function fields, where this group is a well-defined monodromy group), Conjecture 2.1 appears as a very particular case of the Katz-Sarnak philosophy. But much more significantly, the random matrix models can make more precise predictions, such as when trying to estimate the number of curves in a family with rank at least 2. Being stronger, the resulting conjectures are easier to test for numerically. We will now describe a few examples.

Consider the subfamily $\mathcal{E}_{E,+}$ of the the family $\mathcal{E}_E$ of quadratic twists of $E/\mathbf{Q}$ such that $w(E_d) = 1$, and the average $A_{\mathcal{E}_E}(T, V'')$, which means that we count the number of twists with even order $\geqslant 2$. Using the Waldspurger formula for $L(E_d, 1)$, Sarnak conjectured that $A_{\mathcal{E}_{E,+}}(T, V'')$ should be of order $T^{3/4}$ (or equivalently $A_{\mathcal{E}_E}(T)^{3/4}$). Conrey, Keating, Rubinstein and Snaith [CKRS] use random matrix models to predict:

Conjecture 2.3. *There exist constants $c_+(E) > 0$ and $d_+(E) \in \mathbf{R}$, depending only on E, such that*

$$A_{\mathcal{E}_{E,+}}(T, V'') \sim c_+(E) A_{\mathcal{E}_E}(T)^{3/4} (\log T)^{d_+(E)}$$

as $T \to +\infty$.

(In this case, namely the twists with $w(E_d) = 1$, the suspected symmetry group is $SO(2N)$; for odd sign, it is in fact different, namely $SO(2N + 1)$, so it is natural to distinguish between the two cases).

Although $d_+(E)$ can be predicted (not very easily), there is not yet a predicted value of the constant $c_+(E)$, which makes numerical tests of this conjecture not as convincing as it could be. However, another conjecture is proposed in [CKRS] which is easier to test. Namely, fix a prime p where E has good reduction, and split the family $\mathcal{E}_{E,+}$ in $\mathcal{E}_{E,s}$ and $\mathcal{E}_{E,i}$ where $E_d \in \mathcal{E}_{E,s}$ if $\chi_d(p) = 1$ and $E_d \in \mathcal{E}_{E,i}$ if $\chi_d(p) = -1$. Then

---

[6] There is not yet a compelling definition of this notion, beyond the kind of minimal assumptions we have postulated; however, see D. W. Farmer's article in this volume for a stronger set of analytic axioms which is sufficient to confidently make very strong predictions from Random Matrix Theory, but note that his "orthogonality" axiom is not usually sufficient to prove rigorously the expected asymptotic formulas.

Conjecture 2.4. *We have*

$$\lim_{T\to+\infty} \frac{A_{\mathcal{E}_{E,s}}(T,V'')}{A_{\mathcal{E}_{E,i}}(T,V'')} = \sqrt{\frac{p+1-a_E(p)}{p+1+a_E(p)}}.$$

The numerical evidence about this in [CKRS] is quite good. Note the interpretation of the limit as the ratio of $|E(\mathbf{Z}/p\mathbf{Z})|$ and $|E^t(\mathbf{Z}/p\mathbf{Z})|$, where E^t is the quadratic twist of the reduction of E modulo p. An arithmetic explanation for the appearance of such a constant would be quite interesting. This conjecture has been generalized by M. Young to the family of all elliptic curves indexed by height [Yo4].

Random Matrix models have also been used to study the behavior of the rank of the group of points on an elliptic curve $E/\mathbf{Q}$ which are defined in some extension field $K/\mathbf{Q}$. To do this, one uses the general form of the Birch and Swinnerton-Dyer conjecture that predicts that the rank of $E(K)$ be equal to the order of vanishing of the L-function $L_K(E,s)$ of E over the field K. If $K/\mathbf{Q}$ is a finite Galois extension with group G, there is a factorization

$$L_K(E,s) = \prod_\chi L(E\otimes\chi,s)$$

of $L_K(E,s)$ in terms of twists of E by irreducible characters of the group G. Although the analytic continuation and functional equation of these twists are not often known, there are conjectures and partial results if χ is of degree 1 or 2. In particular, the desired properties hold if $K/\mathbf{Q}$ is a cyclic extension of degree k.

David, Fearnley and Kisilevsky [DFK] used this approach to make in particular a conjecture on cyclic extensions $K/\mathbf{Q}$ for which the rank of $E(K)$ can be strictly larger than that of $E(\mathbf{Q})$; under the Birch and Swinnerton-Dyer Conjecture, this is the same as asking for what characters χ one of the values $L(E\otimes\chi,1)$ can vanish. They predict for instance that there should be only finitely many cyclic field of fixed prime degree $k\geqslant 7$ for which $\operatorname{rank} E(K) > \operatorname{rank} E(\mathbf{Q})$.

Another recent development has been the use by Delaunay [D2] of the general conjectures and principles about moments of L-functions of various type (due to Conrey, Farmer, Keating, Rubinstein and Snaith [CFKRS]) together with the conjecture of Birch and Swinnerton-Dyer to predict the leading order asymptotic for moments of all order of the order of the Tate-Shafarevitch groups in a family of quadratic twists. More precisely, one can define the analytic order $S(E)$ using (1.10), for a curve with $w(E) = 1$, pretending that it has rank 0 (so $S(E) = 0$ if $\operatorname{rank} E(\mathbf{Q}) \geqslant 1$), and then predict the asymptotic of $S(E)^k$ by averaging $L(E,1)^k$. Similarly one can do the same with elements of the family with $w(E) = -1$, using the derivative $L'(E,1)$ to define an analytic order $S'(E)$. The expectation from Goldfeld's Conjecture is that summing both predicted heuristics for $S(E_d)^k$ and $S'(E_d)^k$, restricted each respectively to twists with $w(E_d) = 1$ or $w(E_d) = -1$, one should get the correct asymptotic for $\text{III}(E)^k$ as higher rank twists, for which $S(E) = S'(E) = 0$, should contribute less.

2.4 Theoretical results

We will now discuss some of the known theoretical results towards the problems and conjectures discussed in the previous sections. First, we should comment a little bit more on the relation between this kind of average consideration and the B-SD conjecture. Obviously we do not expect to prove the conjecture by this method, even in special cases. One could envision that, if it fails very badly, this could be proved by averaging the rank and the order of vanishing for a family and noting a discrepancy (e.g., if an algebraic family $\mathcal{E}$ built so that $\text{rk}(E) \geqslant 11$ for all $E \in \mathcal{E}$, as in [M], happened to be such that $\text{ord}(E) = 9$ or 10 for $E \in \mathcal{E}$, where one would expect that $A_{\mathcal{E}}(T, L^{(9)})$ would be comparable to $A_{\mathcal{E}}(T)...$), but that is of course highly unlikely for many reasons.

More seriously, even without B-SD it is certainly an interesting problem (requiring no special justification) to study the average of special values of L-functions, and we can hope to provide some evidence for B-SD by giving examples where the rank and the order of vanishing are of comparable size, on average. This is an appealing problem, but (to the author's knowledge), there is no known "non-obvious" family $\mathcal{E}$ for which we can prove unconditionally that

$$\alpha \leqslant \frac{A_{\mathcal{E}}(T, \text{rk})}{A_{\mathcal{E}}(T, \text{ord})} \leqslant \beta$$

for some constants α, $\beta > 0$ and all T large enough (with the convention $0/0 = 1$ which is natural here). Even in the case of $J_0(q)$, q prime, one can prove

$$\text{ord}(J_0(q)) \leqslant 0.1q + o(q)$$

(see [KM2], [KMV]) in addition to (2.6) and (2.7), but there is no known upper bound for $\text{rk}(J_0(q))$ of the right order of magnitude.

Another way that average studies could be useful would be if they yielded some insight into why the B-SD conjecture is true: what exactly makes this local-global principle operate?

The best understood case is that of the family $\mathcal{E}_E$ of twists of a given E. In fact, Heath-Brown has proved very precise unconditional results on the distribution of the order s_2 of the 2-Selmer group of twists of the congruent number curve. Precisely let $\mathcal{E}$ be the family of curves E_d of type

$$E_d : y^2 = x^3 - d^2 x$$

for d odd and squarefree. Heath-Brown proves [H1], [H2] the following results on the distribution of s_2.

Theorem 2.5. *Let $\mathcal{E}$ be the family above.*

(1) For any $k \geqslant 0$ we have

$$A_{\mathcal{E}}(T, s_2^k) \sim d_k A_{\mathcal{E}}(T)$$

as $T \to +\infty$, where

$$d_k = \prod_{j=1}^{k} (1 + 2^j).$$

(2) For any $r \geqslant 0$, let χ_r be the characteristic function of $s_2(E) = 2^{r+2}$. We have

$$A_{\mathcal{E}}(T, \chi_r) \sim \tfrac{1}{2} e_r A_{\mathcal{E}}(T) \tag{2.9}$$

as $T \to +\infty$, where

$$e_r = 2^r \prod_{j=1}^{r} (2^j - 1)^{-1} \prod_{j \geqslant 0} (1 - 2^{-2j-1}).$$

(3) In particular for $T \geqslant 2$ we have

$$\alpha A_{\mathcal{E}}(T) \leqslant A_{\mathcal{E}}(T, \mathrm{rk}) \leqslant \beta A_{\mathcal{E}}(T)$$

for some constants α, $\beta > 0$ and

$$A_{\mathcal{E}}(T, v) \gg A_{\mathcal{E}}(T).$$

(3) For $k \geqslant 0$, let $r_k(E)$ be the characteristic function of the condition $\mathrm{rk}(E) = k$. We have

$$A_{\mathcal{E}}(T, r_k) \leqslant 1.742^{-(k^2-k)/2} A_{\mathcal{E}}(T)$$

for T large enough.

As discussed below, we only know about the order of vanishing ord in this family that

$$(\tfrac{1}{2} + o(1)) A_{\mathcal{E}}(T) \leqslant A_{\mathcal{E}}(T, \mathrm{ord}) \quad \text{and} \quad A_{\mathcal{E}}(T, \mathrm{ord}) = o(T \log T)$$

as $T \to +\infty$. (On GRH, we do get a bounded order of vanishing on average).

We can only say a few words about the strategy of the proof, since it is very different from the other cases we will discuss below, and is independent of L-functions (see Heath-Brown's short note in this volume [H4]). The main fact is (1), from which the rest follows quite directly. For instance, (3) follows using the fact that $4 \mid s_2(E_d)$ since the 2-torsion points are rational, and $s_2(E_d) = 4$ (resp. $= 8$) implies that $\mathrm{rk}(E_d) = 0$ (resp. $= 1$) by the standard exact sequence (1.14) for $\ell = 2$ and the ensuing formula

$$\mathrm{rk}(E_d) = \dim_{\mathbf{Z}/2\mathbf{Z}} \mathrm{Sel}_2(E_d) - \dim_{\mathbf{Z}/2\mathbf{Z}}(\text{Ш}(E_d)[2]) - 2,$$

together with the fact that the last two terms are even (by the existence of Cassels's pairing (1.13)).

By the classical descent theory that we sketched in Section 1.6, the elements of the 2-Selmer group of E_d are identified with quadruplets (D_1, D_2, D_3, D_4) of integers $\geqslant 1$ such that $d = D_1 D_2 D_3 D_4$ and the system

$$D_1 X^2 + D_4 W^2 = D_2 Y^2 \text{ and } D_1 X^2 - D_4 W^2 = D_3 Z^2$$

has solutions modulo p for all primes p (it automatically has real solutions); compare with (1.12). This condition need be checked only at $p \mid d$, and leads to a formula for $s_2(E_d)$ as a fairly involved sum of quadratic residue symbols. Then one needs to take the k-th power of this expression and perform the average: this is a very impressive analytic feat, which can not be summarized here.

Remark 2.6. Although the presence of the Tate-Shafarevitch group means that we can not confirm Conjecture 2.1 using this result, interesting evidence comes from confronting it with the heuristics for the order of $\text{III}(E)$ proposed by Delaunay (see [D1] and his article in this volume [D3]), on the Cohen-Lenstra model. Indeed, his heuristics suggest that for $r \geqslant 0$, the proportion of d for which E_d has rank 0 and $\text{III}(E_d)[2] \simeq (\mathbf{Z}/2\mathbf{Z})^{2r}$ should be

$$f_r = 2^{-r(2r-1)} \prod_{j=1}^{r} \left(1 - \frac{1}{4^j}\right) \prod_{j \geqslant r+1} (1 - 2^{-2j+1})$$

(we use the fact that if $\text{III}(E)$ is finite, then $\text{III}(E)[2] \simeq \text{III}(E)/2\text{III}(E)$, which is the group really considered by Delaunay.)

A simple computation reveals that $f_r = e_{2r}$ in (2.9), therefore all twists with $\dim_{\mathbf{Z}/2\mathbf{Z}} \text{Sel}_2(E_d) = r + 2$ (even) are accounted for by curves of rank 0 if this heuristic is correct. Or, to state it another way: if Conjecture 2.1 holds, the heuristic for the 2-rank of $\text{III}(E_d)$ is correct. Obviously, such consistency is quite convincing. Note that for curves of rank 1 and r odd, one needs to alter a little bit the proposed heuristic of Delaunay (specifically, replace $M_u(f) = M^s_{u/2}(f)$ by $M_u(f) = M^s_u(f)$ in [D1, p. 195, Heuristic Assumption]) to get agreement, but this seems a reasonable change (the $u/2$ does not carry great evidence towards it).

When seeing this, Delaunay also noticed that Heath-Brown proved in [H1] that the average of $s_2(E_d)/4$ is equal to 3, both over curves with even rank and over curves with odd rank (note that the parity conjecture is proved by Monsky in the Appendix to [H2] for this family, so having even or odd rank translates to a congruence modulo 8 for d). Assuming Conjecture 2.1 in this case, this translates to statements on the average of $|\text{III}(E_d)[2]|$, namely it should be $= 3$ for rank 0 and $3/2$ for rank 1. This agrees with [D1, Example 7], again using M_u instead of $M_{u/2}$.

The methods of Heath-Brown have been partly generalized to the family of quadratic twists for a curve E with the 2-torsion points rational [Y2], but not in full generality. One can however study the order of vanishing using analytic techniques. The best results are due to Perelli and Pomykala [PP]. They prove the following:

Theorem 2.7. *Let $E/\mathbf{Q}$ be a fixed elliptic curve and consider the family $\mathcal{E}_E$ of quadratic twists by d coprime with $\mathfrak{f}(E)$. For any $\varepsilon > 0$ we have*

$$A_{\mathcal{E}_E}(T, \mathrm{rk}) \geqslant A_{\mathcal{E}_E}(T, v') \gg A_{\mathcal{E}_E}(T)^{1-\varepsilon} \tag{2.10}$$

for $T \geqslant 2$, and any $\varepsilon > 0$, the implied constant depending only on E and ε. Moreover we have

$$A_{\mathcal{E}_E}(T, \mathrm{ord}) = o(T \log T) \ as \ T \to +\infty. \tag{2.11}$$

Recall that (2.5) gives the trivial upper bound

$$A_{\mathcal{E}}(T, \mathrm{ord}) \ll (A + 1) A_{\mathcal{E}}(T)(\log T)$$

for any family $\mathcal{E}$ and $T \geqslant 2$, the implied constant depending on the family, so the gain in (2.11) is quite small. It seems very hard to improve this however, as it depends on a difficult large-sieve type inequality of Heath-Brown for real characters.

On GRH, Goldfeld [G] has proved that the order of vanishing is bounded on average. Also, Ono [On] currently has the best lower bound on the proportion of twists with $L(E_d) \neq 0$ (hence of twists with $\mathrm{rk}(E_d) = 0$):

$$A_{\mathcal{E}_E}(T, v) \gg A_{\mathcal{E}_E}(T)(\log T)^{c(E)-1}$$

for $T \geqslant 2$ and some constant $c(E) \geqslant 0$, the implied constant depending only on E. In some cases this has been improved to a positive proportion (see e.g. [Y1]).

The method of proof for (2.10) is based on computing the first and second moments ($A_{\mathcal{E}_E}(T, L')$ and $A_{\mathcal{E}_E}(T, |L'|^2)$) of $L'(E_d)$: Cauchy's inequality gives

$$A_{\mathcal{E}_E}(T, v') \geqslant \frac{A_{\mathcal{E}_E}(T, L')^2}{A_{\mathcal{E}_E}(T, |L'|^2)} \tag{2.12}$$

so it suffices to give a lower bound for the first moment $A_{\mathcal{E}_E}(T, L')$ and an upper bound for the second moment $A_{\mathcal{E}_E}(T, |L'|^2)$; this is done (as is the case for (2.11)) using the methods sketched in the next section. Note that the moment conjectures for mollified families of L-functions, if applicable in this case, immediately imply that the order of vanishing of the twists is bounded on average.

Consider now the case of an algebraic family $\mathcal{E}$. Basically the same methods used for quadratic twists are available, but the averaging is much more difficult to perform in general and currently the only non-trivial results are known under the additional assumption of the Generalized Riemann Hypothesis to study the L-functions. Then one can prove the following result:

Theorem 2.8. *Let $\mathcal{E}$ be an algebraic family of elliptic curves over $\mathbf{Q}$. Assume GRH and the Tate conjecture for the elliptic surface associated to $\mathcal{E}/\mathbf{Q}$. Then we have*

$$A_{\mathcal{E}}(T, \mathrm{ord}) \leqslant (\mathrm{rank}\,\mathcal{E}(\mathbf{Q}(t)) + \deg N_{\mathcal{E}} + \tfrac{1}{2})(1 + o(1))A_{\mathcal{E}}(T)$$

as $T \to +\infty$, where $\mathrm{rank}\,\mathcal{E}(\mathbf{Q}(t))$ is the rank of $\mathcal{E}$ as an elliptic curve over $\mathbf{Q}(t)$ and $N_{\mathcal{E}}$ is the conductor polynomial defined by

$$N_{\mathcal{E}} = \prod_{\Delta(t)=0} (X - t) \prod_{c_4(t)=c_6(t)=0} (X - t) \in \mathbf{Z}[X], \tag{2.13}$$

$c_4(t)$ and $c_6(t)$ being the usual invariants for the curves (2.3).

We will describe more precisely below the particular case of Tate's conjecture which is required. The theorem as stated is due to Silverman [S], building on earlier work of Fouvry and Pomykala [FP] and Michel [Mi] which established weaker or slightly different inequalities, all with the conclusion that the average order of vanishing of $\mathcal{E}$ is bounded, hence also the rank on B-SD. (Note however that the generic rank of $\mathcal{E}(\mathbf{Q}(t))$ arises in the bound independently of B-SD).

This average boundedness of the rank was also proved by Brumer [B] for the family $\mathcal{E}_H$ of all elliptic curves ordered by height. The best known result is due to M. Young [Yo1, Yo3]:

Theorem 2.9. *Let $\mathcal{E}_H$ be the family* (2.4)*. Assume GRH for L-functions of elliptic curves. Then we have*

$$A_{\mathcal{E}_H}(T, \mathrm{ord}) \leqslant (\tfrac{25}{14} + o(1))A_{\mathcal{E}_H}(T)$$

as $T \to +\infty$.

Brumer had the constant 2.3 instead of $25/14$, which had been improved to 2 by Heath-Brown [H3]. Young's result is significant because a constant which is strictly smaller than 2 implies that a positive proportion of the curves must have rank either 0 or 1.

2.5   Basic analytic tools

The analytic investigations of the L-functions of elliptic curves are based on two quite general formulas which go back in principle to Riemann and other early investigators of the Riemann zeta function. The first, called somewhat misleadingly "the approximate functional equation", is a convenient expression for L-functions in the critical strip where the Dirichlet series is not convergent. Here is one of many variants, which we state for modular forms, since it is not in any way specific to elliptic curves.

Proposition 2.10. *Let f be a primitive cusp form of weight 2 and conductor q with Fourier coefficients $\lambda_f(n)$ and root number $w(f)$. We have for $X \geqslant 1$*

$$L(f, 1) = \sum_n \frac{\lambda_f(n)}{n} \exp\left(-\frac{2\pi n}{X\sqrt{q}}\right) + w(f) \sum_n \frac{\overline{\lambda_f(n)}}{n} \exp\left(-\frac{2\pi n X}{\sqrt{q}}\right).$$

In particular if $E/\mathbf{Q}$ is an elliptic curve we have for any $X \geqslant 1$

$$L(E, 1) = \sum_n \frac{a_E(n)}{n} \exp\left(-\frac{2\pi n}{X\sqrt{\mathfrak{f}(E)}}\right) + w(E) \sum_n \frac{a_E(n)}{n} \exp\left(-\frac{2\pi n X}{\sqrt{\mathfrak{f}(E)}}\right).$$

For a proof see e.g. [IK, 5.2].

The second principle is of the same type, but applies to the logarithmic derivative of the L-function instead. It is (also misleadingly) called the "explicit formula". To state one of its variants, let f be again a modular form and define $\Lambda_f(n)$ by the Dirichlet series expansion

$$-\frac{L'}{L}(f, s) = \sum_{n \geqslant 1} \Lambda_f(n) n^{-s}$$

for $\sigma > 3/2$, so that Λ_f is supported on prime powers and

$$\Lambda_f(p^k) = (\alpha_p^k + \beta_p^k)(\log p)$$

where $\alpha_p \beta_p = p$ and $\alpha_p + \beta_p = \lambda_f(p)$ (at least for $(p, q) = 1$). In particular note that for $(p, q) = 1$ we have

$$\Lambda_f(p) = a_E(p)(\log p), \quad \Lambda_f(p^2) = (a_E(p^2) - 2p)(\log p). \tag{2.14}$$

Proposition 2.11. *Let f be a primitive cusp form of weight 2 and conductor q with Fourier coefficients $\lambda_f(n)$, and let η be a sufficiently smooth function on $]0, +\infty[$ with compact support such that $\eta(x^{-1}) = \eta(x)$. We have*

$$2 \sum_{n \geqslant 1} \frac{\Lambda_f(n)}{n} \eta(n) = \eta(1) \log q - \sum_\rho \hat{\eta}(\rho - 1)$$

$$+ \frac{1}{2i\pi} \int_{(1/2)} \left(\frac{\Gamma'}{\Gamma}(s) + \frac{\Gamma'}{\Gamma}(1 - s)\right) \hat{\eta}(s) ds$$

where ρ runs over zeros of $L(f, s)$ and $\hat{\eta}$ is the Mellin transform of η. In particular if $E/\mathbf{Q}$ is an elliptic curve we have

$$2 \sum_{n \geqslant 1} \frac{\Lambda_E(n)}{n} \eta(n) = \eta(1) \log \mathfrak{f}(E) - \operatorname{ord}(E)\hat{\eta}(0) - \sum_{\rho \neq 1} \hat{\eta}(\rho - 1)$$

$$+ \frac{1}{2i\pi} \int_{(1/2)} \left(\frac{\Gamma'}{\Gamma}(s) + \frac{\Gamma'}{\Gamma}(1 - s)\right) \hat{\eta}(s) ds.$$

For a proof, see e.g. [IK, 5.5]; note that in the second formula we have isolated the (possible) zero at $s = 1$ from the others.

Before going further, here are a few remarks on these two formulas. In the first one, a standard choice of the parameter X is $X = 1$, in which case the effective length of summation is essentially $n \leqslant \sqrt{\mathfrak{f}(E)}$ for both sums, after which the tails are very small. However, taking $X = \sqrt{\mathfrak{f}(E)}$ is also useful because, although it lengthens the first sum to $n \leqslant \mathfrak{f}(E)$ – which can make it unmanageable –, it gets rid of the second sum involving the root number. The latter can be very much of a problem. There are similar formulas for all moments $L^{(k)}(E)^m$ of all derivatives, with $a_E(n)$ replaced by the coefficients of the corresponding Dirichlet series, and the "length" becoming $\mathfrak{f}(E)^{m/2}$.

The second formula does not involve the root number, and partly for this reason it has been the most commonly used. But because it requires a certain control of the zeros of $L(E, s)$, in applications it has usually been used on the assumption that GRH holds. In this case, we have $\mathrm{Re}(\rho - 1) = 0$ and it is easy to choose a simple test function η for which $\hat{\eta}(s) \geqslant 0$ for all s with $\mathrm{Re}(s) = 0$. Because of the sign, this makes it suitable for *upper bounds* for $\mathrm{ord}(E)$, but not for lower bounds. As we'll see, in applications the support of η is such that n appears up to $\mathfrak{f}(E)^\kappa$ for some $\kappa > 0$ (a small κ can still be useful).

2.6 The Delta-symbol for a family

In any case, for both formulas, if one wishes to use them for families, performing the average over $E \in \mathcal{E}$ yields expressions for $A_{\mathcal{E}}(T, L)$ or $A_{\mathcal{E}}(T, \mathrm{ord})$ in terms of the following fundamental averages, called the Delta-symbols, or twisted Delta-symbols, of the family:

$$\Delta_T(n, m) = A_{\mathcal{E}}(T, a_E(n)a_E(m)) = \sum_{E \in \mathcal{E}(T)} a_E(n)a_E(m), \qquad (2.15)$$

$$\Delta_T^w(n, m) = A_{\mathcal{E}}(T, w(E)a_E(n)a_E(m)) = \sum_{E \in \mathcal{E}(T)} w(E)a_E(n)a_E(m). \qquad (2.16)$$

Precisely, for $A_{\mathcal{E}}(T, L)$, one gets $\Delta_T(n, 1)$ and $\Delta_T^w(n, 1)$, and for $A_{\mathcal{E}}(T, \mathrm{ord})$, one gets combinations of $\Delta_T(p^k, 1)$. The second parameter m is potentially significant if one could involve a mollifier (as has been done with $J_0(q)$); it can be dispensed with using the formula

$$a_E(n)a_E(m) = \sum_{\substack{d|(n,m) \\ (d,\mathfrak{f}(E))=1}} d a_E\left(\frac{nm}{d^2}\right),$$

but this is not necessarily the best arrangement.

The reason of the terminology is the heuristic that, for $n = m$, there is no cancellation in the sum and $\Delta_T(n, m)$ should be large. In fact, recalling the

Hasse bound $|a_E(n)| \leqslant d(n)\sqrt{n}$ and $w(E) = \pm 1$, we see that

$$|\Delta_T(n,m)| \leqslant \tau(n)\tau(m)\sqrt{nm}A_{\mathcal{E}}(T), \qquad (2.17)$$

$$|\Delta_T^w(n,m)| \leqslant \tau(n)\tau(m)\sqrt{nm}A_{\mathcal{E}}(T), \qquad (2.18)$$

and one would expect that $\Delta_T(n,n)$ is roughly of the order of magnitude of the bound above, from the Sato-Tate conjecture. On the other hand, if $n \neq m$ one can expect random sign changes in $a_E(n)a_E(m)$, $E \in \mathcal{E}$, that make the sum smaller than this trivial bound. This leads to expect some kind of approximate orthogonality, at least in certain ranges of n and m. Similar effects are easy to see for the analogue case of Dirichlet characters modulo q, where we have

$$\sum_{\chi \,(\mathrm{mod}\, q)} \chi(n)\overline{\chi(m)} = \varphi(q)\delta_q(n,m),$$

where $\delta_q(n,m) = 1$ if $n \equiv m \,(\mathrm{mod}\, q)$ and $(nm, q) = 1$, and $\delta_q(n,m) = 0$ otherwise, in particular if $n \neq m$ and $n, m \leqslant q$. Similarly, but not so easily, for the variety $J_0(q)$ one has the following easy consequence of the Petersson formula:

$$\sum_f \omega_f \lambda_f(n)\lambda_f(m) = \sqrt{mn}\,\delta(m,n) + O(mn(m,n,q)q^{-3/2})$$

(see e.g. [IK, 14]), where f runs over an orthogonal basis of the space of weight 2 cusp forms of level q, and $\omega_f^{-1} = 4\pi\langle f, f\rangle$, the Petersson norm of f, which is of size about q. Alternately, one can use the Selberg Trace Formula for the Delta symbol in this case (see [V]), with a slightly weaker estimate on the error term.

The simplest case for families of elliptic curves is that of the family $\mathcal{E}_E$ of quadratic twists E_d (with $(d, \mathfrak{f}(E)) = 1$) of a given E. In this case $a_{E_d}(n) = (\frac{d}{n})a_E(n)$ and $w(E_d)$ is given by (2.8) so

$$\Delta_T(n,m) = a_E(n)a_E(m)\sum_{|d| \leqslant T}^{\flat} \left(\frac{d}{nm}\right),$$

$$\Delta_T^w(n,m) = w(E)a_E(n)a_E(m)\sum_{|d| \leqslant T}^{\flat} \left(\frac{-d\,\mathfrak{f}(E)}{nm}\right)$$

where $\sum^{\flat}$ is the sum over the relevant d. Since $d \mapsto (\frac{d}{nm})$ is a Dirichlet character modulo nm which is trivial if and only if nm is a square, it is not hard to derive better individual bounds than (2.17), (2.18). However (partly because of the restrictions on d) to prove an asymptotic formula for, say, $A_{\mathcal{E}_E}(T, L')$, it is necessary to keep the Delta symbol in non-estimated form and perform some transformations in the ensuing sum over n. See [I1] for the details, which yield for instance

$$A_{\mathcal{E}_E}(T, L') = \alpha_1 A_{\mathcal{E}_E}(T)(\log T) + \alpha_2 A_{\mathcal{E}_E}(T) + O(T^{27/28})$$

as $T \to +\infty$, for some constants $\alpha_1 > 0$ and $\alpha_2 \geqslant 0$. This gives the lower bound required in (2.12).

In this situation, it is unimportant (for analytic purposes) that we are dealing with elliptic curves: all the necessary estimates are in fact valid for arbitrary primitive cusp forms of weight 2.

2.7 Sketch of proof of Theorem 2.8

We consider the situation of an algebraic family. In this case one uses the explicit formula for a test function of the type $\eta_\lambda(x) = \eta(x^{\lambda^{-1}})$, the parameter $\lambda \geqslant 1$ allowing to "localize" optimally the sum. The fixed test function η is compactly supported in $[e^{-1}, e]$, and such that $\eta(it) \geqslant 0$ for all $t \in \mathbf{R}$, so the sum over zeros is $\geqslant 0$ under GRH. (For instance, the triangle function $\eta(x) = \max(1 - |\log x|, 0)$ is commonly used.) Since $\hat{\eta}_\lambda(s) = \lambda\hat{\eta}(\lambda s)$, by positivity one derives from Proposition 2.11 that

$$\lambda \operatorname{ord}(E)\hat{\eta}(0) \leqslant \eta(1) \log \mathfrak{f}(E) - 2 \sum_n \frac{\Lambda_E(n)}{n}\eta(n^{\lambda^{-1}}) + (\text{arch}),$$

where (arch) designates the archimedean contribution of the Gamma function, which is easily handled because it is independent of t. Hence

$$\lambda\hat{\eta}(0)A_{\mathcal{E}}(T, \operatorname{ord}) \leqslant \eta(1)A_{\mathcal{E}}(T, \mathfrak{f}) - 2 \sum_n \eta(n^{\lambda^{-1}})A_{\mathcal{E}}\left(T, \frac{\Lambda_E(n)}{n}\right) + O(A_{\mathcal{E}}(T))$$

$$(2.19)$$

for $T \geqslant 2$, the implied constant depending on the family. The average of the conductor is easily handled by (2.2) which holds with $A = \deg \Delta$, where Δ is the discriminant polynomial, or with $A = \deg N_{\mathcal{E}}$ where $N_{\mathcal{E}}$ is the conductor polynomial (2.13):

$$A_{\mathcal{E}}(T, \mathfrak{f}) \leqslant (\deg N_{\mathcal{E}})A_{\mathcal{E}}(T)(\log T)(1 + o(1))$$

as $T \to +\infty$

The last sum can be expressed in terms of the Delta symbols for $n = p^k$. Those are given by

$$\Delta_T(p^k, 1) = \sum_{|t| \leqslant T} a_{E_t}(p^k).$$

The individual Hasse bound is sufficient to treat all $k \geqslant 3$ to show that for $T \geqslant 2$ we have

$$\sum_{\substack{n=p^k \\ k \geqslant 3}} \Lambda_E(n)n^{-1}\eta(n^{\lambda^{-1}}) \ll A_{\mathcal{E}}(T).$$

For the remaining values of k, the main observation is that $t \mapsto a_{E_t}(p^k)$ is periodic of period p, except for some innocuous problems when $p \mid \Delta(t)$. Thus one is led to study the local average

$$\mathcal{A}^k_{\mathcal{E}}(p) = \frac{1}{p} \sum_{\substack{t \,(\mathrm{mod}\,p) \\ \Delta(t)\neq 0}} a_{E_t}(p^k), \qquad (2.20)$$

and apart from boundary terms and those t where $p \mid \Delta(t)$ we have

$$\Delta_T(p^k, 1) \simeq \mathcal{A}^k_{\mathcal{E}}(p) A_{\mathcal{E}}(T). \qquad (2.21)$$

The treatments of Fouvry-Pomykala, Michel and Silverman diverge at this point. Before going further, we remark that this shows that (except if $\mathcal{A}^k_{\mathcal{E}}(p) = 0$) there is in fact no cancellation in $\Delta_T(p^k, 1)$ as T grows, for fixed p and k. (Contrast with the Petersson formula). So improving the results below require a non-trivial treatment of the sum over p afterward, which is quite difficult (Young [Yo1], for instance, did succeed in exploiting this sum over p).

Chronologically, Fouvry and Pomykala used the trivial bound for $k = 2$, with a contribution of size about $\lambda A_{\mathcal{E}}(T)(1 + o(1))$ to (2.19). For $k = 1$ they use the character sum expression (1.8) for $a_E(p)$ (if E has good reduction at $p \geqslant 5$)

$$a_E(p) = - \sum_{x \,(\mathrm{mod}\,p)} \left(\frac{f_t(x)}{p} \right)$$

if E_t is put in Weierstrass form

$$E_t \ : \ y^2 = x^3 - 27c_4(t)x - 54c_6(t) = f_t(x),$$

yielding an expression for $\mathcal{A}_{\mathcal{E}}(p) = \mathcal{A}^1_{\mathcal{E}}(p)$ as a two-variable character sum

$$\mathcal{A}_{\mathcal{E}}(p) = -\frac{1}{p} \sum_{x,t} \sum \left(\frac{x^3 - 27c_4(t)x - 54c_6(t)}{p} \right).$$

One expects, at least generically, square-root cancellation in this sum, i.e. that this expression should be bounded. Fouvry and Pomykala prove this under some genericity assumptions by invoking general bounds of Adolphson and Sperber.

On the other hand, Michel sees the estimation of $\mathcal{A}_{\mathcal{E}}(p)$ as a problem about a one-variable sum of local traces of Frobenius acting on the ℓ-adic sheaf $\mathcal{F}$ of rank 2 whose local traces at a point of $U = \{t \in \mathbf{Z}/p\mathbf{Z} \mid \Delta(t) \neq 0\}$ are $a_{E_t}(p)$, namely

$$\mathcal{F} = R_1 \pi_! \mathbf{Q}_\ell$$

where $\pi : \mathcal{E} \to \mathbf{P}^1$ is the morphism defining the algebraic surface $\mathcal{E}$ (modulo p). Estimating $\mathcal{A}_{\mathcal{E}}(p)$ reduces to the computation of the cohomology groups of $\mathcal{F}$, and Michel treats this with the same kind of arguments that Katz [K] used to prove (for instance) the vertical Sato-Tate distribution for Kloosterman sums. The only assumption on $\mathcal{E}$ that remains necessary (in order that the required monodromy group be as large as possible) is that the family be non-constant, i.e. the polynomial $j(t)$ is not constant (modulo p), which excludes only finitely many p if $j(t) \in \mathbf{Z}[t]$ is not constant, and the estimate obtained is:

Proposition 2.12. *Let $\mathcal{E}$ be a non-constant algebraic family of elliptic curves modulo p. We have*

$$|\mathcal{A}_{\mathcal{E}}^{k}(p)| \leqslant (k+1)(\deg \Delta(t) - 1)p^{(k-1)/2}.$$

and if $k = 1$ one can replace $2(\deg \Delta(t) - 1)$ by $\deg N_{\mathcal{E}}$.

Finally, Silverman handles the case $k = 2$ as Michel did, except that he shows that the bound he obtained remains valid for a constant family (using Rankin-Selberg convolution). For $k = 1$, he does not use character sums to handle $\mathcal{A}_{\mathcal{E}}(p)$ but instead a formula conjectured by Nagao and proved by Rosen and himself [RoS], namely

$$\sum_{p \leqslant x} \mathcal{A}_{\mathcal{E}}(p)(\log p) \sim -\operatorname{rank} \mathcal{E}(\mathbf{Q}(t))x,$$

as $x \to +\infty$, under the assumption that the Tate Conjecture holds for $\mathcal{E}/\mathbf{Q}$, i.e. that the order of the pole at $s = 2$ (on the edge of the region of absolute convergence) of the L-function attached to $H^2(\mathcal{E}/\bar{\mathbf{Q}}, \mathbf{Q}_\ell)$ is equal to the rank of the Néron-Severi group of $\mathcal{E}/\mathbf{Q}$. Thus, summing the contribution of $n = p$ in (2.19) yields a contribution which is

$$-\frac{\lambda}{2}(1 + o(1))(\operatorname{rank} \mathcal{E}(\mathbf{Q}(t)))A_{\mathcal{E}}(T)$$

by summation by parts (using the fact that $p \leqslant e^\lambda$).

The outcome of all this (and a correct treatment of boundary terms) is the estimate

$$\lambda \hat{\eta}(0)A_{\mathcal{E}}(T, \operatorname{ord}) \leqslant (1 + o(1))A_{\mathcal{E}}(T)\Big\{ (\deg N_{\mathcal{E}})(\log T) + \frac{\lambda}{2}(\operatorname{rank} \mathcal{E}(\mathbf{Q}(t)))$$
$$+ \frac{\lambda}{2} + O(1) + O(\lambda e^\lambda)\Big\}.$$

Taking λ slightly smaller than $\log T$ gives the desired average bound.

An interesting point is that the periodicity of $t \mapsto a_{E_t}(p)$ means that in fact the family can be restricted to certain parameters t as long as they are very well-distributed in arithmetic progressions. For instance, one checks easily that using

$$\pi(x; q, a) = \frac{\operatorname{li}(x)}{\varphi(q)} + O(x^{1/2}(\log x))$$

which follows from GRH for Dirichlet L-functions, one can deduce that the average rank is still bounded for the family $\mathcal{E}_p$ which restricts t to prime values:

$$\mathcal{E}_p(T) = \{E_t \mid 1 \leqslant t \leqslant T \text{ and } t \text{ is prime}\}.$$

There are (at least) two interesting problems arising out of this proof. The first is to remove the dependence on GRH; this has been done for quadratic twists, but at the cost of getting only (2.11) instead of boundedness of the average rank. No other instance is known. (Even Brumer's family of curves indexed by height, which seems the most accessible, remains out of reach; see however the recent work of Young [Yo2] concerning the proportion of non-vanishing). It should be noted that in the case of $J_0(q)$, this removal requires much more delicate analysis in the form of density theorems for the possible zeros of L-functions close to the critical line, see [KM2].

The other possibility is to improve on the average rank. The argument shows that the only possible way to do so is to treat non-trivially the sum over the zeros in the explicit formula to get some cancellation with the other terms (especially with the $\Lambda_E(p^2)$ contributions). This is closely related with the issues surrounding the predictions of random matrix theory discussed previously, and has been done quite successfully by Iwaniec, Luo and Sarnak [ILS] for $J_0(q)$ and other families of automorphic L-functions. Some results for algebraic families are due to M. Rubinstein [R2], S. Miller [Mil1] and M. Young [Yo1], confirming the 1-level and 2-level densities (the latter does distinguish between various symmetry groups).

2.8   Some discussion of numerical evidence

Due to the large choice of well-documented and well-implemented algorithms for computing with elliptic curves[7], particularly over $\mathbf{Q}$, many of the conjectures concerning them can be put to the test, and in particular those concerning the rank. This has led to the controversial problem of "excess rank", as a number of experiments with seemingly innocuous families revealed a fairly large proportion of curves with rank $\geqslant 2$ (see e.g. [KZ], [Fe] or [B, 1st paragraph]). Since an occurrence of excess rank for an infinite family (not chosen in a special way to have large rank) would put into question the general Katz-Sarnak philosophy, there is a certain agreement that the data available simply reflects a problem with the size of the sample.

The most convincing theoretical argument against excess rank maybe follows by doing the numerical experiments with the 2-rank of the Selmer group for the congruence number curves, and comparing with the results of Heath-Brown, since those are unconditional. Of course, the Selmer group could have a very different behavior, but as explained in [H2, p. 336], small tests tend to reveal an "excess rank" in this case too, which has to disappear in the long run... Moreover, discussing his proof, Heath-Brown indicates one possible explanation for a very slow convergence towards the limiting distribution: in his arguments, the k-th moment of $s_2(E_d)$ must be averaged over numbers having at least 16^k prime factors before it gets close to the asymptotic value!

Another very simple type of experiments which has not been widely per-

---

formed is that of computing the Delta symbols for some families of elliptic curves. This is much faster, of course, than computing the rank exactly, but it can in theory be quite useful if one compares the results for two families, one of which is – if possible – well-understood. As indicated in Section 2.5, for a number of analytical arguments it is the quasi-orthogonality of $\Delta_T(m, n)$ which is at the heart of a successful average study of special values of L-functions. Even if one family is inaccessible, a behavior of the Delta symbol similar to that of another would be good indication that the rank might also behave in a similar way.

With this in mind, we performed the computation for the most mysterious family $\mathcal{E}_c$, that of elliptic curves indexed by the conductor. For this we used Cremona's table, which currently lists all 845960 elliptic curves over $\mathbf{Q}$ with conductor $< T = 130000$, up to isomorphism. (The table also contains the rank, for which the distribution is: rank 0, 340655 curves, rank 1, 427012 curves, rank 2, 77357 curves, rank 3, 936 curves, and no rank $\geqslant 4$). We limited the computation to $\Delta(p, 1)$ where p is a prime $p \leqslant \sqrt{T}(\log T) < 4246$, since according to the approximate functional equation those are sufficient to recover $L(E, 1)$, and we also computing the twisted Delta-symbols $\Delta_T^w(p, 1)$ as well as the sum $\frac{1}{2}(\Delta_T(p, 1) + \Delta_T^w(p, 1))$.

The following graphs show for instance the Delta symbols for $p = 61$ and $p = 797$; the horizontal axis is the number of curves counted up to x, i.e. $A_{\mathcal{E}_c}(x)$, $x \leqslant T$.

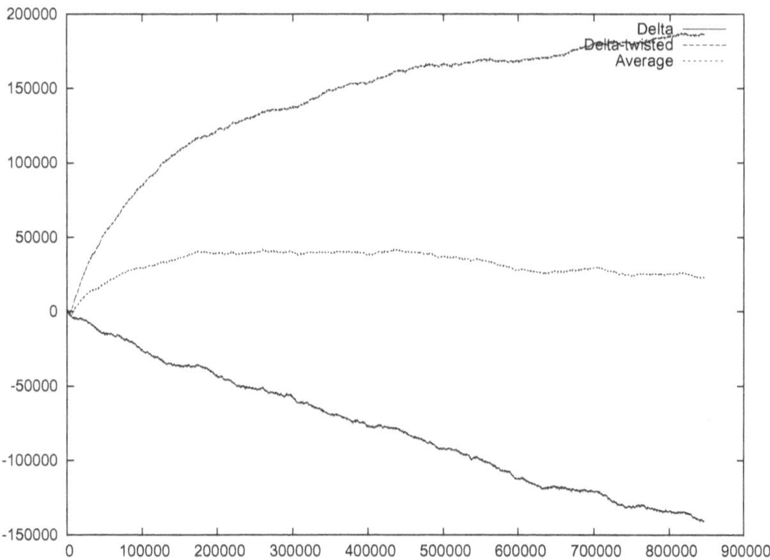

Figure 1: Delta symbols for $p = 61$

Notice that, as far as $\Delta_T(p, 1)$ is concerned, in the first case $(p = 61)$ one has a curve very close to linear, which suggests strongly an effect of periodicity

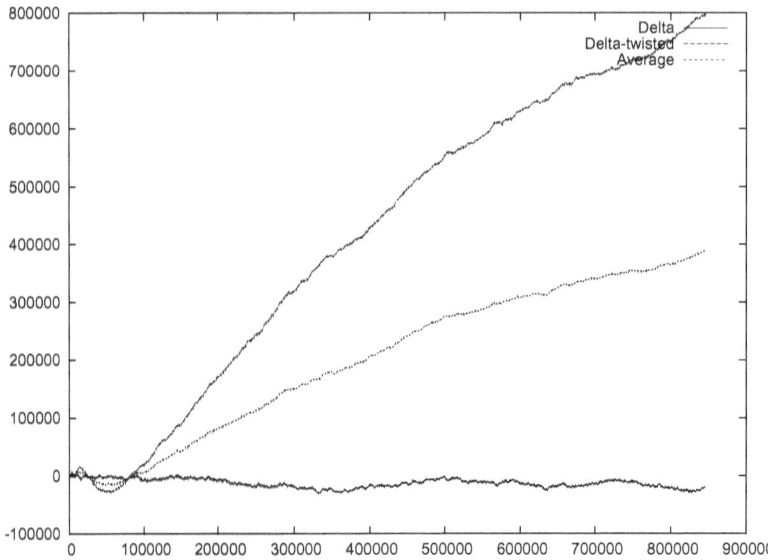

Figure 2: Delta symbols for $p = 797$

like the one that does happen for the Delta symbol of an algebraic family of elliptic curves. This might seem natural and taken to be the reflection of an equidistribution statement for the reductions of curves $E \in \mathcal{E}_c(T)$ modulo p as $T \to +\infty$, among the elliptic curves modulo p, except that the natural linear factor to expect for $\Delta_T(p, 1)$ would be (compare (2.20) and (2.21))

$$\mathcal{A}_c(p) = \frac{1}{p} \sum_{E \,(\mathrm{mod}\, p)} a_E(p) = 0,$$

where the sum is over all elliptic curves modulo p, so quadratic twists with a_p, $-a_p$ cancel out. This does not fit the data available at all!

On the other hand, for $p = 797$, the Delta symbol behaves much more randomly. It should be mentioned that those are the two most common aspects of the graphs for the primes considered, with the first case being most common, and maybe more significantly, with the "linear" curves having negative slopes. Note this means a tendency to have $a_E(p) < 0$, which increases the number of points modulo p, something which is often an indication of a larger rank.

In both cases, however (again this is typical) the introduction of the root number changes the picture completely: in the first case, there seems to be a correlation between $a_E(p)$ negative and $w(E) = -1$ (not too surprising in fact, by the above remark), so that the twisted Delta symbol seems to be increasing (the curve does not look quite as linear). For $p = 797$, the twisted symbol first seems to oscillate, but then has a marked tendency to increase.

All in all, these graphs look very mysterious to me. I think however that the Delta symbol deserves better scrutiny, both numerical and theoretical.

References

[Ba]   A. Baker: *Transcendental number theory*, Cambridge Math. Library, Cambridge Univ. Press (1975).

[B]   A. Brumer: *The average rank of elliptic curves, I*, Invent. math. 109 (1992), 445–472.

[CFKRS]   B. Conrey, D. Farmer, J.P. Keating, M. Rubinstein and N. Snaith: *Integral moments of L-functions*, Proc. Lond. Math. Soc., 91 (2005), 33–104.

[CKRS]   B. Conrey, J.P. Keating, M. Rubinstein and N. Snaith: *On the frequency of vanishing of quadratic twists of modular L-functions*, in Number theory for the millennium, I (Urbana, IL, 2000), 301–315, A K Peters (2002).

[DFK]   C. David, J. Fearnley and H. Kisilevsky: *Vanishing of L-functions of elliptic curves over number fields*, in this volume.

[D1]   C. Delaunay: *Heuristics on Tate-Shafarevitch groups of elliptic curves defined over* $\mathbf{Q}$, Exper. Math. 10 (2001), 191–196.

[D2]   C. Delaunay: *Moments of the orders of Tate-Shafarevitch groups*, International J. of Number Theory, Vol. 1, No. 2 (2005) 243–264.

[D3]   C. Delaunay: *Heuristics on class groups and on Tate-Shafarevich groups*, in this volume.

[Fa]   D. Farmer: *Modeling families of L-functions*, in this volume.

[Fe]   S. Fermigier: *Étude expérimentale du rang de familles de courbes elliptiques sur* $\mathbf{Q}$, Exper. Math. 5 (1996), 119–130.

[FP]   É. Fouvry and J. Pomykala: *Rang des courbes elliptiques et sommes d'exponentielles*, Mh. Math. 116 (1993), 111–125.

[G]   D. Goldfeld: *Conjectures on elliptic curves over quadratic fields*, Number Theory (Carbondale 1979), Springer Lecture Notes Math. 751 (1979), 108–118.

[H1]   D.R. Heath-Brown: *The size of the Selmer group for the congruent number problem, I*, Invent. math. 111 (1993), 171–195.

[H2]   D.R. Heath-Brown: *The size of the Selmer group for the congruent number problem, II*, Invent. math. 118 (1994), 331–370.

[H3]   D.R. Heath-Brown: *The average analytic rank of elliptic curves*, Duke Math. J. 122 (2004), no. 3, 591–623.

[H4]   D.R. Heath-Brown: *A Note on the 2-Part of* III *for the Congruent Number Curves*, in this volume.

[He]   H.A. Helfgott: *Root numbers and the parity problem*, PhD thesis, Princeton University (2003).

[I1]   H. Iwaniec: *On the order of vanishing of modular L-functions at the critical point*, Sém. Théor. Nombres Bordeaux 2 (1990), 365–376.

[I2]   H. Iwaniec: *Topics in classical automorphic forms*, Grad. Studies in Math. 17, A.M.S (1997).

[IK]   H. Iwaniec and E. Kowalski: *Analytic Number Theory*, A.M.S Colloquium Publications, vol 53 (2004).

[ILS]   H. Iwaniec, W. Luo and P. Sarnak: *Low -lying zeros of families of L-functions*, Inst. Hautes Études Sci. Publ. Math. 91 (2001), 55–131.

[K]   N. Katz: *Gauss sums, Kloosterman sums and monodromy groups*, Princeton Univ. Press (1988).

[KS1]   N. Katz and P. Sarnak: *Random matrices, Frobenius eigenvalues, and monodromy*, A.M.S Colloquium Publications, vol. 45 (1999).

[KS2]   N. Katz and P. Sarnak: *Zeroes of zeta functions and symmetry*, Bull. Amer. Math. Soc. 36 (1999), 1–26.

[K]   N. Koblitz: *Introduction to elliptic curves and modular forms*, G.T.M 97.

[KM1]   E. Kowalski and P. Michel: *A lower bound for the rank of $J_0(q)$*, Acta Arith. 94 (2000), 303–343.

[KM2]   E. Kowalski and P. Michel: *The analytic rank of $J_0(q)$ and zeros of automorphic L-functions*, Duke Math. J. 100 (1999), 503–542.

[KMV]   E. Kowalski, P. Michel and J. VanderKam: *Non-vanishing of high derivatives of automorphic L-functions at the center of the critical strip*, J. reine angew. Math. 526 (2000), 1–34.

[KZ]   G. Kramarz and D. Zagier: *Numerical investigations related to the L-series of certain elliptic curves*, J. Indian Math. Soc., New Ser. 52 (1987), 51–60.

[Ma]   B. Mazur: *On the passage from local to global in number theory*, Bull. A.M.S 29 (1993), 14–50.

[M]   J-F. Mestre: *Formules explicites et minorations de conducteurs de variétés algébriques*, Compositio Math. 58 (1986), 209–232.

[Mi]   P. Michel: *Rang moyen de familles de courbes elliptiques et lois de Sato-Tate*, Mh. Math. 120 (1995), 127–136.

[Mil1]   S. Miller: *1- and 2-Level Densities for Rational Families of Elliptic Curves: Evidence for the Underlying Group Symmetries*, Compositio Math. 140 (2004), 952–992.

[Mil2]   S. Miller: *Investigations of zeros near the central point of elliptic curve L-functions*, preprint (2005), `arXiv:math.NT/0508150`

[N]   J. Nekovàr: *On the parity of ranks of Selmer groups, II*, C. R. Acad. Sci. Paris Sér. I Math. 332 (2001), 99–104.

[O]   A. Odlyzko: *The 10^{20}-th zero of the Riemann zeta function and 70 million of its neighbors*, ATT Bell Laboratories preprint, 1989.

[On]   K. Ono: *Nonvanishing of quadratic twists of modular L-functions and applications to elliptic curves*, J. reine angew. Math. 533 (2001), 81–97.

[Pari]   PARI/GP, version 2.2.?, Bordeaux, 2005, `http://pari.math.u-bordeaux.fr/`.

[PP]   A. Perelli and J. Pomykala: *Averages over twisted elliptic L-functions*, Acta Arith. 80 (1997), 149–163.

[RoS]   M. Rosen and J. Silverman: *On the rank of an elliptic surface*, Invent. math. 133 (1998), 43–67.

[R1]   M. Rubinstein: *Evidence for a spectral interpretation of the zeros of L-functions*, PhD thesis, Princeton University (1998).

[R2]   M. Rubinstein: *Low-lying zeros of L-functions and random matrix theory*, Duke Math. J. 109 (2001), 147–181.

[RS]   Z. Rudnick and P. Sarnak: *Zeros of principal L-functions and random matrix theory*, Duke Math. J. 81 (1996), 269–322.

[AEC]   J. Silverman: *The arithmetic of elliptic curves*, Springer G.T.M 106.

[S]   J. Silverman: *The average rank of an algebraic family of elliptic curves*, J. reine angew. Math. 504 (1998), 227–236.

[Sn]   N. Snaith: *Derivatives of random matrix characteristic polynomials with applications to elliptic curves*, preprint (2005), `arXiv:math.NT/0508256`.

[Sw]   P. Swinnerton-Dyer: *2-Descent Through the Ages*, in this volume.

[U1]   D. Ulmer: *Elliptic curves with large rank over function fields*, Ann. of Math. (2) 155 (2002), 295–315.

[U2]   D. Ulmer: *Functions fields and random matrices*, in this volume.

[V]   J. VanderKam: *The rank of quotients of $J_0(N)$*, Duke Math. J. 97 (1999), 545–577.

[Yo1]   M. Young: *Low-lying zeros of families of elliptic curves*, Journal of the A. M. S., 19 (1) (2006), 205–250.

[Yo2]   M. Young: *On the non-vanishing of elliptic curve L-functions at the central point*, to appear in Proc. London Math. Soc.

[Yo3]   M. Young: *Analytic number theory and ranks of elliptic curves*, in this volume.

[Yo4]   M. Young: *Moments of the critical values of families of elliptic curves, with applications*, preprint (2005).

[Y1]   G. Yu: *Rank* 0 *Quadratic Twists of a Family of Elliptic Curves*, Compositio Mathematica 135 (2003), 331–356.

[Y2]   G. Yu: *On the quadratic twists of a family of elliptic curves*, preprint.

Université Bordeaux I - A2X
351, cours de la Libération
33405 Talence Cedex
France

emmanuel.kowalski@math.u-bordeaux1.fr

Modeling families of L-functions

David W Farmer[*]

Abstract

We discuss the idea of a "family of L-functions" and describe various methods which have been used to make predictions about L-function families. The methods involve a mixture of random matrix theory and heuristics from number theory. Particular attention is paid to families of elliptic curve L-functions. We describe two random matrix models for elliptic curve families: the Independent Model and the Interaction Model.

1   Introduction

Using ensembles of random matrices to model the statistical properties of a family of L-functions has led to a wealth of interesting conjectures and results in number theory. In this paper we survey recent results in the hopes of conveying our best current answers to these questions:

1. What is a family of L-functions?

2. How do we model a family of L-functions?

3. What properties of the family can the model predict?

In the remainder of this section we briefly review some commonly studied families and describe some of the properties which have been modeled using ideas from random matrix theory. In Section 2 we provide a definition of "family of L-functions" which has been successful in permitting precise conjectures, and we briefly describe how to model such a family. In Section 3 we discuss families of elliptic curve L-functions and show that there is an additional subtlety which requires us to slightly broaden the class of random matrix models we use. Then in Section 4 we discuss how to go beyond the leading-order terms which random matrix theory can model, and how one can avoid using random matrix theory when modeling a family of L-functions.

If you only care about elliptic curves and their L-functions you can safely skip to Section 3.

I thank Brian Conrey, Nina Snaith, Matt Young, and Steven J. Miller for many helpful conversations.

[*]Work supported by the American Institute of Mathematics and by the Focused Research Group grant (0244660) from the NSF. This paper is an expanded version of a talk given at the workshop "Clay Mathematics Institute Special week on Ranks of Elliptic Curves and Random Matrix Theory" held at the Isaac Newton Institute, February 2004.

1.1   A quick history of families

The idea that collectively the zeros of a single L-function behave in a manner that can be modeled statistically started with Montgomery's [Mon] work on the pair correlation of zeros of the zeta-function. Combined with the large-scale numerical calculations of Odlyzko [Odl], this provided convincing evidence that, to leading order, the local statistics of the zeros of the zeta-function, suitably rescaled, were the same as those of large random unitary matrices.

A similar collective behavior was noted long ago and is termed the "q-analogue" for Dirichlet L-functions. That is, results about all Dirichlet L-functions mod q look just like results for the Riemann ζ-function in t-aspect. For example, the formulas for moments of $|\zeta(\frac{1}{2}+it)|$ and $|L(s,\chi)|$ are identical but for replacing t by q. Another early example of collective behavior is the pair correlation of zeros of quadratic Dirichlet L-functions [OS1]. Clearly something interesting is going on.

The idea of a family of L-functions with an associated symmetry type began with the work of Katz and Sarnak [KSa]. They consider families of function field L-functions, where in this case a "family" of L-functions is the set of L-functions associated to a set of curves having certain properties. Here the collection of curves must be "natural" in the sense that the monodromy group of the family ties it all together. They show that to leading order the statistics of the (normalized) zeros of these L-functions, when averaged over the family, are the same as the statistics of the (normalized) eigenvalues of random matrices chosen from a classical compact group. Here the matrices are chosen uniformly with respect to Haar measure, and the size of the matrices scales with the conductor of the L-function.

For global L-functions there does not (yet?) exist an analogue of "monodromy", but it still has been found that to naturally occurring families one can associate a classical group of matrices. The zeros of the L-functions have, to leading order, the same statistics as the eigenvalues of a randomly chosen matrix from the group. And the appropriately rescaled critical values of the L-functions have, to leading order, the same distribution as the "critical values" of the characteristic polynomials of the matrices from the group, chosen uniformly with respect to Haar measure. The correspondence between the matrices and the L-functions involves equating the eigenvalue spacing with the zero spacing, or equivalently setting the matrix size equal to the conductor of the L-function. See Section 2.4 for more details.

After the work of Katz and Sarnak there quickly appeared many examples of L-functions families behaving in a manner predicted by random matrix theory. Some of the families considered were: L-functions associated to holomorphic cusp forms (in either weight or level aspect); Dirichlet L-functions (either all or quadratic); and various twists or symmetric powers of L-functions. Low-lying zeros were considered by Iwaniec, Luo, and Sarnak [ILS], Rubinstein [Rub], Özlük and Snyder [OS2], and others. Moments were considered by Iwaniec and Sarnak [IS], Kowalski, Michel, and VanderKam [KMV],

Soundararajan [S], and others.

There is an important distinction between the predictions for zeros of L-functions and the predictions for moments. For the zeros there is a natural way to normalize: rescale so that the average spacing is 1. This rescaling involves the conductor of the L-function and the degree of the characteristic polynomial, and this is the source of the principle that one chooses the size of the random matrix to equal the (logarithm of the) conductor of the L-function. With these normalizations one obtains accurate predictions for the leading-order behavior of statistics of zeros of L-functions. However, in addition to its zeros a polynomial is determined by an overall scale factor. So one may hope to use the characteristic polynomials to model the L-functions, but there will be a correction factor that does not come from random matrix theory. The use of the characteristic polynomial to model the L-function was begun by Keating and Snaith [KS1]. The first situation which was well understood is the moments of L-functions, for which there are explicit predictions for the arithmetic scale factor. See [CG, CF, KS2].

2 What is a family?

There does not yet exist an adequate definition of "family of L-functions". An attempt is made in [CFKRS] to define a family axiomatically, and we will describe that definition here. In that definition the axioms are chosen so that it is possible to produce a plausible conjecture for the critical moments of the family.

Complete details and many examples are in [CFKRS], so we will just highlight the key features of that definition of a family. The main idea is that one starts with a fixed L-function and a family of "characters", and the family of L-functions is produced by twisting the fixed L-function by the family of characters. Note that here the term *character* is used to cover more general classes of functions than just Dirichlet characters.

2.1 *L*-functions

We wish to define a "family of L-functions", so first we have to give the definition we will use for "L-function". The definition of an L-function which we give below is slightly different than what is known as the "Selberg class," but it is conjectured that the two are in fact equal.

Let $s = \sigma + it$ with σ and t real. An *L-function* is a Dirichlet series

$$L(s) = \sum_{n=1}^{\infty} \frac{a_n}{n^s}, \tag{2.1}$$

with $a_n = O_\varepsilon(n^\varepsilon)$ for every $\varepsilon > 0$, which has three additional properties.

Analytic continuation: $L(s)$ continues to a meromorphic function of finite order with at most finitely many poles, and all poles are located on the $\sigma = 1$ line.

Functional equation: There is a number ε with $|\varepsilon| = 1$, and a function $\gamma_L(s)$ of the form

$$\gamma_L(s) = P(s)Q^s \prod_{j=1}^{w} \Gamma(\tfrac{1}{2}s + \mu_j), \tag{2.2}$$

where $Q > 0$, $\Re\mu_j \geq 0$, and P is a polynomial whose only zeros in $\sigma > 0$ are at the poles of $L(s)$, such that

$$\xi_L(s) := \gamma_L(s)L(s) \tag{2.3}$$

is entire, and

$$\xi_L(s) = \varepsilon\overline{\xi_L}(1 - s), \tag{2.4}$$

where $\overline{\xi_L}(s) := \overline{\xi_L(\bar{s})}$ and $\bar{s}$ denotes the complex conjugate of s .

The number w is called the *degree* of the L-function. That number will also appear in the Euler product.

Euler product: For $\sigma > 1$ we have

$$L(s) = \prod_{p} \prod_{j=1}^{w} (1 - \gamma_{p,j}p^{-s})^{-1}, \tag{2.5}$$

where the product is over the primes p, and each $|\gamma_{p,j}|$ equals 1 or 0.

Note that $L(s) \equiv 1$ is the only constant L-function, the set of L-functions is closed under products, and if $L(s)$ is an L-function then so is $L(s + iy)$ for any real y. An L-function is called *primitive* if it cannot be written as a nontrivial product of L-functions. Throughout this paper we assume all L-functions are primitive, although we usually omit the word "primitive."

Conductor: Associated to an L-function is its *conductor*, a number which measures the "size" of the L-function. The paper [CFKRS] introduced a refined notion of conductor which, to leading order, is the logarithm of the usual notion of conductor. The refined conductor is necessary in order to have any hope of conjecturing the full main term in a general mean value of the L-function. Write the functional equation in asymmetric form:

$$L(s) = \varepsilon X_L(s)\overline{L}(1 - s), \tag{2.6}$$

where $X_L(s) = \dfrac{\overline{\gamma_L}(1 - s)}{\gamma_L(s)}$. Then the refined conductor of $L(s)$, denoted $c(L)$, is given by $c(L) = |X'_L(\tfrac{1}{2})|$.

2.2 Families of characters

By a family of characters we mean a collection of arithmetic functions $\mathcal{F}$, where each $f \in \mathcal{F}$ is a sequence $f(1) = 1$, $f(2) = a_{2,f}$, $f(3) = a_{3,f}, \ldots$ whose generating function

$$L_f(s) = \sum_{n=1}^{\infty} \frac{a_{n,f}}{n^s} = \prod_{p} \prod_{j=1}^{v} (1 - \beta_{p,j}p^{-s})^{-1} \tag{2.7}$$

is a (primitive) L-function such that the collection $\{L_f : f \in \mathcal{F}\}$ has some nice properties. If we order the L-functions L_f by conductor $c(f)$, then the data $\{Q; \mu_1, \ldots, \mu_w\}$ in the functional equation of L_f should be monotonic functions of the conductor, and the counting function $M(X) := \#\{f \in \mathcal{F} \mid c(f) \leq X\}$ should be nice. The final condition on the family of characters is the existence of an orthogonality relation among the $f \in \mathcal{F}$. Specifically, we require that if $m_1, \ldots, m_k$ are integers then the average

$$\delta_\ell(m_1, \ldots, m_k) := \lim_{X \to \infty} M(X)^{-1} \sum_{\substack{f \in \mathcal{F} \\ c(f) \leq X}} f(m_1) \ldots f(m_\ell)\overline{f(m_{\ell+1}) \ldots f(m_k)} \tag{2.8}$$

exist and be multiplicative. That is, if $(m_1 m_2 \ldots m_k, n_1 n_2 \ldots n_k) = 1$, then

$$\delta_\ell(m_1 n_1, m_2 n_2, \ldots, m_k n_k) = \delta_\ell(m_1, \ldots, m_k)\delta_\ell(n_1, \ldots, n_k). \tag{2.9}$$

See Section 3.1 of [CFKRS] for more details.

2.3 Families of L-functions

Now we create a family of L-functions by starting with a fixed L-function

$$L_g(s) = \sum_{n=1}^{\infty} \frac{a_{n,g}}{n^s} = \prod_p \prod_{j=1}^{w} (1 - \gamma_{p,j} p^{-s})^{-1}. \tag{2.10}$$

Then the elements of our L-function family $\mathcal{L}(\mathcal{F})$ are the Rankin-Selberg convolutions

$$\mathcal{L}(s, f) = L_{f \times g}(s) = \prod_p \prod_{i=1}^{v} \prod_{j=1}^{w} (1 - \beta_{p,i}\gamma_{p,j} p^{-s})^{-1}$$
$$= \sum_{n=1}^{\infty} \frac{a_{n,f \times g}}{n^s}. \tag{2.11}$$

(There may be some issues with the local factors at the bad primes). Note that if $w = 1$ or $v = 1$ then

$$\mathcal{L}(s, f) = \sum_{n=1}^{\infty} \frac{a_{n,f} a_{n,g}}{n^s}. \tag{2.12}$$

And in particular if L_g is the Riemann zeta-function, then $\mathcal{L}(s, f) = L_f(s)$.

The point of this definition of "family" is that the axioms provide the necessary ingredients to apply the recipe in [CFKRS] to conjecture the full main term in the shifted Kth moment

$$M(X)^{-1} \sum_{c(f) \leq X} \prod_{1 \leq k \leq K} L(\tfrac{1}{2} + \alpha_k), \tag{2.13}$$

or more generally a shifted ratio [CFZ]

$$M(X)^{-1} \sum_{c(f) \le X} \prod_{1 \le k \le K} \frac{L(\frac{1}{2} + \alpha_k)}{L(\frac{1}{2} + \delta_k)}. \qquad (2.14)$$

Having such a mean value is sufficient to conjecture just about anything you would like to know about the zeros and the value distribution of the L-function. See [CS] for examples.

Note that some families are unions of increasingly large pieces having the same conductor. Examples are the Dirichlet L-functions and the L-functions associated to holomorphic cusp forms (in either weight or level aspect). For those families it is believed that the heuristics for moments will produce a reasonable conjecture for the average over a fixed (large) conductor.

Although this definition of "family" is useful for certain applications, it lacks the concreteness of the function field case. In particular, there does not yet exist an analogue of monodromy for such a family, and computing the symmetry type of the family is not straightforward. We discuss this in the next section.

2.4 Modeling a family of L-functions

Given a family of L-functions one can ask questions about its value distribution or about the distribution of its zeros. In most cases current technology is not sufficient to answer the interesting questions, so the next hope is to find a plausible conjecture. Only recently have such conjectures been found, and the new ingredient is to use random matrices to model the family of L-functions.

The idea is to associate a classical compact group, $U(N)$, $Sp(2N)$, $O(N)$, $SO(2N)$, or $SO(2N+1)$, to the family. The local statistics of the eigenvalues should agree, to leading order, with the corresponding local statistics of the zeros of the L-functions. And, to leading order and after compensating by an arithmetic constant, the value distribution of the characteristic polynomial

$$\Lambda(z) = \Lambda_A(z) = \det(I - A^* z) = \prod_{n=1}^{N} \left(1 - z e^{-i\theta_n}\right) \qquad (2.15)$$

near the point $z = 1$ should agree with the value distribution of the L-functions near the critical point. Here A is an $N \times N$ unitary matrix A and A^* is the Hermitian conjugate of A, so the eigenvalues of A lie on the unit circle and are denoted by $e^{i\theta_n}$.

In the above correspondence the size of the matrix is set equal to the conductor of the L-function. (Actually, to an integer close to the conductor, but to leading order such discrepancies do not matter). To see why this is a natural choice, consider the functional equation satisfied by the characteristic polynomial:

$$\Lambda_A(z) = (-1)^N \det(A) z^N \Lambda_{A^*}(z^{-1}). \qquad (2.16)$$

If we identify $(-1)^N \det(A)$ with ε and z^N with $X_L(s)$ then we have a perfect correspondence between the functional equations of $\Lambda_A(z)$ and $L(s)$, the unit circle playing the role of the critical line and $z = 1$ the critical point. Just as for L-functions, we define the conductor as $\frac{d}{dz} z^N$ evaluated at the critical point, so N is the conductor. Note that identifying conductors is equivalent to equating the average spacing between the zeros. Values near the critical point are modeled using the correspondence $\Lambda(e^{-z}) \leftrightarrow L(\frac{1}{2} + z)$.

It remains to identify the matrix group which corresponds to the family. From the functional equation it is almost possible to determine the group: the only ambiguity is to distinguish between $SO(2N)$ and $Sp(2N)$. At one time it was thought that this case could be easily resolved because $SO(2N)$ families always arise as "half" of a larger family, the other half being modeled by $SO(2N + 1)$. On the other hand $Sp(2N)$ families do not have such a "partner". A counterexample to that hope is described in [MD2]. But even if that approach were viable, it is unsatisfactory because it relies on the fact that the symmetry type can be found among a small list of possibilities. Fortunately, there are other methods. One possibility is to compute the 1- and 2-level densities of the family. This usually can be done rigorously for functions with small support, and this is sufficient to distinguish among the classical compact groups. But again we are relying on the fact that the symmetry type can be found on a short list. That objection can be overcome if one can conjecture the level densities in the full range, but that can be quite difficult in practice. Another possibility is to use the recipe in [CFKRS] to conjecture the moments of the family. This unambiguously identifies the group, and it also can tell you if the family is not modeled by one of those groups. Unfortunately, it is not clear that the recipe in [CFKRS] can be applied to all interesting families of elliptic curve L-functions, such as the family $\mathcal{F}_3$ given in (3.6).

2.5 Summary of modeling

Just to be pedantic, we note the following answers to the questions posed at the beginning of Section 1:

1. A family of L-functions is a set of L-functions, ordered by conductor, which is built in a particular way from a family of characters. The counting function of the family should be nice, and the data in the functional equation should be monotonic functions of the conductor.

2. A family is modeled by associating to it a classical compact matrix group. The specific compact group can usually be determined by computing the level densities of the low-lying zeros of the family, or by conjecturing the moments of the family. The size of the matrices scales with the (logarithmic) conductor of the L-functions.

3. To leading order the rescaled local zero statistics of the family are the same as the rescaled local eigenvalue statistics of the group. To leading

order the critical moments of the family equal the critical moments of the characteristic polynomials, up to a multiplicative arithmetic constant. For the family we average over the L-functions of conductor less than X, and then let $X \to \infty$. For the matrix groups the averages are with respect to Haar measure.

The modeling described above will produce leading order asymptotics. To make more precise predictions requires other methods, which are described in Section 4.

3   Elliptic curve families

For the remainder of the paper, E is an elliptic curve over $\mathbb{Q}$, with root number $w_E = \pm 1$, and $L(s, E)$ is the L-function associated to E normalized so that $s = \frac{1}{2}$ is the critical point. If F is any family of elliptic curves we write $F = F^+ \cup F^-$ where F^+ or F^-, respectively, are the curves $E \in F$ with $w_E = +1$ or $w_E = -1$. We write $E_{a,b}$ for the curve $y^2 = x^3 + ax + b$.

Most of the information in this section can be found in recent papers by Steven J. Miller and Eduaro Dueñez [M, MD1, MD2], Nina Snaith [Sn1, Sn2] and Matthew Young [Y1, Y2, Y3]. The author of this paper is just trying to convey the current understanding of the relationship between families of elliptic curve L-functions and random matrix theory: he makes no claim to any of the ideas presented here.

3.1   Families with a given rank

The following question is not well posed:

What is the correct random matrix model for the L-functions of a family of elliptic curves having a prescribed rank r?

The question is not well posed because there are (at least) two reasonable models, both of which seem to be appropriate for certain families of elliptic curves. We will first examine the simplest case of rank $r = 1$.

Consider the following families of rational elliptic curves:

$$\mathcal{F}_1(X) = \{E_{a,b} \; : \; |a| \leq X^{\frac{1}{3}}, \; |b| \leq X^{\frac{1}{2}}\} \tag{3.1}$$

and

$$\mathcal{F}_2(X) = \{E_{a,b^2} \; : \; |a| \leq X^{\frac{1}{3}}, \; b^2 \leq X^{\frac{1}{2}}\}. \tag{3.2}$$

Note that the point $(0, b)$ on E_{a,b^2} almost always has infinite order, so almost all of the curves in $\mathcal{F}_2$ have rank at least 1.

Let's consider $\mathcal{F}_1^-$ and $\mathcal{F}_2^-$. In both families we have $L(\frac{1}{2}, E) = 0$ because $w_E = -1$. However, for $E \in \mathcal{F}_2^-$ we could have said $L(\frac{1}{2}, E) = 0$ because $rank(E) \geq 1$. The fact that the zero at $L(\frac{1}{2}, E)$ for $E \in \mathcal{F}_2$ was *constructed*, instead of just arising from parity considerations, has a profound influence on the behavior of the L-function near the critical point.

To understand the influence of the critical zero, we first consider the distribution of $L'(\frac{1}{2}, E)$, which assuming standard conjectures is nonzero for almost all curves in our $w_E = -1$ families. We have the following conjectures from [Y2]:

$$\frac{1}{|\mathcal{F}_1^-(X)|} \sum_{E \in \mathcal{F}_1^-(X)} L'(\tfrac{1}{2}, E)^k \sim c_1(k)(\log X)^{k(k+1)/2} \qquad (3.3)$$

while

$$\frac{1}{|\mathcal{F}_2^-(X)|} \sum_{E \in \mathcal{F}_2^-(X)} L'(\tfrac{1}{2}, E)^k \sim c_2(k)(\log X)^{k(k-1)/2} \qquad (3.4)$$

As the formulas show, the behavior at the critical point is different for the two families, even though both families could be described as "a rank 1 family of elliptic curves." In particular, we see that the derivative $L'(\frac{1}{2}, E)$ tends to be smaller for $E \in \mathcal{F}_2^-$. This can be explained by the tendency for the low-lying zeros of $L(s, E)$ to be closer to the critical point for $E \in \mathcal{F}_2^-$. That is, $\mathcal{F}_2^-$ should have more low-lying zeros, which will cause the L-function to stay small near the critical point, and so its derivative will also be small. To make this idea precise we consider the 1-level density of the zeros.

Let $0 < \gamma_{E,1} \le \gamma_{E,2} \le \gamma_{E,3} \le \cdots$ denote the imaginary parts of the zeros of $L(s, E)$ in the upper half of the critical strip. Note that we have omitted the zero(s) at the critical point. The *one-level density* of the family $F(X)$ is defined to be the function W_1 which satisfies

$$\frac{1}{|F(X)|} \sum_{E \in F(X)} \sum_j \phi(\gamma_{E,j}) \sim \int \phi(t) W_1(t) dt, \qquad (3.5)$$

as $X \to \infty$, for nice functions ϕ. That is, W_1 measures the density of the zeros of the family.

The observation about the relative size of $L'(\frac{1}{2}, E)$ can be restated as: the one-level density for the family $\mathcal{F}_2^-$ should be more concentrated near 0 than the one-level density for the family $\mathcal{F}_1^-$. By using random matrix theory and some other ideas we explain below, it is possible to produce a precise conjecture for the one-level densities of these families. These are given in Figure 3.1. The functions are rescaled so that the average spacing between zeros is 1. In the next section we explain where those conjectures came from.

3.2   Two models for two kinds of families

The plots in Figure 3.1 are familiar. The plot on the left is the rescaled one-level density of the eigenvalues of matrices from the group $SO(2N + 1)$, in the limit as $N \to \infty$. The plot on the right is the rescaled one-level density of the eigenvalues of matrices from the group $SO(2N)$, in the limit as $N \to \infty$. Given those plot, what are our models for the two families?

                             D. W. Farmer

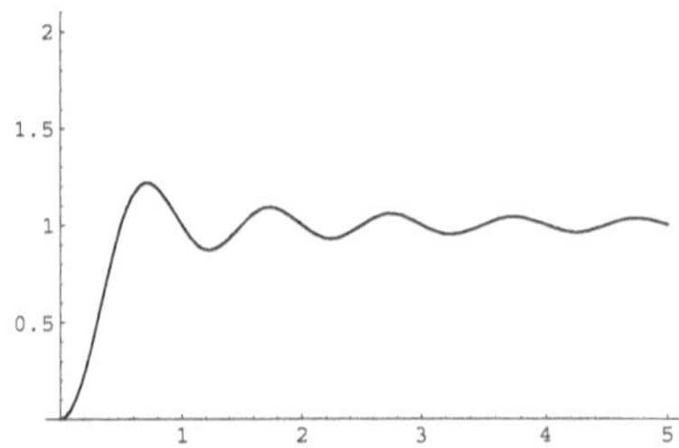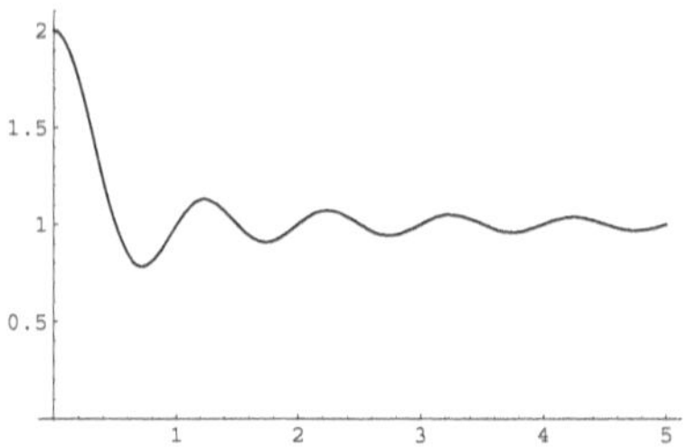

Figure 3.1: Conjectured one-level density of the noncritical L-function zeros of the family $\mathcal{F}_1^-$ (left) and $\mathcal{F}_2^-$ (right).

Consider the following ways to make a polynomial $f(z)$ which has real coefficients, all its zeros on the unit circle, and (almost surely a simple) zero at $z = 1$:

- The characteristic polynomial of a matrix in $SO(2N + 1)$

- $(z - 1)$ times the characteristic polynomial of a matrix in $SO(2N)$

It should be clear that those two examples will have the one-level densities pictured in Figure 3.1. These examples are the simplest cases of the two most commonly studied higher rank families of elliptic curve L-functions, which we now describe.

Suppose E_T is a curve $y^2 = x^3 + a(T)x + b(T)$ of rank r over $\mathbb{Q}(T)$. Consider the following two families of rational elliptic curves:

$$\mathcal{F}_3(X) = \{E_{a,b} \ : \ |a| \leq X^{\frac{1}{2}}, \ |b| \leq X^{\frac{1}{3}}, \ \text{rank}(E) \geq r\} \tag{3.6}$$

and

$$\mathcal{F}_4(X) = \{E_t \ : \ |a(t)| \leq X^{\frac{1}{2}}, \ |b(t)| \leq X^{\frac{1}{3}}, \ t \in \mathbb{N}\}. \tag{3.7}$$

As in our rank 1 example, we have $\mathcal{F}_4(X) \subset \mathcal{F}_3(X)$. Again we consider the subfamilies according to the sign of w_E. Let F be either of the above rank r families. If r is odd then almost all the curves in F^- have rank r, and almost all the curves in F^+ have rank $r + 1$. If r is even then the F^+ curves have rank r and the F^- curves have rank $r + 1$. It is conjectured that $\mathcal{F}_4^+$ and $\mathcal{F}_4^-$ are approximately the same size provided E has at least one place of multiplicative reduction. See [H]. If r is even then it is possible that $\mathcal{F}_3^+$ and $\mathcal{F}_3^-$ are approximately the same size.

We describe the two models which are believed to correspond to these families. The names for these models was coined by Steven J. Miller.

3.3 Selecting to have zeros

The Interaction Model

We are modeling a family that arises by restricting a much larger family to a subfamily having at least r zeros at the critical point. A matrix model for this family can be described as follows: start with $SO(M)$ where $M = 2N$ or $2N + 1$ depending on whether we are modeling $\mathcal{F}_3^+$ or $\mathcal{F}_3^-$, and restrict to those matrices having 1 as an eigenvalue of multiplicity at least r. That is, you take matrices in $SO(M)$ and drag r zeros to the critical point.

There are some bad things about this model. First, it is a set of matrices, but it is not a group. And while it is a perfectly well-defined set, it is a measure zero subset of $SO(M)$, so there is no canonical way to restrict Haar measure to it.

One solution, which has been analyzed by Snaith [Sn1] and Dueñez [D, MD1] is to first restrict to those matrices which have r eigenvalues in $[-\varepsilon, \varepsilon]$, and then let $\varepsilon \to 0$. The resulting measure is the same as one obtains by taking Haar measure on $SO(M)$, formally substituting $\theta = 0$ for r of the eigenvalues, and then omitting those terms which vanish identically. With the eigenvalues given by $e^{i\theta_j}$, the induced measure on that set is

$$C(M, r) \prod_{j=1}^{M} (1 - \cos\theta_j)^r \prod_{1 \leq j < k \leq M} (\cos\theta_j - \cos\theta_k)^2 d\theta_1 \cdots d\theta_M \qquad (3.8)$$

where $C(M, r)$ is a normalization constant.

It is instructive to look at the one-level density for such matrices. The one-level density is given by [Sn1, MD1]

$$\frac{\pi^2}{2}\theta \left(J_{r-\frac{3}{2}}^2(\theta\pi) + J_{r-\frac{1}{2}}^2(\theta\pi) - \frac{2r-1}{\theta\pi} J_{r-\frac{1}{2}}(\theta\pi) J_{r-\frac{3}{2}}(\theta\pi) \right). \qquad (3.9)$$

There is some numerical evidence [MD1] that this model is accurate. For this model it is possible to compute the critical moments of the characteristic polynomials [Sn1], but it does not seem that all the ingredients are available to use the recipe in [CFKRS] to conjecture the moments of the family.

Note that when $r = 0$ we recover $SO(2N)$ and when $r = 1$ we have $SO(2N + 1)$.

3.4 Imposing zeros

The Independent Model

We are modeling a family that has an rth order zero at the origin which arises from an explicit construction. A model for this situation can be found by assuming that the extra critical zeros are just inserted at the critical point, and all the other zeros ignore them. That is, start with a matrix in $SO(M)$ where $M = 2N - r$ or $2N + 1 - r$ depending on the parity of r and the sign of w_E. Then the polynomial which models the L-function is the characteristic

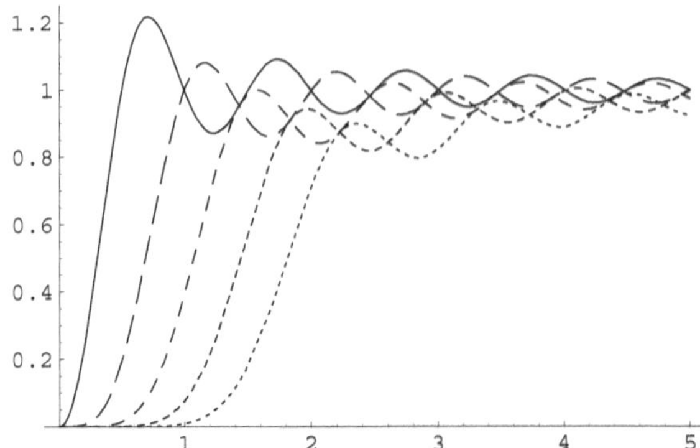

Figure 3.2: One-level density for $SO(M)$, restricted to have exactly r eigenvalues at $\theta = 0$, for $r = 1, 2, 3, 4, 5$.

polynomial of the matrix, multiplied by $(x - 1)^r$. The one-level density only depends on the parity of r and the sign of w_E, and will be one of the functions shown in Figure 3.1.

One can phrase the model strictly in terms of matrices by saying that the model is given by the group

$$\begin{pmatrix} I_{r \times r} & \\ & SO(M) \end{pmatrix} \tag{3.10}$$

where $I_{r \times r}$ is the $r \times r$ identity matrix and $M = 2N - r$ or $2N + 1 - r$ is chosen according to the parity of r and the sign of the functional equation.

There is numerical evidence [MD1], level density calculations [M, MD1, Si, Y1] and conjectures for moments [Y2] that this model gives accurate predictions for some specific families.

3.5   Some issues

It is worth repeating that the above models, even if they are correct, are only intended to capture leading-term asymptotics. Computer experiments [MD1] find that the low lying zeros of the family $\mathcal{F}_4$ exhibit some anomalous behavior which presumably will disappear when larger examples are computed. It is entirely possible that family $\mathcal{F}_3$ is a union of families of the form $\mathcal{F}_4$, and this may contribute to a bias in those numerics. In families of type $\mathcal{F}_4$ the generators of the set of rational points have very small height, and this may also introduce a bias. That is, the heights of the generators are on the order of the logarithm of the conductor, while it is more typical to have the heights as large as a power of the conductor. See Silverman [Si2], Chapter 10, for a discussion of heights of generators.

It is possible that more accurate predictions of the one-level density (using methods described in the next section) will show better agreement with the data. Those methods are also able to give extremely precise predictions for the moments of the L-functions [Y2], and these give support to the models.

It is not clear that these families of elliptic curves give rise to families of L-functions as described in Section 2. For the purpose of conjecturing the moments of the family (which is why that definition of 'family' was developed), the key property is the orthogonality relation (2.8). For specific families of type $\mathcal{F}_4$ it should be possible to evaluate such sums. The result is likely to be quite complicated, as in [Y2]. A subtle problem is that the parameter X in the elliptic curve families is approximately the discriminant, not the conductor. By Szpiro's conjecture the logarithm of the discriminant is within a factor of 6 of the logarithm of the conductor, so it is possible that ordering by discriminant is almost as good as ordering by conductor. It seems reasonable to model by setting N, the size of the matrix, equal to $\log X$, since what else would you choose? If that choice is correct it suggests that X is close to the discriminant most of the time. This has been shown for some families [Y1].

One can cook up an elliptic curve family which presumably is a hybrid of the models described here: take a rank r family of type $\mathcal{F}_4$ and restrict to those curves having rank at least $r+2$. If one makes the reasonable assumption that the "extra" zeros created by this process do not interact with the original r zeros imposed at the critical point, then one can use methods similar to [CKRS] to predict how many curves are in the restricted family. I am not advocating a reckless proliferation of elliptic curve models, but am merely noting that even if the two models described here are correct and can be refined to predict lower order terms, they may not cover all families of interest.

In Figure 3.3 one can see that if r is large then you are unlikely to find noncritical zeros close to the critical point. The name "repulsion" has been given to this phenomenon. The logic behind the name is that the "lowest zero" is further from the critical point than it would be if there were not a multiple critical zero. Unfortunately, the "lowest zero" is not a well defined object. If you drag the lowest zero to the critical point then the other zeros follow it toward the critical point, and at the moment you increase the order of the critical zero there becomes a new "lowest zero". If there is an rth order critical zero and you count zeros correctly, then the "lowest zero" is actually the $(r+1)$st zero, and it is likely to be *closer* to the critical point than a typical $(r+1)$st zero. In other words, the word "attraction" more accurately describes the situation! There is no reason to change the current terminology, but keep in mind that in the model where one restricts to those matrices having multiple eigenvalues at 1, the other eigenvalues have actually moved *toward* the critical point.

4 Refined modeling

Random matrix theory is useful for making leading-order asymptotic predictions about families of L-function. To understand the finer behavior of the family one must use heuristic techniques from number theory.

There are two main refinements to the leading term asymptotics. First, L-function families are ordered by conductor, and we use the conductor to determine the appropriate size matrices for our model. There will be a discrepancy between the limiting behavior for large matrices and the behavior for finite size matrices. For some quantities, such as the nearest neighbor spacing of the zeros/eigenvalues, it is difficult to see the difference between the asymptotics of the distribution and the distribution for moderate size matrices. For other quantities, such as the value distribution, for any computable range there is a notable difference between the limiting quantity and the values that can be computed. See Keating and Snaith [KS1], for a spectacular example concerning the value distribution of $\Re \log \zeta(\frac{1}{2} + it)$. By a theorem of Selberg that is Gaussian in the limit $t \to \infty$, but for finite t it differs from a Gaussian in the same way as the characteristic polynomial of an appropriately sized random unitary matrix.

The second issue is the fact that there are lower order terms, and the lower order terms for L-functions involve arithmetic factors, while the lower order terms of random matrices do not. Thus, the general shape of expressions from random matrix theory can reveal what to expect for L-functions, but the arithmetic "correction terms" must be determined in some other manner. For zero spacings there are no arithmetic corrections in the leading order terms. For moments of L-functions the leading order correction terms are fairly straightforward to determine [CG, CF, KS2]. Just about everything else is quite subtle and one needs sophisticated number-theoretic methods in order to make sensible conjectures.

For moments of L-functions such conjectures are covered in detail in [CFKRS]. Matt Young [Y2] used these heuristics to compute the full main term for various families of elliptic curve L-functions. Those conjectures give (3.3) and (3.4) as special cases. Thus, the heuristics appear to correctly handle a variety of interesting families. (For the family $\mathcal{F}_3$ in (3.6) our current understanding of the distribution of the coefficients a_p does not seem adequate to conjecture the moments of the family. Even finding the leading order arithmetic factor seems difficult in this case. However, random matrix calculations [Sn1] predict the general shape of the moments.)

For quantities involving zero statistics, moments are insufficient and one needs averages of ratios of the L-functions. This is addressed in [CFZ]. For example, the expected value of the ratio

$$\frac{L(\frac{1}{2} + \alpha, f)}{L(\frac{1}{2} + \beta, f)} \tag{4.1}$$

is sufficient to determine the one-level density of the family $L(s, f)$, including the lower order correction terms due to arithmetic effects. See [CS] for many examples. It should be possible to use these methods to conjecture the ratio (4.1) for various higher rank elliptic curve families, and thus give a precise conjecture for the one-level density. As of this writing this has not been done, but probably it will have been done by the time this paper appears in print.

In summary, by choosing the matrix group and the size of the matrices appropriately, and with the appropriate arithmetic correction factor, a family of L-functions can be modeled by the characteristic polynomials of a collection of matrices. The example of elliptic curve L-functions where the elliptic curves are selected to have large rank shows that the collection of matrices may not be a group. In order to capture the lower order terms one must use heuristics from number theory which do not explicitly involve random matrix theory. Those heuristics also recover the leading order behavior which previously required random matrix theory.

References

[CG]   J. B. Conrey and A. Ghosh, *Mean values of the Riemann zeta-function*, Mathematika **31** (1984) pp. 159–161

[CF]   J. B. Conrey and D. W. Farmer, *Mean values of L-functions and symmetry*, Internat. Math. Res. Notices **17** (2000) pp. 883–908

[CFKRS]   J.B. Conrey, D.W. Farmer, J. Keating, M. Rubinstein, and N.C. Snaith, *Integral moments of L-functions*, Proc. Lond. Math. Soc. **91** (2005) pp. 33–104, arxiv.org/abs/math.NT/0206018

[CKRS]   J.B. Conrey, J. Keating, M. Rubinstein, and N.C. Snaith, *On the frequency of vanishing of quadratic twists of modular L-functions.* in "Number Theory for the Millennium I"; MA Bennett et al., eds, A.K. Peters, Natick, 2002. math.NT/0012043

[CFZ]   J. B. Conrey, D. W. Farmer, and M. R. Zirnbauer, *Howe pairs, supersymmetry, and ratios of random characteristic polynomials for the classical compact groups*, preprint

[CS]   J.B. Conrey and N.C. Snaith, *Applications of the L-functions ratios conjectures*, to be published in Proc. Lond. Math. Soc., math.NT/0509480

[D]   E. Dueñez, *Random matrix ensembles associated to compact symmetric spaces*, Commun. Math. Phys, **244**(1), (2004) pp. 29–61, math-ph/0111005

[H]   H. Helfgott, *On the distribution of root numbers in families of elliptic curves*, preprint, math.NT/0408141

[ILS]   H. Iwaniec, W. Luo, and P. Sarnak, *Low lying zeros of families of L-functions*, Inst. Hautes Études Sci. Publ. Math. No. 91 (2000), pp. 55–131 (2001)

[IS]   H. Iwaniec and P. Sarnak, *The non-vanishing of central values of automorphic L-functions and Landau-Siegel zeros*, Israel J. Math. **120** (2000), part A, pp. 155–177

[KSa]   N. M. Katz and P. Sarnak, Random matrices, Frobenius eigenvalues, and monodromy, AMS Colloquium Publications, 45 AMS, Providence, RI (1999)

[KS1]   J. P. Keating and N. C. Snaith, *Random matrix theory and $\zeta(\frac{1}{2} + it)$*, Comm. Math. Phys. **214** (2000) pp. 57–89

[KS2]   J. P. Keating and N. C. Snaith, *Random matrix theory and L-functions at $s = \frac{1}{2}$*, Comm. Math. Phys. **214** (2000) pp. 91–110

[KMV]   E. Kowalski, P. Michel, and J. VanderKam, *Mollification of the fourth moment of automorphic L-functions and arithmetic applications*, Invent. Math. **142** (2000), no. 1, 95–151

[M]   S. J. Miller, *1- and 2-Level Densities for Families of Elliptic Curves: Evidence for the Underlying Group Symmetries*, Compositio Mathematica **140** (2004), no. 4, pp. 952–992. math.NT/0310159

[MD1]   S. J. Miller, Investigations of Zeros Near the Central Point of Elliptic Curve L-Functions, with an appendix by E. Dueñez. math.NT/0508150

[MD2]   S. J. Miller and E. Dueñez, *The Low Lying Zeros of a GL(4) and a GL(6) family of L-functions*, preprint, math.NT/0506462

[Mon]   H.L. Montgomery, *The pair correlation of zeros of the Riemann zeta-function*, Proc. Symp. Pure Math. **24** (1973) pp. 181–93

[Odl]   A. Odlyzko, *The 10^{20}th zero of the Riemann zeta-function and 70 million of its neighbors.*, preprint (1989)

[OS1]   A. Özlük and C. Snyder, *Small Zeroes of Quadratic L -Functions*, Bull. Aust. Math. Soc. **47** (1993) pp. 307–319

[OS2]   A. Özlük and C. Snyder, *On the distribution of the nontrivial zeros of quadratic L-functions close to the real axis*, Acta Arith. 91 (1999), no. 3, 209–228.

[Rub]   M. Rubinstein *Low-lying zeros of L-functions and random matrix theory*, Duke Math. J. **109** (2001), no. 1, pp. 147–181

[Si]   J. Silverman, *The average rank of an algebraic family of elliptic curves*, J. reine angew. Math. **504** (1998), 227–236

[Si2]   J. Silverman, "The Arithmetic of Elliptic Curves", GTM 106, Springer-Verlag, New York, 1994

[Sn1]   N.C. Snaith *Derivatives of random matrix characteristic polynomials with applications to elliptic curves*, J. Phys. A **38** No. 48 (2005) pp. 10345-10360, math.NT/0508256

[Sn2]   N.C. Snaith *The derivative of SO(2N + 1) characteristic polynomials and rank 3 elliptic curves*, in this volume

[S]   Soundararajan, *Nonvanishing of quadratic Dirichlet L-functions at $s = \frac{1}{2}$*, Ann. of Math. (2) **152** (2000), no. 2, pp. 447–488

[Y1]   M. Young, *Lower-Order Terms of the 1-Level Density of Families of Elliptic Curves*, Int. Math. Res. Not. **10** (2005), pp. 587–633, math.NT/0408359

[Y2]   M. Young, *Moments of the critical values of families of elliptic curves, with applications*, preprint

[Y3]   M. Young, *Low-lying zeros of families of elliptic curves*, J. AMS **19** No. 1 (2006) pp. 205-250, math.NT/0406330

American Institute of Mathematics
360 Portage Ave.
Palo Alto, CA 94306
farmer@aimath.org

Analytic number theory and ranks of elliptic curves

Matthew P. Young

Abstract

We discuss recent applications of analytic number theory to the study of ranks of elliptic curves.

1   Introduction

This article is meant to be a sampling of techniques and interesting results on the (analytic) ranks of elliptic curves. The main result discussed is an upper bound on the average rank of the family of all elliptic curves. The bound obtained is less than 2, which implies (by work of Kolyvagin [Kol]) that a positive proportion of elliptic curves have finite Tate-Shafarevich group and algebraic rank equal to analytic rank. The synergy here between algebraic and analytic methods is extremely pleasant.

We also discuss the problem of showing that a large number of elliptic curve L-functions do not vanish at the central point. Many of the techniques used in bounding the average rank are useful in this direction.

Our exposition is meant to be somewhat colloquial. The interested reader should consult [Y1] and [Y2] for all technical details.

In this volume E. Kowalski has given a broad overview of what is known on ranks of elliptic curves in families. We shall refer to his article [Kow] for general background knowledge on elliptic curves. We have attempted to minimize overlap with his article without loss of coherence of this paper.

We shall assume the Generalized Riemann Hypothesis throughout this article.

1.1   Acknowledgements

I would like to thank Henryk Iwaniec for supporting my last-minute decision to attend the workshops. I also thank the organizers of the Newton Institute program for inviting me to attend.

1.2   Notation

To be definite we set our notation here. We shall differ slightly from Kowalski in that we normalize our L-functions to have central point $1/2$.

We suppose E is an elliptic curve over $\mathbb{Q}$ with conductor N and L-function

$$L(s, E) = \prod_p \left(1 - \frac{\lambda_E(p)}{p^s} + \frac{\psi_N(p)}{p^{2s}} \right)^{-1},$$

where here ψ_N is the principal Dirichlet character $(\mathrm{mod}\ N)$. With our normalization $\sqrt{p}\lambda(p) \in \mathbb{Z}$ and the Hasse bound is

$$|\lambda(p)| \leqslant 2,$$
$$|\lambda(n)| \leqslant d(n).$$

If E has minimal Weierstrass equation $y^2 = g(x)$ then for $p \neq 2$ we have

$$\lambda_E(p) = -\frac{1}{\sqrt{p}} \sum_{x\,(\mathrm{mod}\ p)} \left(\frac{g(x)}{p} \right). \tag{1.1}$$

2 Bounding the average rank

The prototypical result of interest in this article is Theorem 2.9 of [Kow], namely Brumer's upper bound of 2.3 for the average rank of the family of all elliptic curves [B]. In what follows we will show how Theorem 2.9 is proved and discuss how improvements may be made. We concentrate on a handful of families for which the best results are known and which exhibit interesting behavior.

2.1 The tools

The main tool for studying the zeros of an L-function is the explicit formula. We refer to §2.5 of [Kow] for a thorough discussion of the explicit formula and the approximate functional equation. We prefer to use the 'Fourier' form of the explicit formula (as opposed to the form of Proposition 2.11 of [Kow]). See Theorem 5.12 of [IK] for our preferred form. In addition, we make a preliminary cleaning by trivially estimating various terms.

Proposition 2.1. *Let ϕ be an even Schwartz-class function whose Fourier transform has compact support (so ϕ extends to an entire function). Let E be an elliptic curve with conductor N and L-function $L(s, E)$ with nontrivial zeros $1/2 + i\gamma$. Then for any $X > 1$ we have*

$$\sum_\gamma \phi\left(\gamma \frac{\log X}{2\pi} \right) = \frac{1}{2}\phi(0) + \widehat{\phi}(0)\frac{\log N}{\log X} - 2\sum_p \frac{\lambda_E(p)}{p^{1/2}} \frac{\log p}{\log X} \widehat{\phi}\left(\frac{\log p}{\log X} \right)$$

$$- 2\sum_p \frac{\lambda_E(p^2)}{p} \frac{\log p}{\log X} \widehat{\phi}\left(\frac{2\log p}{\log X} \right) + O\left(\frac{1}{\log X} \right),$$

the implied constant depending on ϕ only.

We see that taking X close to N scales the zeros properly when the curves vary over a family. It is useful to have freedom in choosing X rather than taking $X = N$ since the conductor may have irregular behavior. In all cases we shall assume $N \ll X$.

Let $D(E; \phi)$ be the quantity on the left-hand side of Proposition 2.1 and likewise set

$$P_k(E; \phi) = -2 \sum_p \frac{\lambda_E(p^k)}{p^{k/2}} \frac{\log p}{\log X} \widehat{\phi}\left(\frac{k \log p}{\log X}\right)$$

for $k = 1, 2$.

The quantity $D(E; \phi)$ can be interpreted as the density of zeros near the central point. The philosophy of Katz and Sarnak ([KS1], [KS2]) makes predictions on the asymptotic behavior of $D(E; \phi)$ as E varies over a family. By comparison with the function field case as well as with the family of all weight 2 level q newforms (see [ILS]) we expect a family of elliptic curves to have orthogonal symmetry. Let $\mathcal{F}_X$ be a family of elliptic curves with parameter X as in §2.1 of [Kow]. Then the natural prediction is that

$$\frac{1}{|\mathcal{F}_X|} \sum_{E \in \mathcal{F}_X} D(E; \phi) \sim \frac{1}{2}\phi(0) + \widehat{\phi}(0),$$

as $X \to \infty$ for any ϕ satisfying the conditions of Proposition 2.1. In practice there will be a restriction on the size of the support of $\widehat{\phi}$ of the type $\text{supp}(\widehat{\phi}) \subset (-\nu, \nu)$ for some $\nu > 0$ (with an abuse of language we will say that $\widehat{\phi}$ has support up to ν). The $\frac{1}{2}\phi(0)$ term should be interpreted as corresponding to zeros that occur at the central point. The factor $1/2$ ostensibly appears because half of all elliptic curves should vanish by virtue of the sign in the functional equation. See [He1] for recent work indicating how the root number behaves in families (one part of his work is reducing the problem to a classical conjecture of Chowla on the behavior of the Möbius function at polynomial values, strongly indicating that the root number is equidistributed between $+1$ and -1 in 'typical' families). The fact that the coefficient of $\phi(0)$ is not larger than $1/2$ predicts that vanishings to order 2 or higher are rare (i.e. that such curves do not constitute a positive percentage of elliptic curves). The $\widehat{\phi}(0)$ term is interpreted as capturing the aggregate of zeros of the L-functions in the family. It is highly desired to reduce this interference by taking ϕ with $\phi(0)$ large and $\widehat{\phi}(0)$ small, because then one can extract more quantitative information about the central zeros.

On the assumption of GRH for $L(s, E)$ we have $\gamma \in \mathbb{R}$ and we may take ϕ non-negative with $\phi(0) = 1$ so that the left-hand side in Proposition 2.1 is at least $\text{ord}_{s=\frac{1}{2}} L(s, E)$. This observation is the starting point for obtaining an upper bound on the average rank of a family of elliptic curves (assuming GRH). The goal then is to obtain an upper bound on the average of the right-hand side of Proposition 2.1. There is motivation for obtaining the asymptotic value of $D(E; \phi)$ because it tests the Katz-Sarnak predictions. Proving an upper bound can sometimes be a simpler task than getting the asymptotic due to difficulties with the size of the conductor.

There are three terms to average in the explicit formula; we address each in turn. A trivial estimation shows $P_2(E; \phi) \ll 1$; virtually any cancellation in this sum shows that $P_2(E; \phi) = o(1)$. The GRH for the symmetric-square L-function attached to E shows

$$P_2(E; \phi) \ll \frac{\log \log N}{\log X}.$$

It is possible to treat this term unconditionally by virtue of averaging over the family; we refer to Proposition 2.12 and surrounding discussions of [Kow] for a more extensive exposition of such results.

For the application of bounding the average rank we require a bound of the type

$$\sum_E \frac{\log N}{\log X} \leqslant \sum_E (1 + o(1)). \tag{2.1}$$

In practice it is exceedingly easy to ensure this condition because the conductor can be trivially bounded from above (by the absolute value of the discriminant for example). Obtaining an asymptotic formula for the average of the logarithm of the conductors is a much more difficult task. Usually this amounts to controlling the frequency of large square divisors of a polynomial. Current technology (the square-free sieve) is strong enough for polynomials of degree 3 or less [He2].

The most difficult term to treat is $P_1(E; \phi)$ (from now on we set $P(E; \phi) = P_1(E; \phi)$). The GRH for $L(s, E)$ implies

$$P(E; \phi) \ll X^\varepsilon,$$

so it is necessary to beat the Riemann hypothesis on average! In general we obtain results of the type

$$\sum_E P(E; \phi) = \sum_E (c + o(1)) \tag{2.2}$$

for an integer c. Here c should be interpreted as a kind of 'forced' rank; this terminology is vague, however if the family is given by specializations of a fixed elliptic surface then c would be the rank of the surface. See [N] and [RS] for further explanation and partial results.

Assuming we have obtained the previous inequalities, we therefore have

$$\frac{1}{|\mathcal{F}_X|} \sum_{E \in \mathcal{F}_X} \mathrm{ord}_{s=1/2} L(s, E) \leqslant \frac{1}{2} + c + \widehat{\phi}(0) + o(1)$$

as $X \to \infty$. To obtain the best upper bound for the average rank it is necessary to choose ϕ with $\widehat{\phi}(0)$ minimized subject to $\phi(0) = 1$ and supp $\widehat{\phi} \subset [-\nu, \nu]$. Taking the Fourier pair

$$\phi(x) = \left(\frac{\sin \pi \nu x}{\pi \nu x} \right)^2, \qquad \widehat{\phi}(y) = \frac{1}{\nu} \left(1 - \frac{|y|}{\nu} \right), \qquad |y| < \nu$$

gives optimal results. See [ILS], Appendix A for a derivation. Technically speaking, ϕ is not Schwartz-class but by taking suitable approximations this issue is easily rectified. The conclusion is that we obtain the upper bound on the average rank r of

$$r \leqslant \frac{1}{2} + c + \frac{1}{\nu} + \varepsilon. \tag{2.3}$$

Clearly, obtaining (2.2) with large ν is extremely desirable for then the upper bound on r improves.

2.2 The setup

To be concrete, let us list some examples of families of elliptic curves. In all cases we have $E_{a,b} : y^2 = g_{a,b}(x)$ where $g_{a,b}(x)$ is a cubic polynomial in x. We also locate $a \asymp A$ and $b \asymp B$ for certain values of A and B depending on X. Precisely, we set

$$
\begin{aligned}
\mathcal{F}_1 &: g_{a,b} = x^3 + ax + b & A &= X^{1/3} & B &= X^{1/2} \\
\mathcal{F}_2 &: g_{a,b} = x^3 + ax + b^2 & A &= X^{1/3} & B &= X^{1/4} \\
\mathcal{F}_3 &: g_{a,b} = x(x^2 + ax - b) & A &= X^{1/4} & B &= X^{1/2} \\
\mathcal{F}_4 &: g_{a,b} = x(x - a)(x + b) & A &= X^{1/3} & B &= X^{1/3}.
\end{aligned}
$$

The various restrictions on the sizes of a and b with respect to X are imposed to maximize the number of curves in the family subject to the constraint (2.1). For example, $E_{a,b} \in \mathcal{F}_1$ has discriminant $\Delta = -16(4a^3 + 27b^2)$ and $N|\Delta$. Having $a \asymp X^{1/3}$, $b \asymp X^{1/2}$ gives $|\Delta| \asymp X$. The upper bound (2.1) is therefore trivial. The asymptotic is not overly difficult for this family; see [Y1], Lemma 5.1 for a proof.

From an analytic point of view it is best to have families as large as possible. The families $\mathcal{F}_i$ are quite large and are well-suited for averaging. The family $\mathcal{F}_1$ is essentially the family of all elliptic curves, $\mathcal{F}_2$ is a large family consisting of positive (algebraic) rank curves, $\mathcal{F}_3$ has torsion group $\mathbb{Z}/2\mathbb{Z}$, and $\mathcal{F}_4$ has torsion group $\mathbb{Z}/2\mathbb{Z} \times \mathbb{Z}/2\mathbb{Z}$. Certain curves in the family $\mathcal{F}_4$ have a distinguished position in the proof of Fermat's Last Theorem. Another interesting feature of $\mathcal{F}_4$ is that Helfgott [He1] has shown that the root number is equidistributed in this family.

Families of quadratic twists have been very popular for many researchers. These families are conducive for study because of their simple nature in many regards, with many quantities being controlled by a Dirichlet character (i.e. the root number and the $\lambda(p)$'s). Quadratic twists also have different reduction than 'typical' elliptic curves, because almost all prime divisors of the conductors of the twisted curves are of additive reduction. A drawback of quadratic twist families is that they have small size (having $\asymp X^{1/2}$ curves of conductor $\asymp X$). We advocate for more study of the above families $\mathcal{F}_i$ as well as other such variants.

Often it can be good to make slight restrictions on the various families in order to avoid quadratic twists or to ensure that the Weierstrass equations are minimal. For instance, by taking a and b such that there is no prime p such that $p^2 | a$ and $p^3 | b$ we achieve both goals with $\mathcal{F}_1$. For technical reasons we omit this restriction; introducing it involves only minor alterations.

In order to aid in the analysis we introduce a smooth, compactly supported cutoff function $w : \mathbb{R}^2 \to \mathbb{R}$ satisfying $w(x, y) = 0$ whenever $x \leqslant 0$ or $y \leqslant 0$. The following quantity $\mathcal{P}(\mathcal{F}, \phi)$ is our central object of interest

$$\mathcal{P}(\mathcal{F}_i, \phi) = -2 \sum_p \sum_a \sum_b \frac{\lambda_{a,b}(p)}{p^{1/2}} \frac{\log p}{\log X} \widehat{\phi}\left(\frac{\log p}{\log X}\right) w\left(\frac{a}{A}, \frac{b}{B}\right). \qquad (2.4)$$

Here we have used the shorthand $\lambda_{a,b}(p) = \lambda_{E_{a,b}}(p)$ and also suppressed the dependence of $\lambda_{a,b}(p)$ on the family $\mathcal{F}_i$. The concrete formula (1.1) allows the use of classical techniques of analytic number theory for the study of (2.4). The compact support of $\widehat{\phi}$ means that a restriction of the type $p \leqslant X^\nu$ holds in (2.4). An important challenge is to estimate the sum with ν large.

In what follows we shall treat the sum (2.4) with increasing degrees of sophistication.

2.3   Completing the sum

We begin by applying Poisson summation in a and b $(\mathrm{mod}\ p)$ to the sum (2.4) (noting of course that $\lambda_{a,b}(p)$ is periodic in a and b $(\mathrm{mod}\ p)$). We have

$$\sum_a \sum_b \lambda_{a,b}(p) w\left(\frac{a}{A}, \frac{b}{B}\right) \qquad (2.5)$$

$$= \frac{AB}{p^2} \sum_h \sum_k \widehat{w}\left(\frac{hA}{p}, \frac{kB}{p}\right) \sum \sum_{\alpha, \beta \,(\mathrm{mod}\ p)} \lambda_{\alpha,\beta}(p) e\left(\frac{\alpha h + \beta k}{p}\right).$$

By the rapid decay of $\widehat{w}$ we may assume $h \ll (p/A)^{1+\varepsilon}$ and $k \ll (p/B)^{1+\varepsilon}$. Therefore, if $p \ll \min(A, B)$ the only terms that can possibly contribute anything come from $h = k = 0$. It is possible to show that

$$\sum \sum_{\alpha, \beta \,(\mathrm{mod}\ p)} \lambda_{\alpha,\beta}(p) = \begin{cases} 0, & i = 1, 3, 4 \\ -p(p-1)p^{-1/2}, & i = 2. \end{cases}$$

Hence we see

$$\mathcal{P}(\mathcal{F}_i, \phi) = \begin{cases} o(AB), & i = 1, 3, 4 \\ AB\widehat{w}(0,0)\phi(0) + o(AB), & i = 2, \end{cases} \qquad (2.6)$$

provided p is sufficiently small with respect to A and B. The additional main term for $i = 2$ arises from the fact that the curve $y^2 = x^3 + ax + b^2$ always has the point $(0, b)$, which is almost always of infinite order by a simple use of the Lutz-Nagell criterion [Si].

We shall consider these initial bounds as trivial. The challenge is to obtain the same result for p larger with respect to A and B. One can make immediate improvements by completing the sum over a or b only, corresponding to whichever of A and B is larger. This minor alteration allows one to take p as large as $\max(A, B)$ rather than $\min(A, B)$.

To progress further we compute the complete sum on the right hand side of (2.5) for arbitrary h and k. We have

$$\sum_{\alpha,\beta \,(\mathrm{mod}\ p)} \lambda_{\alpha,\beta}(p) e\left(\frac{\alpha h + \beta k}{p}\right) \tag{2.7}$$

$$= \begin{cases} -\varepsilon_p p\left(\frac{k}{p}\right) e\left(\frac{-h^3\overline{k}^2}{p}\right), & i = 1 \\ -p^{3/2}\delta_p(h)\delta_p(k) - \varepsilon_p p\left(\frac{-h}{p}\right) e\left(\frac{k^4\overline{h}^3\overline{2}^6}{p}\right) + p^{1/2}, & i = 2 \\ -\varepsilon_p p\left(\frac{h}{p}\right) e\left(\frac{h^2\overline{k}}{p}\right), & i = 3 \\ -\varepsilon_p p\left(\frac{hk(h-k)}{p}\right), & i = 4, \end{cases}$$

where ε_p is the sign in the Gauss sum (i.e. $\varepsilon_p = 1$ if $p \equiv 1 \pmod 4$, $\varepsilon_p = i$, $p \equiv 3 \pmod 4$) and $\delta_p(n)$ is the indicator function of $p|n$. See Appendix A for these computations.

The upshot of these computations is that we have obtained savings in the nonzero frequencies (i.e. those terms with $h, k \neq 0$). In addition, the explicit form of the above sums allows for additional savings in the summations over h, k, and p. If we apply (2.7) to $\mathcal{P}(\mathcal{F}, \phi)$ and do not exploit any such cancellation then we obtain results that are slightly less trivial than before. The quality of these results rests entirely upon the fact that there is square-root cancellation in (2.7) (after accounting for the main term for $i = 2$ of course).

For the record, the results obtained at this point are that $\mathcal{P}(\mathcal{F}_i, \phi)$ is given by (2.6) provided

$$\begin{cases} \nu \leqslant 5/9, & i = 1 \\ \nu \leqslant 7/18, & i = 2 \\ \nu \leqslant 1/2, & i = 3 \\ \nu \leqslant 4/9, & i = 4. \end{cases}$$

This bound for $i = 1$ is originally due to Brumer [B]. Note that this is the limit of the power of algebraic geometry because to go further requires varying primes and/or estimating very short sums. We do not mean that algebraic geometry cannot help in the further study of these sums, but rather that any application would be ingenious (Burgess' method is an outstanding example). The technology of analytic number theory is suited for treating such sums.

It is wise to consider philosophically whether the application of Poisson summation in (2.5) is advantageous. When p is too large with respect to A

and B then completing the sum modulo p leads to a dual sum that is longer than the original sum (the dual sum is the summation over h and k). The conventional wisdom says that Poisson summation should be applied when the dual sum becomes shorter. Nevertheless, with the above four families it appears always to be advantageous to complete the sums. The reason is that the expression in (2.7) is in closed form that allows for extra savings. We have access to $\sum_a \sum_b \lambda_{a,b}(p)$ only by 'opening' the sum $\lambda_{a,b}(p)$ as in (1.1). Opening this sum is costly because it is necessary to recover the Hasse bound.

Notice that the expressions in (2.7) for $i = 1$, 2, and 3 are remarkably similar (ignoring the main term for $i = 2$ of course). It should not be surprising that a method of estimation for one of these families should also treat the others, but of course the quality of the results may depend on the family. The case $i = 4$ is of a very different character than the other families and unsurprisingly it requires different techniques of estimations. One can attempt to study any family of elliptic curves using these techniques. It is desirable to find large families where the analog of (2.7) can be explicitly computed so that additional savings are obtainable.

In the following sections we discuss the cases $i = 1$ and $i = 4$ in greater depth. For the record, the best results obtained so far are

$$
\nu < \begin{cases}
7/9, & i = 1 \\
23/48, & i = 2 \\
2/3, & i = 3 \\
2/3, & i = 4.
\end{cases}
\tag{2.8}
$$

Actually, the case $i = 3$ has not been worked out in full detail but it is likely that the method sketched in the following section does give the result $2/3$. Notice that any improvement in these results for $i = 3$ or $i = 4$ shows that a positive proportion of the corresponding families with prescribed torsion have rank $\leqslant 1$. The quality of these results should be gauged against the size of the family; the respective families have $X^{5/6}$, $X^{7/12}$, $X^{3/4}$, and $X^{2/3}$ curves of conductor $\leqslant X$.

2.4 Sketching the method for the family of all elliptic curves

Recall that we are interested in estimating the following sum

$$
\mathcal{P}(\mathcal{F}_1, \phi) = 2\frac{AB}{\log X} \sum_p \sum_h \sum_k \varepsilon_p \left(\frac{k}{p}\right) e\left(\frac{-h^3\overline{k}^2}{p}\right) \frac{\log p}{p^{3/2}} \tag{2.9}
$$
$$
\times \widehat{\phi}\left(\frac{\log p}{\log X}\right) \widehat{w}\left(\frac{hA}{p}, \frac{kB}{p}\right).
$$

The requirement is $\mathcal{P}(\mathcal{F}_1, \phi) = o(AB)$. Heath-Brown [H-B1] has obtained this bound for support $\nu < 2/3$. Since his method is of interest we briefly sketch the arguments. To begin, write

$$|\mathcal{P}(\mathcal{F}_1, \phi)| \ll \frac{AB}{\log X} \sum_k \sum_p \left| \sum_h e\left(\frac{-h^3\overline{k}^2}{p}\right) \frac{\log p}{p^{3/2}} \widehat{\phi}\left(\frac{\log p}{\log X}\right) \widehat{w}\left(\frac{hA}{p}, \frac{kB}{p}\right) \right|.$$

Next extend the summation over p to all integers m coprime with k. Using Cauchy's inequality to reverse the order of summation, we are led to bound an exponential sum of the type

$$\sum_k \sum_m \sum_{h_1} \sum_{h_2} e\left(\frac{-(h_1^3 - h_2^3)\overline{k}^2}{m}\right) g(h_1, h_2, k, m),$$

where g is a certain test function that locates the variables in appropriate ranges.

An application of the reciprocity law

$$\frac{\overline{u}}{v} + \frac{\overline{v}}{u} \equiv \frac{1}{uv} \pmod 1, \tag{2.10}$$

where $(u, v) = 1$, $u\overline{u} \equiv 1 \pmod v$, and $v\overline{v} \equiv 1 \pmod u$ effectively reduces the modulus in the exponential to k^2 instead of m. The sum to estimate becomes

$$\sum_k \sum_m \sum_{h_1} \sum_{h_2} e\left(\frac{(h_1^3 - h_2^3)\overline{m}}{k^2}\right) g_1(h_1, h_2, k, m),$$

where

$$g_1(h_1, h_2, k, m) = e\left(\frac{-h_1^3 + h_2^3}{mk^2}\right) g(h_1, h_2, k, m).$$

In the typical range where $m \approx P$, $h_1 \approx h_2 \approx P/A$, and $k \approx P/B$ we have

$$\frac{-h_1^3 + h_2^3}{mk^2} \ll 1$$

so that the exponential of this quantity has small derivatives and can be safely absorbed into the test function g as our notation indicates.

Finally, apply Poisson summation in $m \pmod{k^2}$. We treat the nonzero frequencies trivially and consider the completed sum

$$C = \sum_{x \,(\mathrm{mod}\ k^2)} e\left(\frac{u\overline{x}}{k^2}\right),$$

where $u = h_1^3 - h_2^3$. Here C is a Ramanujan sum that will be large only when u and k^2 have a large greatest common divisor. Using fairly intricate elementary arguments it is possible to bound the frequency of such occurences. Notice that it is obvious from the expression (2.9) that some analysis of the greatest common divisor of h^3 and k^2 is necessary to obtain cancellation due to oscillation of the exponential factor. It is believed that no perverse behavior occurs due to a frequently large greatest common divisor; nevertheless, in practice it requires technical arguments to handle this difficulty. This completes our sketch of Heath-Brown's method. Note that the Riemann Hypothesis is not used in the estimations of $\mathcal{P}(\mathcal{F}_1, \phi)$ although of course it is used in the application of bounding the average rank of the family.

This value $\nu = 2/3$ gives the bound of 2 for the average rank. Notice that any improvement on the value $2/3$ shows that the average rank is less than 2, and hence that a positive proportion of elliptic curves have order of vanishing $\leqslant 1$. By the work of Kolyvagin this shows that a positive proportion of elliptic curves have finite Tate-Shafarevich group and algebraic rank equal to analytic rank! For these reasons, any support $\nu > 2/3$ is a natural goal for any family of elliptic curves. The support range of $2/3$ also appears to be a barrier in the estimations; note that $\nu < 2/3$ is allowable for $i = 3, 4$ in (2.8) as well as in Heath-Brown's result for $i = 1$. With the family of all elliptic curves it appears that there are many natural methods of estimation that lead to a support value of $2/3$. It is these examples that give anecdotal evidence that there is an innate barrier in extending the support beyond $2/3$.

The author has shown that $\mathcal{P}(\mathcal{F}_1) = o(AB)$ holds for any $\nu < 7/9$. This gives the upper bound of $25/14 = 1.78...$ for the average rank. We shall sketch some of the main ideas behind the proof of this result. See the original paper [Y1] for the full technical details.

Rather than considering (2.9) we study the following variant

$$S(H, K, P) = \sum_p \sum_h \sum_{\substack{k \\ (h,k)=1}} \left(\frac{k}{p}\right) e\left(\frac{-h^3\overline{k}^2}{p}\right) g\left(\frac{h}{H}, \frac{k}{K}, \frac{p}{P}\right),$$

where g is a smooth, compactly-supported function satisfying $g(x, y, z) = 0$ if $x \leqslant 0$, $y \leqslant 0$, or $z \leqslant 0$. Here P is an arbitrary parameter and $H = P/A$, $K = P/B$. The sum $S(H, K, P)$ differs from (2.9) in two essential ways. The first difference is that h and k are fixed to be close to their maximal sizes (in (2.9) we have the restriction $h \ll (P/A)^{1+\varepsilon}$, $k \ll (P/B)^{1+\varepsilon}$). Different techniques of estimation are necessary for small values of h and/or k. Nevertheless, the main barrier to obtaining larger values of ν comes from the estimations with $H = P/A$ and $K = P/B$. The second difference is that we have imposed the restriction $(h, k) = 1$. When h and k are close to their maximal sizes then the coprimality restriction is for simplicity only; when the variables are smaller some technical difficulties do arise but they do not pose an essential barrier. The factor ε_p is safely ignored because it only depends on $p \pmod 4$. The sum $S(H, K, P)$ represents a 'pure' form of (2.9) unobscured by technical details. The necessary bound is $S(H, K, P) \ll P^{3/2-\varepsilon}$.

As in the work of Heath-Brown we use the reciprocity law (2.10) to give

$$S(H, K, P) = \sum_p \sum_h \sum_{\substack{k \\ (h,k)=1}} \left(\frac{k}{p}\right) e\left(\frac{h^3 \bar{p}}{k^2}\right) g_1(h, k, p),$$

where

$$g_1(h, k, p) = e\left(\frac{h^3}{pk^2}\right) g\left(\frac{h}{H}, \frac{k}{K}, \frac{p}{P}\right).$$

Again notice that $H^3 = PK^2$ (using $A^3 = B^2 = X$) so we may safely absorb the exponential factor into the test function.

To separate the variables we use the expansion of additive characters into multiplicative characters, namely the formula

$$e\left(\frac{a}{n}\right) = \frac{1}{\varphi(n)} \sum_{\chi \,(\mathrm{mod}\ n)} \tau(\chi)\bar{\chi}(a),$$

valid for $(a, n) = 1$. Assuming $(h, k) = 1$ allows for a simple application of this formula; in the general situation one can impose coprimality by factoring out greatest common divisors. We have

$$S(H, K, P) = \sum_k \frac{1}{\varphi(k^2)} \sum_{\chi \,(\mathrm{mod}\ k^2)} \tau(\chi) \sum_p \sum_h \chi(p) \left(\frac{k}{p}\right) \bar{\chi}^3(h) g_1(h, k, p).$$

Because the variables h and p are separated we may use the Riemann hypothesis for Dirichlet L-functions to obtain square root cancellation in the summations over h and p, as long as the characters are nonprincipal. The bound we obtain from the nonprincipal characters is

$$S(H, K, P) \ll H^{1/2} P^{1/2} K^2 X^\varepsilon,$$

which is $\ll P^{3/2-\varepsilon}$ as long as $P \ll X^{7/9-\varepsilon}$. This reveals the limit of the method. The loss of savings from the nonprincipal characters is made up for by their rarity as well as from savings in the size of the Gauss sums of principal characters ($\tau(\chi_0)$ becomes a Ramanujan sum when χ_0 is principal).

It is important to realize that the above method does not work for all ranges of H, K, and P. For instance, when k is small the exponential factor

$$e\left(\frac{h^3}{pk^2}\right)$$

has large derivatives that cannot be safely ignored. The reciprocity law changing the modulus from p to k^2 becomes more advantageous the smaller k becomes; the tradeoff is that the exponential factor 'correction' becomes more difficult to treat. See Lemma 5.8 of [Y1] for the desired bound on the summation over h and p for general k.

Many different methods of estimation come into play in different ranges; for instance, Weyl's method is used to estimate sums of the type

$$\sum_h e\left(\frac{h^3 m}{n}\right).$$

Also notice the extensive use of reciprocity laws, namely (2.10) as well as quadratic reciprocity to view (k/p) as a character with argument p.

2.5 The family of curves with torsion group $\mathbb{Z}/2\mathbb{Z} \times \mathbb{Z}/2\mathbb{Z}$

In this section we investigate $\mathcal{F}_4$ as we did the family of all elliptic curves in Section 2.4.

The analogous 'pure' sum is the following

$$S(H, K, P) = \sum_h \sum_k \sum_p \left(\frac{hk(h+k)}{p}\right) g\left(\frac{h}{H}, \frac{k}{K}, \frac{p}{P}\right),$$

where g is as in the previous section and $H = K = P/A$. As before, the necessary bound is $S(H, K, P) \ll P^{3/2-\varepsilon}$. Clearly this sum S is very different from the corresponding sum for $\mathcal{F}_1$.

So far the best known method allows P as large as $X^{2/3-\varepsilon}$, which gives the upper bound of 2 for the average rank. It is of great interest to make any improvement to this result no matter how small the increment. Notice that when $P = X^{2/3}$ we have $H = K = P^{1/2}$.

We sketch the argument giving support $2/3 - \varepsilon$. We use two different methods depending on whether $hk(h+k)$ is a square. In the typical case where $hk(h+k)$ is not a square we may use the Riemann hypothesis for Dirichlet L-functions to obtain square-root cancellation in the summation over p. When $hk(h+k)$ is a square there is no cancellation in the summation over p but there is a lot of extra savings due to the rarity of such h and k. If $(h, k) = 1$ we must have that each of h, k, and $h+k$ are squares and are therefore led to the classical problem of counting primitive Pythagorean triples! The general case leads to the problem of counting solutions to

$$l_1 x^2 + l_2 y^2 = l_3 z^2$$

uniformly in l_1, l_2, and l_3. See Lemma 8.4 of [Y1] for a full treatment of the problem of counting how often $hk(h+k)$ is a square.

The sum $S(H, K, P)$ displays beautiful symmetry and exhibits the interactions between addition and multiplication.

2.6 Open problems and directions for improvement

Any improvements on the constants in (2.8) would be a major achievement. The cases $i = 3$ and $i = 4$ are especially enticing because the structure of the sums seem less complicated than for the family of all elliptic curves. The applications of obtaining average rank strictly less than 2 provide a lot of motivation. The family $\mathcal{F}_2$ has been studied by Iwaniec (work in progress) in an attack on the Landau-Siegel zero. There are various barriers in his work, but one of them essentially amounts to obtaining $\nu > 1/2$ in (2.8).

It would be good to have more examples of large families of elliptic curves where the analog of (2.7) may be computed explicitly. Notice that the best known results for $\mathcal{F}_3$ and $\mathcal{F}_4$ are the same even though the former family is larger than the latter ($\mathcal{F}_3$ has $X^{3/4}$ curves of conductor $\leqslant X$ whereas $\mathcal{F}_4$ has $X^{2/3}$ curves of conductor $\leqslant X$). Sometimes the particular structure of a family can provide for surprisingly good results.

One may be interested in removing the assumption of the generalized Riemann hypothesis from the work. Recall that we required the Riemann hypothesis in our application of the explicit formula in order to ignore any zeros not at the critical point. It should be possible using zero-density estimates to unconditionally handle zeros that are too far from the critical line. Kowalski and Michel [KM1], [KM2] have carried out this procedure for weight 2 level q newforms, obtaining the bound of 6.5 for the average order of vanishing. Kowalski, Michel, and VanderKam [KMVdK] improved this constant to less than 1.2 by studying the central values of derivatives of the family of L-functions. Unfortunately there is little hope of applying their methods to the family of all elliptic curves because of difficulties with the root number in applications of the approximate functional equation.

Along similar lines, one might try to remove the use of GRH for Dirichlet L-functions from the treatment of the family of all elliptic curves and obtain support larger than 2/3.

3 Nonvanishing results

The question of how many L-functions in a certain family vanish at the central point is of great interest and has a variety of applications. Many people have studied this problem for different families, including Dirichlet L-functions ([So], [IS1]), weight k Hecke L-functions of level N ([IS2], [KMVdK]), and quadratic twists of a fixed elliptic curve ([PP], [H-B1]), to name a handful of examples.

In this section we discuss the nonvanishing question for the family of all elliptic curves; in many ways the techniques are similar for the other families studied in the previous section, though there are many new difficulties.

3.1 Methods and results

To begin, we discuss the classical analytic method for proving nonvanishing results for a general family of L-functions. There are essentially two ingredients, namely showing bounds of the type

$$\sum_{f\in\mathcal{F}} L(1/2, f) \geqslant \mathcal{A}, \tag{3.1}$$

and

$$\sum_{f\in\mathcal{F}} L^2(1/2, f) \leqslant \mathcal{B}. \tag{3.2}$$

From these two inequalities and a simple application of Cauchy's inequality we immediately obtain

$$\sum_{\substack{f\in\mathcal{F} \\ L(1/2,f)\neq 0}} 1 \geqslant \frac{\mathcal{A}^2}{\mathcal{B}}.$$

It turns out that $L(1/2, f)$ occasionally takes large enough values so that the best possible value of $\mathcal{B}$ is of the order $\mathcal{A} \log \mathcal{A}$, so that this method barely fails to prove a positive proportion of central values do not vanish. By introducing a mollifier $M(f)$ (an approximation to $L(1/2, f)^{-1}$), it becomes possible to show

$$\sum_{f\in\mathcal{F}} L(1/2, f)M(f) \gg \mathcal{A} \tag{3.1'}$$

and

$$\sum_{f\in\mathcal{F}} L^2(1/2, f)M^2(f) \ll \mathcal{A}. \tag{3.2'}$$

and prove that a positive proportion of central values are nonzero. The existence of such a mollifier is not *a priori* obvious and picking a mollifier that optimizes the implied constants can be a tricky problem.

The main barrier to proving a nonvanishing result for the family of all elliptic curves is the lower bound (3.1). The fundamental difficulty to proving such a lower bound is the complete lack of knowledge of the distribution of the root number. So for no one has ruled out the possibility that the root number is -1 for almost all elliptic curves, which obviously causes the lower bound to be unapproachable. It is therefore necessary to make some sort of hypothesis on the distribution of the root number. A feasible goal is to minimize the severity of this hypothesis.

The analytic tool to access $L(1/2, E)$ is of course the approximate functional equation (see Proposition 2.10 of [Kow]), which states

$$L(1/2, E) = \sum_{n=1}^{\infty} \frac{\lambda_E(n)}{\sqrt{n}} g\left(\frac{n}{U}\right) + \epsilon_E \sum_{n=1}^{\infty} \frac{\lambda_E(n)}{\sqrt{n}} g\left(\frac{n}{V}\right),$$

where $UV = N$, ϵ_E is the root number, and g is a certain smooth function with rapid decay (for example, $g(x) = \exp(-2\pi x)$ is a popular choice).

Thus it is desirable to estimate sums of the type

$$\sum_{E\in\mathcal{F}} \sum_{n} \frac{\lambda_E(n)}{\sqrt{n}} g\left(\frac{n}{U}\right). \tag{3.3}$$

Notice the resemblance with (2.4). This new sum is more difficult to study than (2.4) because a workable formula for $\lambda(n)$ is not available for general n. For n prime we have (1.1), but for higher powers of p we must use the Hecke relations to determine a formula. For instance (assuming $(p, N) = 1$)

$$\lambda(p^2) = \lambda^2(p) - 1,$$
$$\lambda(p^3) = \lambda^3(p) - 2\lambda(p),$$

and in general $\lambda(p^k)$ is a polynomial in $\lambda(p)$ of degree k (it is a Tchebyshev polynomial of the second kind). By treating the squarefree and squarefull parts of n separately it is possible to minimize this difficulty. For instance, in [Y2] the following is shown.

Lemma 3.1. *For an integer n let $(n)_2$ be the squarefull part of n, that is, the product of prime powers exactly dividing n to order 2 or higher. Let $\phi(x)$ be a smooth, compactly supported function vanishing for $x > 2$ and satisfying $\phi(x) + \phi(x^{-1}) = 1$. Then on GRH we have*

$$\sum_n \frac{\lambda_E(n)}{\sqrt{n}} g\left(\frac{n}{U}\right) = \sum_n \frac{\lambda_E(n)}{\sqrt{n}} g\left(\frac{n}{U}\right) \phi\left(\frac{(n)_2}{X^\varepsilon}\right) + O(X^{-\delta}),$$

where $\delta > 0$ depends on $\varepsilon > 0$ only, and where $N \ll X$.

This result reduces the problem to estimating (3.3) except where n runs over almost squarefree integers. When n is squarefree we do have the formula

$$\lambda_E(n) = \mu(n)\frac{1}{\sqrt{n}} \sum_{x \,(\mathrm{mod}\ n)} \left(\frac{g(x)}{n}\right)$$

so it is possible to proceed using similar techiques to those in Section 2. Notice that the presence of the Möbius function here means that estimating (3.3) with n squarefree is equivalent in practice to estimating (2.4).

Using the previous lemma and the techniques of Section 2 it is possible to show

Theorem 3.2. *Let $\nu < 7/9$ and set $U = X^\nu$. Let $\mathcal{F}_1$ be the family of all elliptic curves as in Section 2.2 with corresponding smoothing function w. Then for some $c > 0$ we have*

$$\sum_{E_{a,b} \in \mathcal{F}_1} \sum_n \frac{\lambda_E(n)}{\sqrt{n}} g\left(\frac{n}{U}\right) w\left(\frac{a}{A}, \frac{b}{B}\right) \sim c|\mathcal{F}_1| \tag{3.4}$$

as $X \to \infty$.

The important feature of this result is the large allowable size for U with respect to X. The proof of the result follows the method sketched in Section 2.4 but there are many new technical difficulties arising from the presence of higher prime powers dividing n (no matter how small they may be). The requirement $\nu < 7/9$ is essentially equivalent to (2.8). A large part of this work has been to apply the methods used to prove (2.8) to the proof of Theorem 3.2. Similar results most certainly hold for the other families $\mathcal{F}_i$, $i = 2, 3, 4$ using similar methods although the details have not been carried out. The study of these families of elliptic curves it not conducive to general theories.

The constant c is a certain arithmetical factor that can be expressed as an absolutely convergent Euler product involving the average values of $\lambda(p^k)$.

To obtain a lower bound of the type (3.1) it is then necessary to control

$$R = \sum_E \epsilon_E \sum_n \frac{\lambda_E(n)}{\sqrt{n}} g\left(\frac{n}{V_E}\right),$$

where $UV_E = N$. By taking U very large it shortens this sum. Nevertheless, it is hopeless to attack this sum using the methods of harmonic analysis as in Section 2. The reason is that the root number is fundamentally incompatible with harmonic analysis. To elaborate, one may show under the restriction $4a^3 + 27b^2$ squarefree that the root number of the curve $y^2 = x^3 + ax + b$ is given by

$$\epsilon_E = \mu(4a^3 + 27b^2)\left(\frac{a}{3b}\right)\chi_4(b)\epsilon_2(-1)^{a+1},$$

where χ_4 is the primitive Dirichlet character modulo 4 and ϵ_2 is the local root number at 2. There is no nice formula for ϵ_2 because there are many cases to consider [Ha]. Bounding R is very difficult because it requires strong cancellation arising from the variation in sign of the Möbius function evaluated at the thin sequence of values of $4a^3 + 27b^2$. While there has been tremendous recent progress on the polynomials $x^2 + y^4$ and $x^3 + 2y^3$ ([FI], [H-B2], respectively), equidistribution is not yet known for our required polynomial. Furthermore, it is required that there be a lot of cancellation (power savings) in the sum; even in the simpler case of $\sum_{n \leqslant Y} \mu(n)$ the Riemann hypothesis (or at least a zero free region for $\mathrm{Re}\, s > 1 - \delta$ for some $\delta > 0$) is required to obtain power-savings in Y. It is not clear how GRH would help for the sequence of values of $4a^3 + 27b^2$ because it is required to identify an appropriate L-function associated to this sequence.

Based on the expected variation of the sign of the root number we make the following

Conjecture 3.3. *Let $\nu < 7/9$ and set $V_E = X^{-\nu}N$. Then*

$$\sum_E \epsilon_E \sum_n \frac{\lambda_E(n)}{\sqrt{n}} g\left(\frac{n}{V_E}\right) \ll (AB)^{1/2+\varepsilon}.$$

Using Theorem 3.2 and this Conjecture we have the lower bound (3.1) with $\mathcal{A} \gg AB$.

The upper bound (3.2) can be treated in a number of ways. It may be possible to obtain unconditional results because the difficulties with the root number can perhaps be avoided by positivity arguments. This is an interesting direction for further research. Since we already have been assuming GRH we instantly have (via Lindelöf) the upper bound (3.2) with $\mathcal{B} \ll ABX^{\varepsilon}$.

3.2   Future progress

To make progress on this problem it is necessary to understand the distribution of the root number. A starting point would be to show that

$$\sum_{a \leqslant A} \sum_{b \leqslant B} \mu(4a^3 + 27b^2) = o(AB),$$

which already appears to be worthy challenge.

Perhaps a better approach would be to use the approximate functional equation with $U \geqslant N$ so that the effect of the root number becomes implicit in the long sum of Dirichlet coefficients. It is not clear how the root number is captured by this sum (the Möbius function of the discriminant is nowhere to be seen when studying the sums of Dirichlet coefficients). Any work showing such a connection, even conjecturally, would be of interest.

In a related, yet different, direction, it would be good to have quantitative vanishing results. Based on the conjectured rarity of rank 2 and higher curves, statistically all vanishing central values should arise from $\epsilon_E = -1$. Perhaps it is possible to show $\epsilon_E = -1$ for many elliptic curves without necessarily showing that the root number is equidistributed. Helfgott's result on the equidistribution of the root number for the family $y^2 = x(x - a)(x + b)$ shows that at least $X^{2/3}$ curves of conductor $\ll X$ vanish at the central point.

A   Computing the complete character sums

This appendix is devoted to the rather pleasant task of computing the sums (2.7). We let $T_i(h, k, p)$ be the sum on the left hand side of (2.7).

A.1   The family of all elliptic curves

Here we carry out the case $i = 1$.

By definition,

$$T_1(h, k; p) = -\frac{1}{\sqrt{p}} \sum_{x \,(\mathrm{mod}\ p)} \sum_{\alpha \,(\mathrm{mod}\ p)} \sum_{\beta \,(\mathrm{mod}\ p)} \left(\frac{x^3 + \alpha x + \beta}{p}\right) e\left(\frac{\alpha h + \beta k}{p}\right).$$

The change of variables $\beta \to \beta - x^3 - \alpha x$ gives

$$T_1(h,k;p) = -\frac{1}{\sqrt{p}} \sum_{x \,(\mathrm{mod}\, p)} e\left(\frac{-x^3 k}{p}\right) \sum_{\alpha \,(\mathrm{mod}\, p)} e\left(\frac{\alpha(h-xk)}{p}\right)$$

$$\times \sum_{\beta \,(\mathrm{mod}\, p)} \left(\frac{\beta}{p}\right) e\left(\frac{\beta k}{p}\right)$$

$$= -\varepsilon_p p \left(\frac{k}{p}\right) e\left(\frac{-h^3 \bar{k}^2}{p}\right),$$

as claimed.

A.2 The large positive rank family

This is the case $i = 2$. We have

$$T_2(h,k;p) = -\frac{1}{\sqrt{p}} \sum_{x \,(\mathrm{mod}\, p)} \sum_{\alpha \,(\mathrm{mod}\, p)} \sum_{\beta \,(\mathrm{mod}\, p)} \left(\frac{x^3 + \alpha x + \beta^2}{p}\right) e\left(\frac{\alpha h + \beta k}{p}\right)$$

$$= -\frac{1}{\sqrt{p}} \sum_{\alpha} \sum_{\beta} \left(\frac{\beta^2}{p}\right) e\left(\frac{\alpha h + \beta k}{p}\right)$$

$$- \frac{1}{\sqrt{p}} \sum_{x \neq 0} \sum_{\alpha} \sum_{\beta} \left(\frac{x^3 + \alpha x + \beta^2}{p}\right) e\left(\frac{\alpha h + \beta k}{p}\right)$$

$$= T_0 + T_2',$$

say. We easily have

$$T_0 = -p^{1/2} \delta_p(h)(p\delta_p(k) - 1)$$

where $\delta_p(n)$ is the characteristic function of $p|n$.

The sum T_2' is, after the linear change of variables $\alpha \to \alpha - x^2 - \beta^2 \bar{x}$, given by

$$T_2' = -\frac{1}{\sqrt{p}} \sum_{x \neq 0} \left(\frac{x}{p}\right) \sum_{\beta} e\left(\frac{-h\beta^2 \bar{x} + \beta k - hx^2}{p}\right) \sum_{\alpha} \left(\frac{\alpha}{p}\right) e\left(\frac{\alpha h}{p}\right)$$

$$= -\varepsilon_p \left(\frac{h}{p}\right) \sum_{x \neq 0} \left(\frac{x}{p}\right) e\left(\frac{-hx^2}{p}\right) \sum_{\beta} e\left(\frac{-hx\beta^2 + xk\beta}{p}\right) \quad (\text{from } \beta \to x\beta).$$

To evaluate the summation over β we apply the formula

$$\sum_{x \,(\mathrm{mod}\, p)} e\left(\frac{ax^2 + bx}{p}\right) = \begin{cases} \varepsilon_p \sqrt{p}\left(\frac{a}{p}\right) e\left(\frac{-\bar{a}b^2\bar{4}}{p}\right) & \text{if } (a,p) = 1, \\ p & \text{if } a \equiv b \equiv 0 \ (\mathrm{mod}\, p), \\ 0 & \text{otherwise.} \end{cases} \quad (\text{A.1})$$

We obtain

$$T_2' = -\varepsilon_p^2 p^{1/2} \left(\frac{h^2}{p}\right)\left(\frac{-1}{p}\right) \sum_{x \neq 0} e\left(\frac{-hx^2 + k^2 \bar{h}\bar{4}x}{p}\right).$$

Applying (A.1) again we obtain

$$T_2' = -p^{1/2} \left(\frac{h^2}{p}\right) \left(\varepsilon_p \sqrt{p} \left(\frac{-h}{p}\right) e\left(\frac{\bar{h}^3 k^4 \bar{2}^6}{p}\right) - 1\right).$$

Gathering terms and simplifying finishes the calculation.

A.3 The family with 2-torsion

Here we do the case $i = 3$. Using the change of variables $\beta \to -\beta + \alpha x + x^2$ we have

$$
\begin{aligned}
T_3 &= -\frac{1}{\sqrt{p}} \sum_x \sum_\alpha \sum_\beta \left(\frac{x(x^2 + \alpha x - \beta)}{p}\right) e\left(\frac{\alpha h + \beta k}{p}\right) \\
&= -\frac{1}{\sqrt{p}} \sum_x \left(\frac{x}{p}\right) e\left(\frac{x^2 k}{p}\right) \sum_\alpha e\left(\frac{\alpha(h + xk)}{p}\right) \sum_\beta \left(\frac{\beta}{p}\right) e\left(\frac{-\beta k}{p}\right) \\
&= -\varepsilon_p p \left(\frac{h}{p}\right) e\left(\frac{-h^2 \bar{k}}{p}\right),
\end{aligned}
$$

as desired.

A.4 The family with torsion group $\mathbb{Z}/2\mathbb{Z} \times \mathbb{Z}/2\mathbb{Z}$

Using the changes of variables $\alpha \to \alpha + x$ and $\beta \to \beta - x$, we get

$$
\begin{aligned}
T_4 &= -\frac{1}{\sqrt{p}} \sum_x \sum_\alpha \sum_\beta \left(\frac{x(x - \alpha)(x + \beta)}{p}\right) e\left(\frac{\alpha h + \beta k}{p}\right) \\
&= -\frac{1}{\sqrt{p}} \sum_{x, \alpha, \beta \,(\mathrm{mod}\, p)} \left(\frac{x \alpha \beta}{p}\right) e\left(\frac{-\alpha h + \beta k + x(h - k)}{p}\right) \\
&= -\varepsilon_p^3 p \left(\frac{-hk(h - k)}{p}\right) = -p\varepsilon_p \left(\frac{hk(h - k)}{p}\right),
\end{aligned}
$$

which completes the calculation.

References

[B] A. Brumer, *The average rank of elliptic curves. I*, Invent. Math. 109 (1992), no. 3, 445–472.

[FI] J. Friedlander and H. Iwaniec, *The polynomial $X^2 + Y^4$ captures its primes*, Ann. of Math. (2) 148 (1998), no. 3, 945–1040.

[Ha] E. Halberstadt, *Signes locaux des courbes elliptiques en 2 et 3*, C. R. Acad. Sci. Paris Sér. I Math. 326 (1998), no. 9, 1047–1052.

[H-B1] D. R. Heath-Brown, *The average analytic rank of elliptic curves*, Duke Math J., 122 (2004), no. 3, 591-623.

[H-B2] D. R. Heath-Brown, *Primes represented by $x^3 + 2y^3$*, Acta Math. 186 (2001), no. 1, 1–84.

[He1] H. Helfgott, *On the behavior of root numbers in families of elliptic curves*, http://www.arxiv.org/abs/math.NT/0408141.

[He2] H. Helfott, *On the square-free sieve*, Acta Arith. 115 (2004), 349-402.

[IK] H. Iwaniec and E. Kowalski, *Analytic Number Theory*, American Mathematical Society Colloquium Publications, 53. American Mathematical Society, Providence, RI, 2004. xii+615 pp.

[ILS] H. Iwaniec, W. Luo and P. Sarnak, *Low lying zeros of families of L-functions*, Inst. Hautes Études Sci. Publ. Math. no. 91 (2001), 55–131 .

[IS1] H. Iwaniec and P. Sarnak, *Dirichlet L-functions at the central point*, Number theory in progress, Vol. 2 (Zakopane-Kościelisko, 1997), 941–952.

[IS2] H. Iwaniec and P. Sarnak, *The non-vanishing of central values of automorphic L-functions and Landau-Siegel zeros*, Israel J. Math. 120 (2000), part A, 155–177.

[KS1] N. Katz and P. Sarnak, *Random Matrices, Frobenius Eigenvalues, and Monodromy.* American Mathematical Society Colloquium Publications, 45. American Mathematical Society, Providence, RI, 1999.

[KS2] N. Katz and P. Sarnak, *Zeroes of zeta functions and symmetry*, Bull. Amer. Math. Soc. 36 (1999), 1–26.

[Kol] V. Kolyvagin, *Euler systems*, The Grothendieck Festschrift, Vol. II, 435–483, Progr. Math., 87, Birkhäuser Boston, Boston, MA, 1990.

[Kow] E. Kowalski, *Elliptic curves, rank in families and random matrices*, in this volume.

[KM1] E. Kowalski and P. Michel, *The analytic rank of $J_0(q)$ and zeros of automorphic L-functions*, Duke Math. J. 100 (1999), no. 3, 503–542.

[KM2] E. Kowalski and P. Michel, *Explicit upper bound for the (analytic) rank of $J_0(q)$*, Israel J. Math. 120 (2000), part A, 179–204.

[KMVdK] E. Kowalski, P. Michel, and J. VanderKam, *Non-vanishing of high derivatives of automorphic L-functions at the center of the critical strip*, J. Reine Angew. Math. 526 (2000), 1–34.

[PP] A. Perelli and J. Pomykała, *Averages of twisted elliptic L-functions*, Acta Arith. 80 (1997), no. 2, 149–163.

[N] K. Nagao, $\mathbb{Q}(T)$-*rank of elliptic curves and certain limit coming from the local points*, with an appendix by Nobuhiko Ishida, Tsuneo Ishikawa and the author, Manuscripta Math. 92 (1997), no. 1, 13–32.

[RS] M. Rosen and J. Silverman, *On the rank of an elliptic surface*, Invent. Math. 133 (1998), no. 1, 43–67.

[Si] J. Silverman, *The Arithmetic of Elliptic Curves*, Springer-Verlag, New York, 1986.

[So] K. Soundararajan, *Nonvanishing of quadratic Dirichlet L-functions at $s = \frac{1}{2}$*, Ann. of Math. (2) 152 (2000), no. 2, 447–488.

[W] A. Wiles, *Modular elliptic curves and Fermat's last theorem*, Ann. of Math. (2) 141 (1995), no. 3, 443–551.

[Y1] M. Young, *Low-lying zeros of families of elliptic curves*, J. Amer. Math. Soc. 19 (2006), no. 1, 205–250.

[Y2] M. Young, *On the nonvanishing of elliptic curve L-functions at the central point*, to appear in Proc. London Math. Soc.

American Institute of Mathematics,
360 Portage Ave.,
Palo Alto, CA 94306-2244, USA

myoung@aimath.org

The derivative of $SO(2N+1)$ characteristic polynomials and rank 3 elliptic curves

N. C. Snaith

Abstract

We calculate the value distribution of the first derivative of characteristic polynomials of matrices from $SO(2N+1)$ at the point 1, the symmetry point on the unit circle of the eigenvalues of these matrices. The connection between the values of random matrix characteristic polynomials and values of the L-functions of families of elliptic curves implies that this calculation in random matrix theory is relevant to the problem of predicting the frequency of rank three curves within these families, since the Birch and Swinnerton-Dyer conjecture relates the value of an L-function and its derivatives to the rank of the associated elliptic curve. This article is based on a talk given at the Isaac Newton Institute for Mathematical Sciences during the "Clay Mathematics Institute Special Week on Ranks of Elliptic Curves and Random Matrix Theory".

1 Introduction

1.1 Random matrix theory and number theory

The connection between random matrix theory and number theory began with the work of Montgomery [27] when he conjectured that the distribution of the complex zeros of the Riemann zeta function follows the same statistics as the eigenvalues of a random matrix chosen from $U(N)$ generated uniformly with respect to Haar measure. This conjecture is supported by numerical evidence [28] and also by further work [16, 31, 2, 3] suggesting that the same conjecture is true for more general L-functions. For all these L-functions there is a Generalized Riemann Hypothesis that the non-trivial zeros lie on a vertical line in the complex plane. The conjectures mentioned above concern the statistics of the zeros high on this critical line.

The philosophy of Katz and Sarnak [19, 20] extended the connection with random matrix theory by proposing that rather than averaging over many zeros of a given L-function, if the zeros near to the point where the critical line crosses the real axis are averaged over a family of naturally connected L-functions, then they will be found to follow the statistics of the eigenvalues of one of the three classical compact groups of random matrices: $U(N)$, $O(N)$ or $USp(2N)$, where again the statistics are computed with respect to the probability measure given by Haar measure. There is numerical evidence for this conjecture as well [30], and strong support is given to it by the rigorous work of Katz and Sarnak [19] in the case of function field zeta functions.

For a review of applications of random matrix theory to questions in number theory see, for example, [5] or [23].

There has been a series of papers, starting with [22] and continuing with [6, 18, 17, 21, 8, 7], examining how random matrix theory can be used to predict the distribution of values of the Riemann zeta function and other L-functions, either averaged over an interval high on the critical line, or over a family at the critical point where the critical line crosses the real axis. For large values of the natural asymptotic parameter, for example the variable ordering the L-functions within the family, the moments of L-functions are conjectured to split into a product of an arithmetic contribution, determined by the family being averaged over, and a component derived from a random matrix calculation - the corresponding moment of the characteristic polynomial of the matrices in one of the three groups $U(N)$, $O(N)$ or $USp(2N)$. The asymptotic parameter on the random matrix side is the dimension of the matrix N and a natural equivalence can be made between the two.

In the following section we review the results of Conrey, Keating, Rubinstein and Snaith [9] which use the random matrix prediction for the leading order behaviour of moments of L-functions mentioned above to conjecture the frequency of zeros at the critical point among L-functions in a family corresponding to quadratic twists of an elliptic curve. With the Birch and Swinnerton-Dyer conjecture, this result predicts the frequency of curves of rank two or greater occurring in this family. This work makes use of a discretization formula [36, 32, 24] relating L-values at the critical point to Fourier coefficients of half-integral weight forms. See also David, Fearnley and Kisilevsky [12, 13] for a similar use of random matrix theory to predict frequency of vanishing at the critical point amongst families of elliptic curve L-functions twisted by cubic and higher order characters.

The second half of this article presents the random matrix calculation of the distribution of values of the derivative of characteristic polynomials of matrices from $SO(2N + 1)$ with Haar measure. We find, at equation (2.20), that moment of the derivative grows like

$$\mathcal{M}(N, s) := \int_{SO(2N+1)} |\Lambda'_U(1)|^s dU_{Haar} \sim (2\pi)^{s/2} 2^{-s^2/2} \frac{G(3/2)}{G(3/2 + s)} N^{s^2/2 + s/2},$$

$$(1.1)$$

where $\Lambda_U(e^{i\theta})$ is the characteristic polynomial of $U \in SO(2N + 1)$ and $G(z)$ is the Barnes double gamma function (see 2.11). We also show that the probability that $|\Lambda'_U(1)| < X$ over $U \in SO(2N + 1)$ is, for small X, given by (see

(2.9) and the sentence following)

$$\tfrac{2}{3} X^{\frac{3}{2}} f(N), \tag{1.2}$$

for the function f given at (2.10). With a discretization formula for the derivatives of elliptic curve L-functions at the critical point similar to that of the L-values themselves, (1.2) could be used to predict the frequency of curves of rank 3 or higher within a family of quadratic twists. Unfortunately such a formula for the derivative of L-functions at the critical point is not yet known, but is being investigated [11], where numerical support is also given to the validity of the model presented here.

We believe that the model presented in Section 2 should apply to the L-functions with odd functional equation selected from families of quadratic twists of elliptic curves for the property that they have a zero of at least order one at the critical point. However, there has been some very interesting theoretical work computing the one-level densities of zeros of families of L-functions selected in a different way (through parametric families of elliptic curves constructed so that the rank is at least 1) that implies that in that case the zero at the critical point does not affect the position of nearby zeros (see [26] and [39]). This does not seem to be a contradiction with the model proposed here (where the zero at the critical point does repel the other close by zeros), as the zero statistics are examined over collections of L-functions selected in very different ways.

The result presented at (1.1) has already been applied by Delaunay in [14] in order to predict the moments of the orders of Tate-Shafarevich groups and regulators of elliptic curves with odd rank belonging to a family of quadratic twists.

This work has been extended [33] to considering subsets of matrices from $SO(N)$ that are constrained to have n eigenvalues equal to 1, and investigating the first non-zero derivative of the characteristic polynomial at that point. When $n = 1$ this specializes to the result in the present paper.

1.2 Random matrix theory and elliptic curves

We review here the results of [9] which apply random matrix theory to predicting the frequency of vanishing at the critical point of the L-functions in the family of elliptic curves described below; or equivalently, assuming the Birch and Swinnerton-Dyer conjecture, the frequency of rank 2 or higher curves occurring in the family of elliptic curves. The motivation for the random matrix calculations presented in Section 2 is that they may be used similarly to examine rank 3 curves, see [11].

We consider an L-function (defined by the Dirichlet series and Euler product below when Re $s > 3/2$)

$$L_E(s) = \sum_{n=1}^{\infty} \frac{a_n}{n^s} = \prod_p \mathcal{L}_p(1/p^s) = \prod_{p|\Delta} \left(1 - a_p p^{-s}\right)^{-1} \prod_{p\nmid\Delta} \left(1 - a_p p^{-s} + p^{1-2s}\right)^{-1},$$

$$(1.3)$$

that is associated to an elliptic curve E over $\mathbb{Q}$

$$E : \; y^2 = x^3 + Ax + B. \tag{1.4}$$

The coefficients a_p, for prime p, are determined by $a_p = p + 1 - \#E(\mathbb{F}_p)$, where $\#E(\mathbb{F}_p)$ counts the number of pairs x, y, with $0 \leqslant x, y \leqslant p - 1$, such that $y^2 \equiv x^3 + Ax + B \; (\bmod \; p)$, plus one for the point at infinity. Δ is the discriminant of the cubic $x^3 + Ax + B$. For an extremely clear introduction to elliptic curves in the context discussed here, see the review paper by Rubin and Silverberg [29].

A family of quadratic twists of this elliptic curve is formed by

$$E_d : \; dy^2 = x^3 + Ax + B \tag{1.5}$$

for integer d that are fundamental discriminants, and the corresponding family of L-functions, ordered by $|d|$, are

$$L_E(s, \chi_d) = \sum_{n=1}^{\infty} \frac{a_n \chi_d(n)}{n^s}, \tag{1.6}$$

where the characters $\chi_d(n)$ are the Kronecker symbol:

$$\chi_d(n) = \left(\frac{d}{n}\right). \tag{1.7}$$

The original L-function, $L_E(s)$, and its twisted companions, $L_E(s, \chi_d)$, have an analytic continuation with the following functional equations [38, 34, 4]

$$\left(\frac{2\pi}{\sqrt{Q}}\right)^{-s} \Gamma(s) L_E(s) = w_E \left(\frac{2\pi}{\sqrt{Q}}\right)^{s-2} \Gamma(2-s) L_E(2-s) \tag{1.8}$$

(where Q is the conductor of the curve E and the sign of the functional equation, w_E, takes the value ± 1) and, if $(d, Q) = 1$,

$$\left(\frac{2\pi}{\sqrt{Q}|d|}\right)^{-s} \Gamma(s) L_E(s, \chi_d) = \chi_d(-Q) w_E \left(\frac{2\pi}{\sqrt{Q}|d|}\right)^{s-2} \Gamma(2-s) L_E(2-s, \chi_d).$$

$$(1.9)$$

See also [10] where there is a similar but more detailed discussion of these L-functions.

The sign of the functional equation for the twisted L-function $L_E(s, \chi_d)$ is $\chi_d(-Q) w_E$ and is either $+1$ or -1. We consider the family of L-functions

$$\mathcal{F}_{E+} = \{L_E(s, \chi_d) : \; \chi_d(-Q) w_E = +1\}. \tag{1.10}$$

By the philosophy of Katz and Sarnak it is expected that the zeros near the critical point of such a family have statistics like eigenvalues near 1 of matrices from $SO(2N)$ with Haar measure. These eigenvalues occur in complex conjugate pairs and an eigenvalue at one must have even multiplicity. The even functional equation of the L-functions forces the same symmetry on their zeros lying, by the Generalized Riemann Hypothesis, on the line Re $s = 1$.

In contrast, the low-lying zeros of the L-functions in the family

$$\mathcal{F}_{E^-} = \{L_E(s, \chi_d) : \chi_d(-Q)w_E = -1\} \tag{1.11}$$

display the same statistics as $SO(2N+1)$ eigenvalues near one, since in this case there is always an eigenvalue at one and it has to have odd multiplicity.

Both numerical and analytical evidence have been given already [9], and we review the argument here, that random matrix theory can be used to conjecture the frequency of L-functions vanishing at the critical point in a family such as $\mathcal{F}_{E^+}$. This is particularly important because of the Birch and Swinnerton-Dyer conjecture which asserts that the order of the zero of an elliptic curve L-function at the critical point is equal to the rank of the Mordell-Weil group of the elliptic curve. (See [29] for a discussion of ranks of elliptic curves and for a summary of what is known about the occurrence of ranks of various sizes amongst families of elliptic curves.)

To model values of L-functions from $\mathcal{F}_{E^+}$ near the point $s = 1$, we use the characteristic polynomials of matrices from $SO(2N)$

$$\Lambda_U(e^{i\theta}) = \prod_{n=1}^{N} \left(1 - e^{i(\theta_n - \theta)}\right) \left(1 - e^{i(-\theta_n - \theta)}\right), \tag{1.12}$$

evaluated at the point $\theta = 0$. Here $e^{\pm i\theta_1}, \ldots, e^{\pm i\theta_N}$ are the eigenvalues of the matrix $U \in SO(2N)$.

The moments of $\Lambda_U(1) = \prod_{n=1}^{N} |1 - e^{i\theta_n}|^2$ are easily calculated: using Weyl's expression [37] for Haar measure on the conjugacy classes of $SO(2N)$ and a form of Selberg's integral they are [21]

$$\int_{SO(2N)} \Lambda_U(1)^s dU_{Haar}$$

$$= 2^{2Ns} \prod_{j=1}^{N} \frac{\Gamma(N + j - 1)\Gamma(s + j - 1/2)}{\Gamma(j - 1/2)\Gamma(s + j + N - 1)} \tag{1.13}$$

$$\equiv M_O(N, s).$$

The L-function moments are then conjectured to have the form [6, 21]

$$M_E(T, s) \equiv \frac{1}{T^*} \sum_{\substack{|d| \leqslant T \\ L_E(s, \chi_d) \in \mathcal{F}_{E^+}}} L_E(1, \chi_d)^s$$

$$\sim a_s(E) M_O(N, s) \tag{1.14}$$

for large T. Here $N = \log T$ (from equating the density of zeros near the critical point with the density of the matrix eigenvalues), the sum is over fundamental discriminants d, T^* is the number of terms in the sum and $a_s(E)$ is an Euler product that contains arithmetic information specific to the elliptic curve E and the family of L-functions being averaged over. In practice, it is often a subset of $\mathcal{F}_{E+}$ that is summed over. If, for example, we select those L-functions $L_E(s, \chi_d)$ in $\mathcal{F}_{E+}$ with $d > 0$ and further restricted by a condition on $d \mod Q$, if Q is odd, and on $d \mod 4Q$, if Q is even, then the arithmetic factor would be

$$
\begin{aligned}
a_s(E) = \quad & \prod_{p \nmid Q} \left(1 - p^{-1}\right)^{s(s-1)/2} \left(\frac{p}{p+1}\right) \left(\frac{1}{p} + \frac{1}{2}\left(\mathcal{L}_p(1/p)^s + \mathcal{L}_p(-1/p)^s\right)\right) \\
\times \quad & \prod_{p \mid Q} \left(1 - p^{-1}\right)^{s(s-1)/2} \mathcal{L}_p(a_p/p)^s.
\end{aligned}
\tag{1.15}
$$

See [10] and [8], Section 4.4 for more examples.

Next we consider the distribution of the values of the characteristic polynomials of $SO(2N)$ matrices at the point 1. If $P_O(N, x)dx$ is the probability that the characteristic polynomial of a matrix chosen from $SO(2N)$ with Haar measure has a value between x and $x + dx$, then

$$
\begin{aligned}
P_O(N, x) &= \frac{1}{2\pi i x} \int_{(c)} M_O(N, s) x^{-s} ds \\
&\sim x^{-1/2} h(N)
\end{aligned}
\tag{1.16}
$$

for $x \to 0^+$, since for small x the behaviour is dominated by the pole of $M_O(N, s)$ at $s = -1/2$. Here (c) denotes a path of integration along the vertical line from $c - i\infty$ to $c + i\infty$, $c > 0$.

For large N, $h(N) \sim 2^{-7/8}G(1/2)\pi^{-1/4}N^{3/8}$ (G is the Barnes double gamma function, defined as [1, 35]:

$$
G(1 + z) = (2\pi)^{z/2} e^{-[(1+\gamma)z^2 + z]/2} \prod_{n=1}^{\infty} \left[(1 + z/n)^n e^{-z + z^2/(2n)}\right],
\tag{1.17}
$$

where γ is Euler's constant. See also (2.11) for more properties of this function.) Since the probability that an element of $SO(2N)$ has a characteristic polynomial whose value at 1 is X or smaller is $\int_0^X P_O(N, x)dx$, we find that the the behaviour of this probability for small X and large N is

$$
\lim_{N \to \infty} N^{-3/8} \lim_{X \to 0^+} \left(X^{-1/2} \int_0^X P_O(N, x)dx\right) = 2^{1/8}G(1/2)\pi^{-1/4}.
\tag{1.18}
$$

We see from equation (1.14) that for large d, moments of L-functions are conjectured to be just $a_s(E)$ (the prime product) times the random matrix moment $M_O(N, s)$. If this is true, then we define $P_E(T, x)dx$ as the probability, amongst members of $\mathcal{F}_{E^+}$, that $L_E(1, \chi_d)$, for $|d|$ around e^N, will take a value between x and $x + dx$, giving

$$P_E(T, x) \;=\; \frac{1}{2\pi i x} \int_{(c)} M_E(T, s) x^{-s} ds, \tag{1.19}$$

and an approximation for this probability for small x should be

$$P_E(T, x) \;\sim\; a_{-1/2}(E) x^{-1/2} h(N). \tag{1.20}$$

Here equating densities of zeros gives $N \sim \log T$.

But these L-functions are constrained to take only certain discretized values. The L-values have the form [36, 32, 24]:

$$L_E(1, \chi_d) = \kappa_E \frac{c_E(|d|)^2}{\sqrt{d}}, \tag{1.21}$$

where the $c_E(|d|)$ are integers, the Fourier coefficients of a half-integral weight form.

The argument now is to suppose that if

$$L_E(1, \chi_d) < \frac{\kappa_E}{\sqrt{d}} \tag{1.22}$$

then

$$L_E(1, \chi_d) = 0. \tag{1.23}$$

Thus we integrate (1.20) as we did (1.16) and so predict that

$$\#\{|d| \le T : L_E(1, \chi_d) = 0, L_E(s, \chi_d) \in \mathcal{F}_{E^+}\} \sim \frac{8}{3} \sqrt{\kappa_E} a_{-1/2} \frac{T^*}{T^{1/4}} h(N). \tag{1.24}$$

with $N \sim \log T$. However, the $c(|d|)$ are divisible by some predetermined powers of 2 which change this discretization. To avoid this problem, the conjecture stated in [9] is restricted to prime discriminants.

Conjecture 1.1. (Conrey, Keating, Rubinstein, Snaith):

Let E be an elliptic curve defined over $\mathbb{Q}$. Then there is a constant $c_E \geqslant 0$ such that

$$\sum_{\substack{p \leqslant T \\ L_E(1,\chi_p)=0 \\ L_E(s,\chi_p) \in \mathcal{F}_{E^+}}} 1 \sim c_E T^{3/4} (\log T)^{-5/8}$$

(The conjecture was originally stated in [9] with $c_E > 0$, but in fact numerics have revealed that the constant can be zero [10] for certain families. An explanation of such a case with $c_E = 0$ was given by Delaunay [15].)

With the Birch and Swinnerton-Dyer conjecture, this suggests that out of all the elliptic curves associated with L-functions in $\mathcal{F}_{E+}$ with prime discriminant $p \leqslant T$ (there are of order $T/\log T$ of them) a number of order $T^{3/4}(\log T)^{-5/8}$ should have rank two or greater. The $T^{3/4}$ has been predicted previously by Sarnak using different arguments, but random matrix theory adds more detailed information in the form of the power on the logarithm. For numerical evidence supporting the conjecture, see [9].

In the next section we will calculate the distribution of values of the first derivative at the point 1 of the characteristic polynomials from $SO(2N + 1)$ matrices (with Haar measure). The calculation is similar to that leading to (1.16). If a discretization of values of $L'_E(1, \chi_d)$ were known, in analogy to (1.21), then the following calculation could be used to probe questions of elliptic curves of rank three occurring in families of quadratic twists (the family described in (1.5)).

2   Random matrix calculations

In this section we will calculate the probability of the first derivative of a characteristic polynomial of a random (with respect to Haar measure) $SO(2N+1)$ matrix taking a value less than X at the point 1 on the unit circle. Since the zeros near the critical point of L-functions in the family $\mathcal{F}_{E-}$ are predicted to have statistics like those of the eigenvalues near 1 of a random $SO(2N + 1)$ matrix, it is expected that the probability density of values of the derivative of the characteristic polynomial will model that of the derivative at the critical point of L-functions in this family.

For a matrix $U \in SO(2N + 1)$ the characteristic polynomial looks like

$$\Lambda_U(e^{i\theta}) = (1 - e^{-i\theta}) \prod_{n=1}^{N}(1 - e^{i(\theta_n - \theta)})(1 - e^{i(-\theta_n - \theta)}). \tag{2.1}$$

We will consider the derivative

$$\Lambda'_U(1) \ := \ \frac{d}{d\alpha}\left[(1 - e^{-\alpha})\prod_{n=1}^{N}(1 - e^{i\theta_n - \alpha})(1 - e^{-i\theta_n - \alpha})\right]_{\alpha=0}$$

$$= \ \prod_{n=1}^{N}|1 - e^{i\theta_n}|^2$$

$$= \ 2^N \prod_{n=1}^{N}(1 - \cos\theta_n). \tag{2.2}$$

We will now calculate the moments and value distribution of $\Lambda'_U(1)$ averaged over $SO(2N + 1)$ with respect to Haar measure. Since $\Lambda'_U(1)$ depends only on the eigenvalues of the matrix U, we use the expression

$$C \prod_{n=1}^{N}(1 - \cos\theta_n) \prod_{1 \leqslant j < k \leqslant N}(\cos\theta_j - \cos\theta_k)^2 \tag{2.3}$$

for the measure on conjugacy classes of matrices with the same set of eigenvalues [37]. The normalisation constant is

$$C = 2^{-N^2} \prod_{j=1}^{N} \Gamma(N+j) \ \left(\Gamma(j+1)\Gamma(1/2+j)\Gamma(j-1/2)\right)^{-1}. \tag{2.4}$$

The sth moment of the derivative of the characteristic polynomial is given by

$$\mathcal{M}(N, s) := C \int_0^\pi \cdots \int_0^\pi |\Lambda_U'(1)|^s \prod_{n=1}^{N}(1 - \cos\theta_n)$$

$$\times \prod_{1 \leqslant j < k \leqslant N}(\cos\theta_j - \cos\theta_k)^2 d\theta_1 \cdots d\theta_N$$

$$= C \int_0^\pi \cdots \int_0^\pi 2^{Ns} \prod_{n=1}^{N}(1 - \cos\theta_n)^{1+s} \prod_{1 \leqslant j < k \leqslant N}(\cos\theta_j - \cos\theta_k)^2 d\theta_1 \cdots d\theta_N$$

$$= C2^{Ns} \int_{-1}^1 \cdots \int_{-1}^1 \prod_{n=1}^{N} \frac{(1 - x_n)^{1/2+s}}{(1 + x_n)^{1/2}} \prod_{1 \leqslant j < k \leqslant N}(x_j - x_k)^2 dx_1 \cdots dx_N. \tag{2.5}$$

This can be evaluated using a form of Selberg's integral (for details see [25]):

$$\int_{-1}^1 \cdots \int_{-1}^1 \prod_{1 \leqslant j < l \leqslant n} |(x_j - x_l)|^{2\gamma} \prod_{j=1}^{n}(1 - x_j)^{\alpha-1}(1 + x_j)^{\beta-1} dx_j$$

$$= 2^{\gamma n(n-1)+n(\alpha+\beta-1)} \prod_{j=0}^{n-1} \frac{\Gamma(1 + \gamma + j\gamma)\Gamma(\alpha + j\gamma)\Gamma(\beta + j\gamma)}{\Gamma(1 + \gamma)\Gamma(\alpha + \beta + \gamma(n + j - 1))}, \tag{2.6}$$

if $\mathrm{Re}\,\alpha > 0$, $\mathrm{Re}\,\beta > 0$ and $\mathrm{Re}\,\gamma > -\min\left(\frac{1}{n}, \frac{\mathrm{Re}\,\alpha}{n-1}, \frac{\mathrm{Re}\,\beta}{n-1}\right)$.

We have $\gamma = 1$, $\alpha = 3/2 + s$ and $\beta = 1/2$, so the integral in equation (2.5) is

$$\mathcal{M}(N, s) = C2^{2Ns+N^2} \prod_{j=1}^{N} \frac{\Gamma(j+1)\Gamma(1/2 + s + j)\Gamma(j - 1/2)}{\Gamma(s + N + j)}$$

$$= 2^{2Ns} \prod_{j=1}^{N} \frac{\Gamma(1/2 + s + j)\Gamma(N + j)}{\Gamma(1/2 + j)\Gamma(s + N + j)}. \tag{2.7}$$

102   N. C. Snaith

If the probability that $|\Lambda'_U(1)|$ takes a value between x and $x + dx$ is given by $P(N, x)dx$, then from a standard result in probability (where the contour of integration is a vertical line with real part equal to $c > 0$)

$$
\begin{aligned}
P(N, x) &= \frac{1}{2\pi i x} \int_{(c)} x^{-s} \mathcal{M}(N, s) ds \\
&= \frac{1}{2\pi i x} \int_{(c)} x^{-s} 2^{2Ns} \prod_{j=1}^{N} \frac{\Gamma(1/2 + s + j)\Gamma(N + j)}{\Gamma(1/2 + j)\Gamma(s + N + j)} ds. \quad (2.8)
\end{aligned}
$$

We are particularly interested in the behaviour at small x, and this is dominated by the nearest pole to zero of the integrand: the pole at $s = -3/2$ of $\Gamma(1/2 + s + j)$. Thus for small x

$$
P(N, x) \sim x^{1/2} f(N). \tag{2.9}
$$

The probability that $|\Lambda'_U(1)| < X$ for U chosen from $SO(2N + 1)$ with Haar measure is therefore $\sim \frac{2}{3} X^{\frac{3}{2}} f(N)$ for small X.

The function $f(N)$, derived from (2.8) by a residue calculation, is given by

$$
f(N) = 2^{-3N} \frac{1}{\Gamma(N)} \prod_{j=1}^{N} \frac{\Gamma(j)\Gamma(N + j)}{\Gamma(1/2 + j)\Gamma(N + j - 3/2)}. \tag{2.10}
$$

For large N the behaviour of $f(N)$ can be determined using the Barnes G-function [1, 35]:

$$
G(1 + z) = (2\pi)^{z/2} e^{-[(1+\gamma)z^2 + z]/2} \prod_{n=1}^{\infty} \left[(1 + z/n)^n e^{-z + z^2/(2n)} \right], \tag{2.11}
$$

which has zeros at the negative integers, $-n$, with multiplicity n $(n = 1, 2, 3 \ldots)$. Other properties useful to us are

$$
\begin{aligned}
G(1) &= 1, \tag{2.12} \\
G(z + 1) &= \Gamma(z)\, G(z),
\end{aligned}
$$

and furthermore, for large $|z|$

$$
\log G(z + 1) \sim z^2 (\tfrac{1}{2} \log z - \tfrac{3}{4}) + \tfrac{1}{2} z \log(2\pi) - \tfrac{1}{12} \log z + \zeta'(-1) + O(\tfrac{1}{z}). \tag{2.13}
$$

Thus

$$
\prod_{j=1}^{N} \Gamma(j) = G(N + 1), \tag{2.14}
$$

$$
\prod_{j=1}^{N} \Gamma(N + j) = \frac{\prod_{j=1}^{2N} \Gamma(j)}{\prod_{j=1}^{N} \Gamma(j)} = \frac{G(2N + 1)}{G(N + 1)}, \tag{2.15}
$$

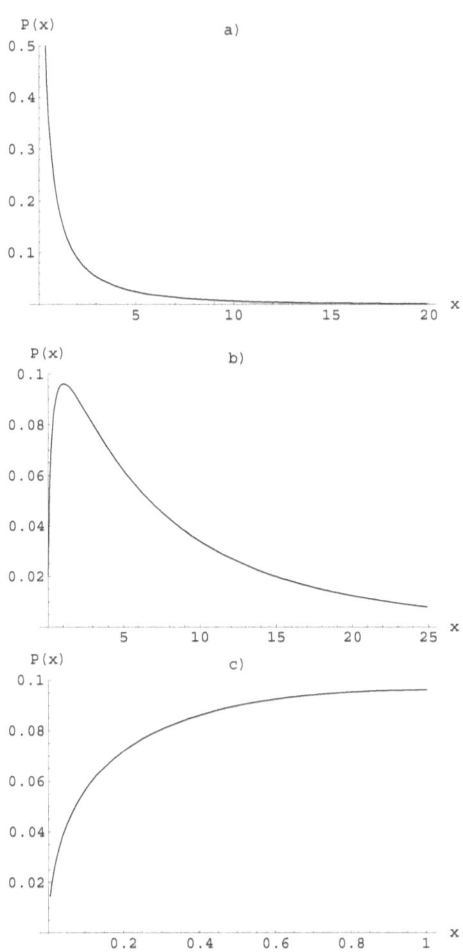

Figure 2.1: Figure a) shows the distribution of values of the characteristic polynomial at the point one of $SO(2N)$ matrices ($P_O(N,x)$ from (1.16)) when $N = 5$. In comparison, figure b) is the value distribution of the derivative of the characteristic polynomial for $SO(2N+1)$ matrices, that is $P(N,x)$ from (2.8), at the point one when $N = 5$. Figure c) shows in more detail the behaviour at the origin of figure b) (see equation (2.9)).

$$\prod_{j=1}^{N} \Gamma(j + 1/2) = \frac{G(3/2 + N)}{G(3/2)} \tag{2.16}$$

and

$$\prod_{j=1}^{N} \Gamma(N + j - 3/2) = \frac{G(2N - 1/2)}{G(N - 1/2)}. \tag{2.17}$$

So we can write

$$f(N) = 2^{-3N} \frac{G(N)G(2N + 1)G(3/2)G(N - 1/2)}{G(N + 1)G(N + 3/2)G(2N - 1/2)} \tag{2.18}$$

and expanding the G-functions for large N gives

$$f(N) \;\sim\; G(3/2) N^{\frac{3}{8}} 2^{-\frac{15}{8}} \pi^{-\frac{3}{4}}. \tag{2.19}$$

Note that $G(3/2) = \Gamma(1/2)G(1/2)$ and $\Gamma(1/2) = \pi^{1/2}$ and [1] $G(1/2) = A^{-3/2}\pi^{-1/4}e^{1/8}2^{1/24}$ with $A = 1.28242713$.

Note also that we can use (2.13) to revisit the moment $\mathcal{M}(N, s)$ and examine that asymptotically for large N. This gives us the large N behaviour of the moments at the point one of the derivative of characteristic polynomials of $SO(2N + 1)$ matrices. We have

$$\mathcal{M}(N, s) \;=\; 2^{2Ns} \frac{G(3/2 + s + N)}{G(3/2 + s)} \frac{G(3/2)}{G(3/2 + N)} \frac{G(2N + 1)}{G(N + 1)} \frac{G(s + N + 1)}{G(s + 2N + 1)}$$

$$\sim\; (2\pi)^{s/2} 2^{-s^2/2} \frac{G(3/2)}{G(3/2 + s)} N^{s^2/2 + s/2}. \tag{2.20}$$

3   Discussion

We have shown that the probability that $|\Lambda'_U(0)| < X$ over $U \in SO(2N + 1)$ with Haar measure is, for small X, given by $\frac{2}{3}X^{\frac{3}{2}}f(N)$. With a better understanding of the values taken by the derivative of L-functions associated to a family of quadratic twists of an elliptic curve this could be used to predict the frequency of rank three curves occurring amongst the members of such a family that have an odd functional equation. Some preliminary numerics have been done to investigate these derivative values, and further work is ongoing [11].

The result (2.20) was applied by Delaunay in [14] to predict moments of the orders of Tate-Shafarevich groups and the regulators of elliptic curves belonging to a family of quadratic twists - a family of type $\mathcal{F}_{E-}$. The Birch and Swinnerton-Dyer conjecture provides a formula for the first non-zero derivative of an L-function in terms of various quantities related to the associated elliptic curve. One of these quantities is the order of the Tate-Shafarevich group, and another, in the case of the first derivative of L-functions with odd functional equation, is the regulator. For families of L-functions with even functional equation the result (1.14) is used by Delaunay to predict the asymptotic form of moments of the order of the Tate-Shafarevich group for the associated family of elliptic curves, and for families with odd functional equation (2.20) was used to conjecture the form of moments of the regulator.

4   Acknowledgments

Thanks to Brian Conrey for his number theoretical insight in recognizing the potential importance of these calculations from the start, to the Royal Society and EPSRC for supporting this research and to the Isaac Newton Institute for Mathematical Sciences for their hospitality during the programme "Random Matrix Approaches in Number Theory" during which this paper was prepared.

References

[1] E.W. Barnes, The theory of the G-function, *Q. J. Math.*, **31**:264–314, 1900.

[2] E.B. Bogomolny and J.P. Keating, Random matrix theory and the Riemann zeros I: three- and four-point correlations, *Nonlinearity*, **8**:1115–1131, 1995.

[3] E.B. Bogomolny and J.P. Keating, Random matrix theory and the Riemann zeros II:n-point correlations, *Nonlinearity*, **9**:911–935, 1996.

[4] C. Breuil, B. Conrad, F. Diamond, and R. Taylor, On the modularity of elliptic curves over $\mathbb{Q}$: Wild 3-adic exercises, *J. Amer. Math. Soc.*, **14**(4):843–939, 2001.

[5] J.B. Conrey, L-functions and random matrices, In *Mathematics Unlimited 2001 and Beyond*; editors, B. Enquist and W. Schmid, pages 331–352. Springer-Verlag, Berlin, 2001, arXiv:math.nt/0005300.

[6] J.B. Conrey and D.W. Farmer, Mean values of L-functions and symmetry, *Int. Math. Res. Notices*, **17**:883–908, 2000, arXiv:math.nt/9912107.

[7] J.B. Conrey, D.W. Farmer, J.P. Keating, M.O. Rubinstein, and N.C. Snaith, Autocorrelation of random matrix polynomials, *Commun. Math. Phys.*, **237**(3):365–395, 2003, arXiv:math-ph/0208007.

[8] J.B. Conrey, D.W. Farmer, J.P. Keating, M.O. Rubinstein, and N.C. Snaith, Integral moments of L-functions, *Proc. Lond. Math. Soc.*, **91**(1):33–104, 2005, arXiv:math.nt/0206018.

[9] J.B. Conrey, J.P. Keating, M.O. Rubinstein, and N.C. Snaith, On the frequency of vanishing of quadratic twists of modular L-functions, In *Number Theory for the Millennium I: Proceedings of the Millennial Conference on Number Theory*; editor, M.A. Bennett et al., pages 301–315. A K Peters, Ltd, Natick, 2002, arXiv:math.nt/0012043.

[10] J.B. Conrey, J.P. Keating, M.O. Rubinstein, and N.C. Snaith, Random matrix theory and the Fourier coefficients of half-integral weight forms, *Experimental Mathematics*, **15**(1), 2006, arXiv:math.nt/0412083.

[11] J.B. Conrey, M.O. Rubinstein, N.C. Snaith, and M. Watkins, Discretisation for odd quadratic twists, in this volume.

[12] C. David, J. Fearnley, and H. Kisilevsky, On the vanishing of twisted L-functions of elliptic curves, *Experimental Mathematics*, **13**(2):185–98, 2004.

[13] C. David, J. Fearnley, and H. Kisilevsky, Vanishing of L-functions of elliptic curves over number fields, in this volume.

[14] C. Delaunay, Moments of the orders of Tate-Shafarevich groups, *International Journal of Number Theory*, **1**(2):243–264, 2005.

[15] C. Delaunay, Note on the frequency of vanishing of L-functions of elliptic curves in a family of quadratic twists, in this volume.

[16] D.A. Hejhal, On the triple correlation of zeros of the zeta function, *Inter. Math. Res. Notices*, **7**:293–302, 1994.

[17] C.P. Hughes, Random matrix theory and discrete moments of the Riemann zeta function, *J. Phys. A-Math. Gen.*, **36**(12):2907–2917, 2003, arXiv:math.nt/0207236.

[18] C.P. Hughes, J.P. Keating, and N. O'Connell, Random matrix theory and the derivative of the Riemann zeta function, *Proc. R. Soc. Lond. A*, **456**:2611–2627, 2000.

[19] N.M. Katz and P. Sarnak, *Random Matrices, Frobenius Eigenvalues and Monodromy*, American Mathematical Society Colloquium Publications, 45. American Mathematical Society, Providence, Rhode Island, 1999.

[20] N.M. Katz and P. Sarnak, Zeros of zeta functions and symmetry, *Bull. Amer. Math. Soc.*, **36**:1–26, 1999.

[21] J.P. Keating and N.C. Snaith, Random matrix theory and L-functions at $s = 1/2$, *Commun. Math. Phys*, **214**:91–110, 2000.

[22] J.P. Keating and N.C. Snaith, Random matrix theory and $\zeta(1/2+it)$, *Commun. Math. Phys.*, **214**:57–89, 2000.

[23] J.P. Keating and N.C. Snaith, Random matrices and L-functions, *J. Phys. A*, **36**(12):2859–81, 2003.

[24] W. Kohnen and D. Zagier, Values of L-series of modular forms at the center of the critical strip, *Invent. Math.*, **64**:175–198, 1981.

[25] M.L. Mehta, *Random Matrices*, Elsevier, Amsterdam, third edition, 2004.

[26] S.J. Miller, One- and two-level densities for rational families of elliptic curves: evidence for the underlying group symmetries, *Compos. Math.*, **140**(4):952–992, 2004.

[27] H.L. Montgomery, The pair correlation of the zeta function, *Proc. Symp. Pure Math*, **24**:181–93, 1973.

[28] A.M. Odlyzko, The 10^{20}th zero of the Riemann zeta function and 70 million of its neighbors, *Preprint*, 1989, http://www.dtc.umn.edu/ odlyzko/unpublished/index.html.

[29] K. Rubin and A. Silverberg, Ranks of elliptic curves, *Bulletin of the American Mathematical Society*, **39**(4):455–74, 2002.

[30] M. Rubinstein, *Evidence for a Spectral Interpretation of Zeros of L-functions*, PhD thesis, Princeton University, 1998.

[31] Z. Rudnick and P. Sarnak, Zeros of principal L-functions and random matrix theory, *Duke Mathematical Journal*, **81**(2):269–322, 1996.

[32] G. Shimura, On modular forms of half integral weight, *Ann. Math.*, **97**(2):440–481, 1973.

[33] N.C. Snaith, Derivatives of random matrix characteristic polynomials with applications to elliptic curves, *J. Phys. A: Math. Gen.*, **38**:10345–10360, 2005, arXiv:math.nt/0508256.

[34] R. Taylor and A. Wiles, Ring-theoretic properties of certain Hecke algebras, *Ann. of Math.*, **141**(3):553–572, 1995.

[35] A. Voros, Spectral functions, special functions and the Selberg zeta function, *Commun. Math. Phys.*, **110**:439–465, 1987.

[36] J.-L. Waldspurger, Sur les coefficients de Fourier des formes modulaires de poids demi-entier, *J. Math. Pures Appl.*, **60**(9):375–484, 1981.

[37] H. Weyl, *Classical Groups*, Princeton University Press, 1946.

[38] A. Wiles, Modular elliptic curves and fermat's last theorem, *Ann. of Math.*, **141**(3):443–551, 1995.

[39] M.P. Young, Low-lying zeros of families of elliptic curves, *J. AMS*, **19**(1):205–250, 2006, arXiv:math.nt/0406330.

School of Mathematics,
University of Bristol,
Bristol BS8 1TW, UK

n.c.snaith@bris.ac.uk

Function fields and random matrices

Douglas Ulmer *

> ... le mathématicien qui étudie ces problèmes a l'impression de déchiffrer une inscription trilingue. Dans la première colonne se trouve la théorie riemannienne des fonctions algébriques au sens classique. La troisième colonne, c'est la théorie arithmétique des nombres algébriques. La colonne du milieu est celle dont la découverte est la plus récente; elle contient la théorie des fonctions algébriques sur un corps de Galois. Ces textes sont l'unique source de nos connaissances sur les langues dans lesquels ils sont écrits; de chaque colonne, nous n'avons bien entendu que des fragments; .... Nous savons qu'il y a des grandes différences de sens d'une colonne à l'autre, mais rien ne nous en avertit à l'avance.
>
> A. Weil, "De la métaphysique aux mathématiques" (1960)

The goal of this survey is to give some insight into how well-distributed sets of matrices in classical groups arise from families of L-functions in the context of the middle column of Weil's trilingual inscription, namely function fields of curves over finite fields. The exposition is informal and no proofs are given; rather, our aim is to illustrate what is true by considering key examples.

In the first section, we give the basic definitions and examples of function fields over finite fields and the connection with algebraic curves over function fields. The language is a throwback to Weil's Foundations, which is quite out of fashion but which gives good insight with a minimum of baggage. This part of the article should be accessible to anyone with even a modest acquaintance with the first and third columns of Weil's trilingual inscription, namely algebraic functions on Riemann surfaces and algebraic number fields.

The rest of the article requires somewhat more sophistication, although not much specific technical knowledge. In the second section, we introduce ζ- and L-functions over finite and function fields and their spectral interpretation. The cohomological apparatus is treated purely as a "black box." In the third section, we discuss families of L-functions over function fields, the main equidistribution theorems, and a small sample of applications to arithmetic. Although we do not give many details, we hope that this overview will illuminate the function field side of the beautiful Katz-Sarnak picture.

In the fourth section we give some pointers to the literature for those readers who would like to learn more of the sophisticated algebraic geometry needed to work in this area.

---

*The author's research is partially supported by grants from the US National Science Foundation.

1   Function fields

In this first section we give a quick overview of function fields and their connection with curves over finite fields. The emphasis is on notions especially pertinent to function fields over finite fields (as opposed to function fields over algebraically closed fields), such as rational prime divisors on curves, places of function fields, and their behavior under extensions of fields and coverings of curves. The section ends with a Cebotarev equidistribution theorem which is a model for later more sophisticated equidistribution statements for matrices in Lie groups.

1.1   Finite fields

If p is a prime number, then $\mathbf{Z}/p\mathbf{Z}$ with the usual operations of addition and multiplication modulo p is a field which we will also denote $\mathbf{F}_p$. If $\mathbf{F}$ is a finite field, then $\mathbf{F}$ contains a subfield isomorphic to $\mathbf{F}_p$ for a uniquely determined p, the *characteristic* of $\mathbf{F}$. (The subfield $\mathbf{F}_p = \mathbf{Z}/p\mathbf{Z}$ is the image of the unique homomorphism of rings $\mathbf{Z} \to \mathbf{F}$ sending 1 to 1.) Since $\mathbf{F}$ is a finite dimensional vector space over its subfield $\mathbf{Z}/p\mathbf{Z}$, the cardinality of $\mathbf{F}$ must be p^f for some positive integer f. Conversely, for each prime p and positive integer f, there is a field with p^f elements, and any two such are (non-canonically) isomorphic. We may construct a field with $q = p^f$ elements by taking the splitting field of the polynomial $x^q - x$ over $\mathbf{F}_p$.

It is old-fashioned but convenient to fix a giant field Ω of characteristic p (say algebraically closed of infinite transcendence degree over $\mathbf{F}_p$) which will contain all fields under discussion. We won't mention Ω below, but all fields of characteristic p discussed are tacitly assumed to be subfields of Ω. Given Ω, we write $\overline{\mathbf{F}}_p$ for the algebraic closure of $\mathbf{F}_p$ in Ω (the set of elements of Ω which are algebraic over $\mathbf{F}_p$) and $\mathbf{F}_q$ for the unique subfield of $\overline{\mathbf{F}}_p$ with cardinality q. Its elements are precisely the q distinct solutions of the equation $x^q - x = 0$. With this notation, $\mathbf{F}_q \subset \mathbf{F}_{q'}$ if and only if q' is a power of q, say $q' = q^k$ in which case $\mathbf{F}_{q'}$ is a Galois extension of $\mathbf{F}_q$ with Galois group cyclic of order k generated by the q-power Frobenius map $\mathrm{Fr}_q(x) = x^q$.

1.2   Function fields over finite fields

We fix a prime p. A *function field F of characteristic p* is a finitely generated field extension of $\mathbf{F}_p$ of transcendence degree 1. The *field of constants* of F is the algebraic closure of $\mathbf{F}_p$ in F, i.e., the set of elements of F which are algebraic over $\mathbf{F}$. Since F is finitely generated, its field of constants is a finite field $\mathbf{F}_q$. When we say "F is a function field over $\mathbf{F}_q$" we always mean that $\mathbf{F}_q$ is the field of constants of F.

Examples:

1. The most basic example is the rational function field $\mathbf{F}_q(x)$ where q is a power of p and x is an indeterminate. More explicitly, the elements of $\mathbf{F}_q(x)$ are ratios of polynomials in x with coefficients in $\mathbf{F}_q$. Its field of constants is $\mathbf{F}_q$.

2. Let q be a power of p, and let F be the field extension of $\mathbf{F}_q$ generated by two elements x and y and satisfying the relation $y^2 = x^3 - 1$. More precisely, let F be the fraction field of $\mathbf{F}_q[x, y]/(y^2 - x^3 + 1)$ or equivalently $F = \mathbf{F}_q(x)[y]/(y^2 - x^3 + 1)$. The field of constants of F is $\mathbf{F}_q$. If $p > 3$, the field F is not isomorphic to the rational function field $\mathbf{F}_q(t)$. (This is a fun exercise. For hints, see [Sha77, p. 7]. Sadly, this point is missing from later editions of Shafarevitch's wonderful book.) The cases $p = 2$ and $p = 3$ are degenerate: F is isomorphic to the rational function field $\mathbf{F}_q(t)$. (If $p = 2$, let $t = (y + 1)/x$ and note that $x = t^2$ and $y = t^3 - 1$. If $p = 3$, let $t = y/(x - 1)$ and note that $x = t^2 + 1$ and $y = t^3$.)

3. Similarly, if $p \neq 2, 5$ and q is a power of p, let F be the function field generated by x and y with relation $y^2 = x^5 - 1$. It can be shown that F has field of constants $\mathbf{F}_q$ and is not isomorphic to either of the examples above.

4. Suppose that $p \equiv 3 \pmod 4$ so that -1 is not a square in $\mathbf{F}_p$. Let F be the function field generated over $\mathbf{F}_p$ by elements x_1, x_2, x_3 with relations $x_1 x_2 = x_3$ and $x_2^2 + x_3^2 = 0$. It is not hard to see that the relations imply that $x_1^2 = -1$ and so $F \cong \mathbf{F}_{p^2}(x_2)$. The moral is that the field of constants of F is not always immediately visible from the defining generators and relations.

If F has constant field $\mathbf{F}_q$, then any element $x \in F \setminus \mathbf{F}_q$ is transcendental over $\mathbf{F}_q$ and so F contains a subfield $\mathbf{F}_q(x)$ isomorphic to the rational function field. Since F has transcendence degree 1, it is algebraic over the subfield $\mathbf{F}_q(x)$.

We can always choose the element $x \in F$ such that F is a finite *separable* extension of $\mathbf{F}_q(x)$. (It suffices to choose x which is not the p-th power of an element of F.) The theorem of the primitive element then guarantees that F is generated over $\mathbf{F}_q(x)$ by a single element y satisfying a separable polynomial over $\mathbf{F}_q(x)$:

$$f(y) = y^n + a_1(x)y^{n-1} + \cdots + a_0(x) = 0 \qquad \text{with } a_i(x) \in \mathbf{F}_q(x).$$

(Separable means that f has distinct roots, or equivalently, f and $\frac{df}{dy}$ are relatively prime in $\mathbf{F}_q(x)[y]$.) This shows that F is $\mathbf{F}_q(x)[y]/(f(y))$.

More symmetrically, we may clear the denominators in the a_i and express the relation between x and y via a two-variable polynomial over $\mathbf{F}_q$:

$$g(x, y) = \sum b_{ij} x^i y^j = 0 \qquad \text{with } b_{ij} \in \mathbf{F}_q.$$

This give us a presentation of F as the fraction field of $\mathbf{F}_q[x,y]/(g(x,y))$. Thus the general function field can be generated over its constant field by two elements satisfying a polynomial relation. Note that this representation is far from unique and it may be more natural in particular cases to give several generators and relations.

1.3   Curves over finite fields

Let $\overline{\mathbf{F}}_p$ be the algebraic closure of $\mathbf{F}_p$ and let $\mathbf{P}^n(\overline{\mathbf{F}}_p)$ denote the projective space of dimension n over $\overline{\mathbf{F}}_p$. Thus elements of $\mathbf{P}^n(\overline{\mathbf{F}}_p)$ are by definition the one-dimensional subspaces of the vector space $\overline{\mathbf{F}}_p^{n+1}$. If $(a_0, \dots, a_n) \in \overline{\mathbf{F}}_p^{n+1} \setminus (0, \dots, 0)$, we write $[a_0 : \cdots : a_n]$ for the element of $\mathbf{P}^n(\overline{\mathbf{F}}_p)$ defined by the subspace spanned by $(a_0, \dots, a_n)$. We let $X_0, \dots, X_n$ denote the standard coordinates on $\overline{\mathbf{F}}_p^{n+1}$; of course the X_i do not give well-defined functions on $\mathbf{P}^n(\overline{\mathbf{F}}_p)$ but the ratio of two homogenous polynomials in the X_i of the same degree gives a well-defined function on the set where the denominator does not vanish. In particular, on the subset $X_0 \neq 0$, the functions $x_i = X_i/X_0$ $(i = 1, \dots, n)$ are a set of coordinates which give a bijection between the set where $X_0 \neq 0$ and the affine space $\mathbf{A}^n(\overline{\mathbf{F}}_p) = \overline{\mathbf{F}}_p^n$.

We put a topology on $\mathbf{P}^n(\overline{\mathbf{F}}_p)$ by declaring that a (Zariski) *closed* subset $Z \subset \mathbf{P}^n(\overline{\mathbf{F}}_p)$ is by definition the set of points where some collection of homogeneous polynomials vanishes. We may always take the set of polynomials to be finite and so a closed set has the form

$$Z = \left\{ [a_0 : \cdots : a_n] \in \mathbf{P}^n(\overline{\mathbf{F}}_p) \,|\, f_1(a_0, \dots, a_n) = \cdots = f_k(a_0, \dots, a_n) = 0 \right\}$$

where $f_1, \dots, f_k \in \overline{\mathbf{F}}_p[X_0, \dots, X_n]$ are homogeneous polynomials. A closed subset Z is said to be *defined over* $\mathbf{F}_q$ if we may take the f_i to have coefficients in $\mathbf{F}_q$.

We will work with the following definition, which is somewhat naive, but suitable for our purposes: A (smooth, projective) *curve $\mathcal{C}$ over $\mathbf{F}_q$* is a closed subset $\mathcal{C} \subset \mathbf{P}^n(\overline{\mathbf{F}}_p)$ defined over $\mathbf{F}_q$, such that:

1. $\mathcal{C}$ is infinite

2. there exist homogeneous polynomials $f_1, \dots, f_k$ vanishing identically on $\mathcal{C}$ such that for every $p \in \mathcal{C}$, the Jacobian matrix $(\frac{\partial f_i}{\partial X_j}(p))$ $(i = 1, \dots, k$ and $j = 0, \dots, n)$ has rank $n - 1$

3. $\mathcal{C}$ is not the union of two proper closed subsets, i.e., if Z_1 and Z_2 are closed subsets and $\mathcal{C} = Z_1 \cup Z_2$ then $\mathcal{C} = Z_1$ or $\mathcal{C} = Z_2$

In the language of algebraic geometry, the first condition implies that $\mathcal{C}$ has positive dimension and the first two conditions imply that it is smooth and of dimension 1. The third condition says that $\mathcal{C}$ is *absolutely irreducible*. If in the third condition we insist that Z_1 and Z_2 be defined over $\mathbf{F}_q$ we arrive at the weaker condition that $\mathcal{C}$ is *irreducible*. Although there are sometimes good reasons to consider irreducible but not absolutely irreducible curves, for simplicity we will not do so except in one example below.

We equip $\mathcal{C}$ with the Zariski topology induced from $\mathbf{P}^n(\overline{\mathbf{F}}_p)$ so that its closed subsets are intersections of $\mathcal{C}$ with closed subsets $Z \subset \mathbf{P}^n(\overline{\mathbf{F}}_p)$.

Warning: in the current literature a curve $\mathcal{C}$ is usually defined in a more sophisticated way. The set we are considering here would be denoted $\mathcal{C}(\overline{\mathbf{F}}_p)$ and called the set of $\overline{\mathbf{F}}_p$-valued points of $\mathcal{C}$.

Examples:

1. $\mathbf{P}^1 = \mathbf{P}^1(\overline{\mathbf{F}}_p)$ is the most basic example. It is defined by the zero polynomial on $\mathbf{P}^1$ (!) or, if that seems too tautological, by the equation $X_2 = 0$ in $\mathbf{P}^2(\overline{\mathbf{F}}_p)$. Either representation makes it clear that $\mathbf{P}^1$ is defined over $\mathbf{F}_p$.

1′. For $p > 2$, let $\mathcal{C}_2$ be the curve in $\mathbf{P}^2(\overline{\mathbf{F}}_p)$ defined over $\mathbf{F}_p$ by the polynomial $X_1^2 + X_2^2 - X_0^2$. Note that restricted to $\mathcal{C}_2 \cap \{X_0 \neq 0\}$, the coordinate functions x_i satisfy $x_1^2 + x_2^2 = 1$.

1″. For any p, let $\mathcal{C}_3$ be the curve in $\mathbf{P}^3(\overline{\mathbf{F}}_p)$ defined over $\mathbf{F}_p$ by the polynomials $X_0 X_2 - X_1^2$, $X_0 X_3 - X_1 X_2$, and $X_1 X_3 = X_2^2$. Note that restricted to $\mathcal{C}_3 \cap \{X_0 \neq 0\}$, the coordinate functions x_i satisfy $x_2 = x_1^2$ and $x_3 = x_1^3$.

2. Assume that $p > 3$ and let $\mathcal{C}_3'$ be the curve in $\mathbf{P}^2(\overline{\mathbf{F}}_p)$ defined over $\mathbf{F}_p$ by the polynomial $X_0 X_2^2 - X_1^3 + X_0^3$. (If $p = 2$ or 3 the second condition in the definition of a curve is not met: the Jacobian matrix is 0 at $[1:0:1]$ if $p = 2$ and at $[1:1:0]$ if $p = 3$.) Note that restricted to $\mathcal{C}_3' \cap \{X_0 \neq 0\}$, the coordinate functions x_i satisfy $x_2^2 = x_1^3 - 1$.

3. Assume that $p \neq 2, 5$ and let $\mathcal{C}_5$ be the closed subset of $\mathbf{P}^3(\overline{\mathbf{F}}_p)$ defined over $\mathbf{F}_p$ by the equation polynomials $X_0 X_2 - X_1^2$, $X_0 X_3^2 - X_1 X_2^2 + X_0^3$, and $X_1 X_3 - X_2^3 + X_0^2 X_1$. Note that restricted to $\mathcal{C}_5 \cap \{X_0 \neq 0\}$, the coordinate functions x_i satisfy $x_2 = x_1^2$ and $x_3^2 = x_1^5 - 1$.

4. Assume that $p \equiv 3 \pmod 4$ so that $-1 \in \mathbf{F}_p$ is not a square. Let $\mathcal{C}_2'$ be defined over $\mathbf{F}_p$ by the three polynomials $X_0^2 + X_1^2$, $X_2^2 + X_3^2$, and $X_0 X_3 - X_1 X_2$. Then $\mathcal{C}_2'$ is irreducible, but it is not absolutely irreducible and so it is not a curve by our definition. Indeed, $\mathcal{C}_2'$ is the union of the two lines $\{X_0 = iX_1, X_2 = iX_3\}$ and $\{X_0 = -iX_1, X_2 = -iX_3\}$ defined over $\mathbf{F}_{p^2}$ where $i \in \mathbf{F}_{p^2}$ satisfies $i^2 = -1$. Note that restricted to $\mathcal{C}_2' \cap \{X_0 \neq 0\}$, the coordinate functions x_i satisfy $x_1^2 = -1$, $x_2^2 + x_3^2 = 0$ and $x_3 = x_1 x_2$.

1.4 Morphisms and rational functions

If $\mathcal{C} \subset \mathbf{P}^m(\overline{\mathbf{F}}_p)$ and $\mathcal{C}' \subset \mathbf{P}^n(\overline{\mathbf{F}}_p)$ are curves defined over $\mathbf{F}_q$, a *morphism* of curves is a map $\phi : \mathcal{C} \to \mathcal{C}'$ with the property that at each point $P \in \mathcal{C}$,

ϕ is represented in an open neighborhood of P by homogenous polynomials. In other words, for each $P \in \mathcal{C}$ there should exist polynomials $f_0, \ldots, f_n \in \overline{\mathbf{F}}_p[X_0, \ldots, X_m]$, all homogeneous of the same degree, such that for all Q in some open neighborhood of P, not all of the f_i vanish at Q and $\phi(Q) = [f_0(Q) : \cdots : f_n(Q)]$. We say that ϕ is *defined over* $\mathbf{F}_q$ if it possible to choose the f_i with coefficients in $\mathbf{F}_q$. An *isomorphism* is a morphism which is bijective and whose inverse is a morphism.

Examples:

1. If f_0 and f_1 are homogeneous polynomials in $\mathbf{F}_q[X_0, X_1]$ of the same degree, not both 0, and with no common factors, then

$$[a_0 : a_1] \mapsto [f_0(a_0, a_1) : f_1(a_0, a_1)]$$

   defines a morphism $\mathbf{P}^1 \to \mathbf{P}^1$. Using that $\mathbf{F}_q[X_0, X_1]$ is a unique factorization domain, one checks that every morphism from $\mathbf{P}^1$ to itself defined over $\mathbf{F}_q$ is of this form.

1'. For $p > 2$, the polynomials $f_0 = X_0^2 + X_1^2$, $f_1 = X_0^2 - X_1^2$, and $f_2 = 2X_0X_1$ define a morphism from $\mathbf{P}^1$ over $\mathbf{F}_p$ to the curve $\mathcal{C}_2$ in Example (1') of Section 1.3. This morphism is an isomorphism with inverse defined by $f_0 = \frac{1}{2}(X_0 + X_1)$ and $f_1 = \frac{1}{2}(X_0 - X_1)$.

1''. For any p, the polynomials $f_0 = X_0^3$, $f_1 = X_0^2 X_1$, $f_2 = X_0 X_1^2$, and $f_3 = X_1^3$ define a morphism ϕ from $\mathbf{P}^1$ over $\mathbf{F}_p$ to the curve $\mathcal{C}_3$ in Example (1'') of Section 1.3. This morphism is an isomorphism with inverse defined on $\{X_0 \neq 0\}$ by $f_0 = X_0$ and $f_1 = X_1$ and on $\{X_3 \neq 0\}$ by $f_0 = X_2$ and $f_1 = X_3$. In this example, it is not possible to define the inverse of ϕ by a single set of polynomials on all of $\mathcal{C}_3$. Note also that the polynomials defining a morphism are in general not at all unique. For example, on $\{X_0 X_3 \neq 0\}$, the inverse of ϕ is defined both by $f_0 = X_0$ and $f_1 = X_1$ and by $f_0 = X_2$ and $f_1 = X_3$.

2. Let $\mathcal{C}_3'$ be as in Example (2) of Section 1.3. We define a morphism $\phi : \mathcal{C}_3' \to \mathbf{P}^1$ by setting $\phi([a_0 : a_1 : a_2]) = [a_0 : a_1]$ on the open set where $a_0 \neq 0$ and $\phi([a_0 : a_1 : a_2]) = [a_1^2 : a_0^2 + a_2^2]$ on the open set where $a_0^2 + a_2^2 \neq 0$. These requirements are compatible since if $a_0 \neq 0$ and $a_0^2 + a_2^2 \neq 0$, then $a_1 \neq 0$ and

$$[a_0 : a_1] = [a_0 a_1^2 : a_1^3] = [a_0 a_1^2 : a_0^3 + a_0 a_2^2] = [a_1^2 : a_0^2 + a_2^2].$$

   If we think of $\mathbf{P}^1(\overline{\mathbf{F}}_p) \setminus \{[0 : 1]\}$ as $\overline{\mathbf{F}}_p$ via $[a_0 : a_1] \mapsto a_1/a_0$, then the morphism ϕ extends the function $x_1 = X_1/X_0$, defined on $\mathcal{C}_3' \cap \{X_0 \neq 0\}$ to a morphism $\mathcal{C}_3' \to \mathbf{P}^1(\overline{\mathbf{F}}_p)$. Again, it is not possible to find a single pair of homogeneous polynomials representing ϕ at all points of $\mathcal{C}_3'$.

$2'$. Let $\mathcal{C}_3'$ be as above. Choose a non-square element $a \in \mathbf{F}_p$ $(p > 3)$ and define $\mathcal{C}_3'' \subset \mathbf{P}^2(\overline{\mathbf{F}}_p)$ by the equation $aX_0X_2^2 - X_1^3 + X_0^3 = 0$. Note that both $\mathcal{C}_3'$ and $\mathcal{C}_3''$ are defined over $\mathbf{F}_p$. Let $b \in \mathbf{F}_{p^2}$ be a square root of a and define a morphism $\phi : \mathcal{C}_3'' \to \mathcal{C}_3'$ by $\phi([a_0 : a_1 : a_2]) = [ba_0 : a_1 : a_2]$. It is clear that ϕ is defined over $\mathbf{F}_{p^2}$ and is an isomorphism. On the other hand, one can show that $\mathcal{C}_3''$ and $\mathcal{C}_3'$ are not isomorphic over $\mathbf{F}_p$. This shows that two curves not isomorphic over their fields of definition may become isomorphic over a larger field. One says that $\mathcal{C}_3''$ is a *twist* of $\mathcal{C}_3'$.

3. If $\mathcal{C} \subset \mathbf{P}^n(\overline{\mathbf{F}}_p)$ is a curve defined over $\mathbf{F}_q$, then there is an important morphism, the q-power Frobenius morphism $\mathrm{Fr}_q : \mathcal{C} \to \mathcal{C}$, defined by $\mathrm{Fr}_q([a_0 : \cdots : a_n]) = [a_0^q : \cdots : a_n^q]$. Note that the fixed points of Fr_q are precisely the points of $\mathcal{C}$ with coordinates in $\mathbf{F}_q$.

4. If $\mathcal{C} \subset \mathbf{P}^n(\overline{\mathbf{F}}_p)$ is a curve and $f_0, \ldots, f_k$ are homogeneous polynomials in $X_0, \ldots X_n$ which do not all vanish identically on $\mathcal{C}$, then the map $\phi : \mathcal{C} \to \mathbf{P}^k(\overline{\mathbf{F}}_p)$ given by

$$\phi([a_0 : \cdots : a_n]) = [f_0(a_0, \ldots, a_n) : \cdots : f_k(a_0, \ldots, a_n)]$$

is well-defined on the non-empty open subset of $\mathcal{C}$ where not all of the f_i vanish. It is an important fact that ϕ can always be extended uniquely to a well-defined morphism on all of $\mathcal{C}$. (NB: This is false for higher dimensional varieties.) In particular, there are globally defined morphisms $x_i : \mathcal{C} \to \mathbf{P}^1$ extending the maps $[a_0 : \cdots : a_n] \mapsto [a_0 : a_i]$ which are *a priori* only defined on $\mathcal{C} \cap \{X_0 \neq 0\}$.

A *rational function* on a curve $\mathcal{C}$ over $\mathbf{F}_q$ is a morphism $\phi : \mathcal{C} \to \mathbf{P}^1$ defined over $\mathbf{F}_q$, except that we rule out the constant morphism with image $\infty = [0 : 1]$. (NB: This is a reasonable definition only for curves, not for higher dimensional varieties.) In a neighborhood of any $P \in \mathcal{C}$, ϕ can be represented by polynomials: $\phi(Q) = [f_0(Q) : f_1(Q)]$ where f_0 and f_1 are homogeneous of the same degree and f_0 does not vanish identically. It is useful to think of ϕ as an $\overline{\mathbf{F}}_p$-valued function (with poles) whose value at Q is $\frac{f_1(Q)}{f_0(Q)}$. We say that ϕ is *regular at* $P \in \mathcal{C}$ if $\phi(P) \neq \infty = [0 : 1]$. If we restrict to an open set where ϕ is regular, i.e., where f_0 does not vanish, then we get a well-defined $\overline{\mathbf{F}}_p$-valued function. If ϕ and ϕ' are two rational functions, we may restrict them to an open set where they both give well-defined $\overline{\mathbf{F}}_p$-valued functions, add or multiply them, and then extend back to rational functions on $\mathcal{C}$. More explicitly, if ϕ and ϕ' are represented on some open set $U \subset \mathcal{C}$ by $[f_0 : f_1]$ and $[f_0' : f_1']$, then $\phi + \phi'$ is represented by $[f_0 f_0' : f_0' f_1 + f_0 f_1']$ and $\phi\phi'$ is represented by $[f_0 f_0', f_1 f_1']$. This gives the set of rational functions the structure of a ring, in fact an algebra over $\mathbf{F}_q$. This algebra turns out to be a field extension of $\mathbf{F}_q$ of transcendence degree 1, i.e., a function field in the sense of the previous subsection. It is denoted $\mathbf{F}_q(\mathcal{C})$.

Note that the ratio f_1/f_0 can be written as a rational function (ratio of polynomials) in $x_1 = X_1/X_0, \ldots, x_n = X_n/X_0$. This shows that if $\mathcal{C} \subset \mathbf{P}^n(\overline{\mathbf{F}}_p)$, then $\mathbf{F}_q(\mathcal{C})$ is generated over $\mathbf{F}_q$ by the rational functions $x_1, \ldots, x_n$. To determine $\mathbf{F}_q(\mathcal{C})$, we need only determine the relations among the x_i.

Examples:

1. As noted above, a rational function on $\mathbf{P}^1$ is given by two homogeneous polynomials f_0 and f_1 of the same degree, with $f_0 \neq 0$. Two rational functions $[f_0 : f_1]$ and $[f_0' : f_1']$ are equal if and only if $f_1/f_0 = f_1'/f_0'$. Thus we see that rational functions on $\mathbf{P}^1$ are equivalent to rational functions (ratios of polynomials) in $x = X_1/X_0$, i.e., $\mathbf{F}_p(\mathbf{P}^1) = \mathbf{F}_p(x)$ and more generally $\mathbf{F}_q(\mathbf{P}^1) = \mathbf{F}_q(x)$.

1'. The function fields of the curves $\mathcal{C}_2$ and $\mathcal{C}_3$ in Examples (1') and (1'') of Section 1.3 are also isomorphic to $\mathbf{F}_p(x)$. One can see this by using the relations among the x_i noted above, or by using the fact (to be explained below) that isomorphic curves have isomorphic function fields.

2. Let $\mathcal{C}_3'$ be as in Example (2) of Section 1.3 and let x_1 be the rational function ϕ of that example (so $x_1([a_0 : a_1 : a_2]) = [a_0 : a_1]$ or $[a_1^2 : a_0^2 + a_2^2]$). Let x_2 be the rational function defined on all of $\mathcal{C}_3'$ by $x_2([a_0 : a_1 : a_2]) = [a_0 : a_2]$. Then x_1 and x_2 generate the field of rational functions on $\mathcal{C}_3'$ over $\mathbf{F}_q$ and they satisfy the relation $x_2^2 = x_1^3 - 1$. In other words, $\mathbf{F}_q(\mathcal{C}_3')$ is the field in Example (2) of Section 1.2.

3. Let $\mathcal{C}_5$ be as in Example (3) of Section 1.3 and define rational functions x_1 and x_3 by

$$
x_1([a_0 : a_1 : a_2 : a_3]) = \begin{cases} [a_0 : a_1] & \text{if } a_0 \neq 0 \\ [a_2^2 : a_0^2 + a_3^2] & \text{if } a_0^2 + a_3^2 \neq 0 \end{cases}
$$

and

$$
x_3([a_0 : a_1 : a_2 : a_3]) = [a_0 : a_3].
$$

(We leave it to the reader to check that these formulas do indeed define rational functions on $\mathcal{C}_5$.) It is not hard to see that x_1 and x_3 generate $\mathbf{F}_q(\mathcal{C}_5)$. The equations defining $\mathcal{C}_5$ imply that $x_3^2 = x_1^5 - 1$ and that all relations among x_1 and x_3 are consequences of this one. Thus $\mathbf{F}_q(\mathcal{C}_5)$ is the function field of Example (3) of Section 1.2.

1.5 The function field/curve dictionary

The examples at the end of the last section illustrate the general fact that if $\mathcal{C}$ is a curve defined over $\mathbf{F}_q$, then the field of rational functions $\mathbf{F}_q(\mathcal{C})$ is a function field, i.e., a finitely generated extension of $\mathbf{F}_p$ of transcendence degree one, with field of constants $\mathbf{F}_q$.

Conversely, it turns out that every function field F with field of constants $\mathbf{F}_q$ is the field of rational functions of a curve defined over $\mathbf{F}_q$ which is uniquely determined up to $\mathbf{F}_q$-isomorphism. We sketch one construction of the curve corresponding to a function field F. As we pointed out above, F may be generated over $\mathbf{F}_q$ by two elements x and y satisfying a single relation

$$0 = g(x, y) = \sum b_{ij} x^i y^j \qquad \text{with } b_{ij} \in \mathbf{F}_q.$$

If g has degree d, we form

$$G(X_0, X_1, X_2) = X_0^d g(X_1/X_0, X_2/X_0) = \sum b_{ij} X_0^{d-i-j} X_1^i X_2^j$$

and consider the closed subset of $\mathbf{P}^2(\overline{\mathbf{F}}_p)$ defined by $G = 0$. This closed subset will be infinite and irreducible, but it will not in general be a curve under our definition, since it may not satisfy the Jacobian condition. If it does, we are finished. If not, the closed set $\{G = 0\}$ has singularities and the classical process of blowing up (see [Ful89, Chap. 7]) gives an algorithm to resolve the singularities and find a smooth curve in some high-dimensional projective space with function field F. By a suitable projection, the curve $\mathcal{C}$ can be embedded in $\mathbf{P}^3(\overline{\mathbf{F}}_p)$. In general we will not be able to find a *plane* curve with function field F. This is the case for example with the function field in Example (3) of Section 1.2 generated by x and y satisfying $y^2 = x^5 - 1$. The simplest curve with this function field is a curve in $\mathbf{P}^3(\overline{\mathbf{F}}_p)$ defined by three equations.

The dictionary between curves and function fields extends to morphisms and field extensions. More precisely, if $\mathcal{C}$ and $\mathcal{C}'$ are two curves defined over $\mathbf{F}_q$ and $\phi : \mathcal{C} \to \mathcal{C}'$ is a non-constant morphism defined over $\mathbf{F}_q$, then composition with ϕ induces a "pull-back" homomorphism of fields $\mathbf{F}_q(\mathcal{C}') \hookrightarrow \mathbf{F}_q(\mathcal{C})$ which is the identity on $\mathbf{F}_q$. Conversely, it can be shown that if F and F' are function fields over $\mathbf{F}_q$ with corresponding curves $\mathcal{C}$ and $\mathcal{C}'$, then a field inclusion $F' \hookrightarrow F$ which is the identity on $\mathbf{F}_q$ is induced by a unique non-constant morphism of curves $\mathcal{C} \to \mathcal{C}'$ which is defined over $\mathbf{F}_q$.

Examples:

1. If $\mathcal{C}$ is a curve over $\mathbf{F}_q$ and x is a non-constant rational function on $\mathcal{C}$, then x is transcendental over $\mathbf{F}_q$. Thus the rational function field $F' = \mathbf{F}_q(x)$ is a subfield of $F = \mathbf{F}_q(\mathcal{C})$. The corresponding morphism $\mathcal{C} \to \mathbf{P}^1$ is the morphism x.

2. Suppose F' is a function field with field of constants $\mathbf{F}_q$ and $\mathcal{C}'$ is the corresponding curve over $\mathbf{F}_q$. If r is a power of q so that $\mathbf{F}_r$ is a finite extension of $\mathbf{F}_q$, then the function field $F = \mathbf{F}_r F'$ corresponds to the same curve $\mathcal{C}'$ viewed over $\mathbf{F}_r$. (Here $\mathbf{F}_r F'$ is the compositum of $\mathbf{F}_r$ and F', i.e., the smallest field containing both $\mathbf{F}_r$ and F'.) In other words, $F = \mathbf{F}_r(\mathcal{C}')$.

3. We say that an extension of function fields F/F' is *geometric* if it is separable and if the field of constants of F and F' are the same. If

$n = [F : F']$ is the degree of the field extension, then the corresponding morphism of curves $\phi : C \to C'$ has degree n in the sense that for all but finitely many $P \in C'$, $\phi^{-1}(P)$ consists of n points.

4. If F/F' is a purely inseparable extension of function fields, say of degree p^m, then $F' = F^{p^m}$, the subfield of p^m-th powers. In terms of suitable equations, the morphism $C \to C'$ acts on points by raising their coordinates to the p^m-th power.

An arbitrary extension can be factored into three like these: Given F/F', let $\mathbf{F}_r$ be the field of constants of F and let F^{sep} be the separable closure of F' in F. Then $\mathbf{F}_r F'/F'$ is a constant field extension, $F^{sep}/\mathbf{F}_r F'$ is geometric, and F/F^{sep} is purely inseparable.

1.6   Points, prime divisors, and places

As we have defined it, a curve C over $\mathbf{F}_q$ is a set of points with coordinates in $\overline{\mathbf{F}}_p$. We would like to have a set which reflects the fact that the equations defining C have coefficients in $\mathbf{F}_q$. The naive thing to look at would be the set of $\mathbf{F}_q$-rational points of C, i.e., those with coordinates in $\mathbf{F}_q$, but this set is too small to be useful—it may even be empty. The classical approach is to consider $\mathbf{F}_q$-rational prime divisors.

A *divisor* on C is a finite, formal, linear combination $\mathfrak{d} = \sum a_P P$ of points of C with integer coefficients. A divisor $\mathfrak{d}$ is called *effective* if $a_P \geq 0$ for all P. The *degree* of $\mathfrak{d}$ is $\deg(\mathfrak{d}) = \sum a_P$. The *support* of $\mathfrak{d}$, written $|\mathfrak{d}|$, is the set of points appearing in $\mathfrak{d}$ with non-zero coefficient.

If $\sigma \in \mathrm{Gal}(\overline{\mathbf{F}}_p/\mathbf{F}_q)$ and $P \in C$, then P^σ is again in C. (Here σ acts on the coordinates of P and the claim follows from the fact that the equations defining C have coefficients in $\mathbf{F}_q$.) We extend this action to divisors by linearity $((\sum a_P P)^\sigma = \sum a_P P^\sigma)$ and we say that a divisor $\mathfrak{d} = \sum a_P P$ is $\mathbf{F}_q$-*rational* if it is fixed by the Galois group, i.e., if $\mathfrak{d}^\sigma = \mathfrak{d}$ for all $\sigma \in \mathrm{Gal}(\overline{\mathbf{F}}_p/\mathbf{F}_q)$.

A *prime divisor* is an effective $\mathbf{F}_q$-rational divisor which is non-zero and cannot be written as the sum of two non-zero $\mathbf{F}_q$-rational effective divisors. (Note that whether or not a divisor is prime depends on the ground field over which we are considering our curve. A better terminology might be $\mathbf{F}_q$-prime, but we will stick with the traditional terminology.) It is not hard to see that the prime divisors of C are in bijection with the orbits of $\mathrm{Gal}(\overline{\mathbf{F}}_p/\mathbf{F}_q)$ acting on C. If $\mathfrak{p}$ is a prime divisor, we define the residue field of $\mathfrak{p}$ to be the field generated over $\mathbf{F}_q$ by the coordinates of any point in the support of $\mathfrak{p}$. If $\mathfrak{p}$ is prime and has degree d, then the residue field at $\mathfrak{p}$ is $\mathbf{F}_{q^d}$.

If P is a point of C and $f \in \mathbf{F}_q(C)$ is a rational function on C, then f has a well-defined order of vanishing or pole at P. One motivation for considering prime divisors is that the order of f at P is the same for all points P in the support of the prime divisor $\mathfrak{p}$ containing P. In other words, the various points in $|\mathfrak{p}|$ cannot be distinguished from one another by the vanishing of $\mathbf{F}_q$-rational functions.

Examples:

1. Let $\mathcal{C} = \mathbf{P}^1$ over $\mathbf{F}_q$. The divisors of degree 1 are simply the points of $\mathbf{P}^1$ with coordinates in $\mathbf{F}_q$. The prime divisors of degree $d > 1$ are in bijection with the irreducible, monic polynomials in $\mathbf{F}_q[x]$ of degree d, a polynomial corresponding to the formal sum of its roots.

2. Let $\mathcal{C}'_3$ be the curve in Example (2) of Section 1.3 over $\mathbf{F}_q$. If $a \in \mathbf{F}_q$ with $a^3 - 1 \neq 0$, consider the points $P = [a : b : 1]$ and $Q = [a : -b : 1]$ where $b \in \overline{\mathbf{F}}_p$ satisfies $b^2 = a$. The divisor $\mathfrak{d} = P + Q$ has degree two and it is prime if and only if $b \notin \mathbf{F}_q$. If $b \in \mathbf{F}_q$, then $\mathfrak{d}$ is the sum of two prime divisors, namely P and Q.

The set of prime divisors on $\mathcal{C}$ is more "arithmetical" than the full set of points on $\mathcal{C}$ (since it takes into account that $\mathcal{C}$ is defined over $\mathbf{F}_q$) and more convenient and flexible than the set of $\mathbf{F}_q$-rational points of $\mathcal{C}$.

Prime divisors play the role of the prime ideals of a number field. More precisely, if $\mathfrak{p} = \sum P_i$ is a prime divisor and if $f \in \mathbf{F}_q(\mathcal{C})$ we say f is regular (resp. vanishes) at $\mathfrak{p}$ if it is regular (resp. vanishes) at one and therefore all of the $P_i \in |\mathfrak{p}|$. The set of $f \in \mathbf{F}_q(\mathcal{C})$ which are regular at $\mathfrak{p}$ is a discrete valuation ring $R_\mathfrak{p}$ with fraction field $\mathbf{F}_q(\mathcal{C})$. The maximal ideal of $R_\mathfrak{p}$ is the set of f which vanish at $\mathfrak{p}$. The residue field at $\mathfrak{p}$ as we defined it above turns out to be $R_\mathfrak{p}$ modulo its maximal ideal. We get a valuation $\operatorname{ord}_\mathfrak{p} : \mathbf{F}_q(\mathcal{C})^\times \to \mathbf{Z}$ in the usual way. It turns out that every non-trivial valuation of $\mathbf{F}_q(\mathcal{C})$ is $\operatorname{ord}_\mathfrak{p}$ for a uniquely determined prime divisor $\mathfrak{p}$. (Therefore, it is possible, although not in my opinion advisable, to eliminate the geometry completely and study function fields via their valuations. What one gains in algebraic purity hardly seems to compensate for the loss of geometric intuition this approach entails.)

Here is one respect in which the analogy between function fields and number fields breaks down ("il y a des grandes différences de sens d'une colonne à l'autre"): in a number field F, there is a canonical Dedekind domain contained in F whose primes give the non-archimedean valuations of F, namely the ring of integers. In a function field, to get a Dedekind domain we fix a non-empty set of prime divisors S and then consider the ring R of functions regular at all primes not in S. The prime ideals of R are then in bijection with the prime divisors of $\mathbf{F}_q(\mathcal{C})$ except those in S, and with the valuations of $\mathbf{F}_q(\mathcal{C})$ except those arising from primes in S. One thinks of the primes in S as the "infinite primes", but there is no canonical choice for the set S.

1.7 The Riemann-Roch theorem

The Riemann-Roch theorem is true for curves over non-algebraically closed fields and the statement is essentially the same as for the case of curves over algebraically closed fields. We give the basics in our context.

Let $\mathcal{C}$ be a curve defined over $\mathbf{F}_q$ with function field $F = \mathbf{F}_q(\mathcal{C})$. For each $P \in \mathcal{C}$ and $0 \neq f \in F$, there is a well-defined order of vanishing or pole of f at P, denoted $\mathrm{ord}_P(f)$. The *divisor of* f is defined as the formal sum $(f) = \sum_P \mathrm{ord}_P(f)$ which is in fact a finite sum. It is not hard to see that (f) is $\mathbf{F}_q$-rational and a basic results says that it has degree 0: $\sum_P \mathrm{ord}_P(f) = 0$.

If $\mathfrak{d}$ is an $\mathbf{F}_q$-rational divisor, we define the Riemann-Roch space $L(\mathfrak{d})$ by

$$L(\mathfrak{d}) = \{f \in F^\times | (f) + \mathfrak{d} \text{ is effective}\} \cup \{0\}.$$

Roughly speaking, $L(\mathfrak{d})$ consists of rational functions whose poles are at worst given by $\mathfrak{d}$. It is clear that $L(\mathfrak{d})$ is an $\mathbf{F}_q$ vector space which turns out to be finite dimensional. Note that $L(\mathfrak{d})$ is obviously zero if $\mathfrak{d}$ has negative degree.

The Riemann-Roch theorem in its most basic form is a formula that often allows one to compute the dimension $l(\mathfrak{d})$ of $L(\mathfrak{d})$. The theorem says that there is a non-negative integer g, the *genus of* $\mathcal{C}$ and a divisor ω of degree $2g - 2$ such that for all divisors $\mathfrak{d}$

$$l(\mathfrak{d}) - l(\omega - \mathfrak{d}) = \deg(\mathfrak{d}) - g + 1.$$

The divisor ω is not unique (if ω works, then so does $\omega + (f)$ for any non-zero f)). Despite this ambiguity, ω is called a *canonical divisor*. It turns out that ω can be calculated as the divisor of a rational 1-form (i.e., a 1-form possibly with poles) on $\mathcal{C}$.

It follows immediately that $l(\mathfrak{d}) \geq \deg(\mathfrak{d}) - g + 1$ with equality if $\deg \mathfrak{d} > 2g - 2$. This gives a large supply of functions with controlled poles.

As an example, note that on $\mathbf{P}^1$ over $\mathbf{F}_q$ the Riemann-Roch space $L(d\infty)$ is just the space of polynomials of degree d, which has dimension $d + 1$. It follows that the genus of $\mathbf{P}^1$ is 0. One can check that the genus of the curve in Example (2) of Section 1.3 is 1 and the genus of the curve in Example (3) is 2.

As another application, which we leave as a simple exercise, the theorem implies a partial converse to the statement that $\mathbf{P}^1$ has genus zero: if $\mathcal{C}$ has genus zero and an $\mathbf{F}_q$-rational divisor of degree 1, then $\mathcal{C}$ is isomorphic to $\mathbf{P}^1$. It turns out that over a finite field $\mathbf{F}_q$ every curve has an $\mathbf{F}_q$-rational divisor of degree one, so this partial converse is in fact a complete converse.

The reader curious about what a number field analog of the Riemann-Roch theorem might be should consult Weil's "Basic Number Theory," [Wei95, Chap. VI].

1.8   Extensions, coverings, and splitting

Let $\mathcal{C}$ and $\mathcal{C}'$ be curves defined over $\mathbf{F}_q$ and let $\phi : \mathcal{C} \to \mathcal{C}'$ be a morphism of curves defined over $\mathbf{F}_q$. We say that ϕ has degree n if $n = [\mathbf{F}_q(\mathcal{C}) : \mathbf{F}_q(\mathcal{C}')]$. Given a point P in $\mathcal{C}$ or $\mathcal{C}'$ we write $\mathbf{F}_q(P)$ for the field generated over $\mathbf{F}_q$ by the coordinates of P. We define an inverse image mapping on divisors. If $P \in \mathcal{C}'$ and if set-theoretically the inverse image of P in $\mathcal{C}$ is $\{Q_1, \ldots, Q_k\}$, then we assign a multiplicity e_i to each Q_i by choosing a rational function f vanishing simply at P and setting e_i = the order of vanishing of the pull-back of f at

Q_i. We then define $\phi^{-1}(P)$ as $\sum e_i Q_i$ and extend to divisors by linearity. It turns out that if $\mathbf{F}_q(\mathcal{C})$ is separable over $\mathbf{F}_q(\mathcal{C}')$ (i.e., if we have a geometric extension of function fields), then for all but finitely many P, all the e_i are 1, and in general for all P, $\sum e_i = n$.

If $\mathfrak{p}$ is a prime divisor of $\mathcal{C}$, then we may decompose $\phi^{-1}(\mathfrak{p})$ into a sum of prime divisors $\mathfrak{q}_1, \ldots, \mathfrak{q}_g$. For each $\mathfrak{q}_i$ we may define the *residue degree* f_i as $\deg \mathfrak{q}_i / \deg \mathfrak{p}$ or equivalently, the degree of the field extension $\mathbf{F}_q(Q)/\mathbf{F}_q(P)$ where P is any point in the support of $\mathfrak{p}$ and Q is any point over P in the support of $\mathfrak{q}_i$. The *ramification index* e_i is the e_i defined above for any point P in the support of $\mathfrak{p}$ and any point Q over P in the support of $\mathfrak{q}_i$. It is a basic fact that for all $\mathfrak{p}$, $\sum_{i=1}^{g} e_i f_i = n$ where $n = [F : F']$.

Examples:

1. Suppose $p > 3$, q is a power of p, F is the fraction field of $\mathbf{F}_q[x, y]/(y^2 - x^3 + 1)$, and $\mathbf{F}' = \mathbf{F}_q(x)$, so that the corresponding morphism of curves $\phi : \mathcal{C} \to \mathcal{C}' = \mathbf{P}^1$ is as in Example (2) in Section 1.4. Suppose that $\mathfrak{p}$ is a prime divisor of degree one corresponding to a finite $\mathbf{F}_q$-rational point P with coordinate $x = a$. If $a^3 - 1 = 0$, then $\phi^{-1}(\mathfrak{p})$ is a single prime $\mathfrak{q}$ with $e = 2$ and $f = 1$; we say $\mathfrak{p}$ is ramified. If $a^3 - 1$ is a non-zero square of $\mathbf{F}_q$, then $\phi^{-1}(\mathfrak{p})$ consists of two primes $\mathfrak{q}_1$ and $\mathfrak{q}_2$, both with $e = 1$ and $f = 1$; we say that $\mathfrak{p}$ splits. Finally, if $a^3 - 1$ is a non-square in $\mathbf{F}_q$, then $\phi^{-1}(\mathfrak{p})$ consists of one prime $\mathfrak{q}$ with $e = 1$ and $f = 2$; we say that $\mathfrak{p}$ is inert.

2. With notation as in the last example, if $\mathfrak{p}$ is a general prime, say $\mathfrak{p} = \sum P_i$, then the behavior of ϕ over each of the points P_i is the same (one ramified point, two points with the same field of coordinates as P_i, or two points with coordinates in a quadratic extension of $\mathbf{F}_q(P_i)$) and so $\phi^{-1}(\mathfrak{p}) = 2\mathfrak{q}$ with $\deg \mathfrak{q} = \deg \mathfrak{p}$ ($\mathfrak{p}$ ramifies), $\phi^{-1}(\mathfrak{p}) = \mathfrak{q}_1 + \mathfrak{q}_2$ with $\deg \mathfrak{q}_i = \deg \mathfrak{p}$ ($\mathfrak{p}$ splits), or $\phi^{-1}(\mathfrak{p}) = \mathfrak{q}$ with $\deg \mathfrak{q} = 2 \deg \mathfrak{p}$ ($\mathfrak{p}$ is inert).

3. If $\mathcal{C}$ is defined over $\mathbf{F}_q$ and $r = q^n$, then we may consider the splitting of $\mathbf{F}_q$-rational prime divisors into $\mathbf{F}_r$-rational prime divisors. This splitting is determined purely in terms of degrees: an $\mathbf{F}_q$-rational prime $\mathfrak{p}$ of degree d splits into $\gcd(n, d)$ $\mathbf{F}_r$-rational primes, each with $e = 1$ and $f = n/\gcd(d, n)$.

4. If $\mathcal{C} \to \mathcal{C}'$ is a morphism of curves defined over $\mathbf{F}_q$ and is purely insepa-rable of degree p^m, then every prime $\mathfrak{p}$ of $\mathcal{C}'$ pulls back to a single prime $\mathfrak{q}$ of $\mathcal{C}$ with $e = p^m$ and $f = 1$.

In the case of a morphism $\mathcal{C} \to \mathcal{C}'$ corresponding to a geometric extension F/F' which is Galois, it is easy to see that for a fixed prime $\mathfrak{p}$ of $\mathcal{C}'$, the ramification and residue degrees e_i and f_i are all the same, in other words, $\mathfrak{p}$ splits into g primes, all with ramification index e and residue degree f, and we have

$efg = n = [F : F']$. Only finitely many $\mathfrak{p}$ have $e > 1$ and one can make very precise statements about the distribution of primes having allowable values of f and g. See Section 1.10 below.

1.9   Frobenius elements

Let F' be a function field with constant field $\mathbf{F}_q$ and let F be a finite Galois extension of F' with Galois group G; for simplicity we assume the extension F/F' is geometric, i.e., the field of constants of F is $\mathbf{F}_q$. Let $\phi : \mathcal{C} \to \mathcal{C}'$ be the corresponding morphism of curves over $\mathbf{F}_q$. Fix a finite extension $\mathbf{F}_r$ of $\mathbf{F}_q$ and a point of $P \in \mathcal{C}'$ rational over $\mathbf{F}_r$. We may view P as an $\mathbf{F}_r$-rational prime divisor. Suppose that $\mathfrak{p}_1, \ldots, \mathfrak{p}_g$ are the $\mathbf{F}_r$-rational primes of $\mathcal{C}$ over P, so that as divisors $\phi^{-1}(P) = e\mathfrak{p}_1 + \cdots + e\mathfrak{p}_g$ where e is the ramification index. The Galois group G acts (transitively in fact) on the set of $\mathfrak{p}_i$ and we let $D_{\mathfrak{p}_i} \subset G$ denote the stabilizer of $\mathfrak{p}_i$, the *decomposition group at* $\mathfrak{p}_i$. Then $D_{\mathfrak{p}_i}$ acts on the residue field at $\mathfrak{p}_i$ and so we have a homomorphism $D_{\mathfrak{p}_i} \to \mathrm{Gal}(\mathbf{F}_{r'}/\mathbf{F}_r)$ where $\mathbf{F}_{r'} = \mathbf{F}_r(\mathfrak{p}_i) = \mathbf{F}_r(Q)$ for any $Q \in |\mathfrak{p}_i|$. This homomorphism is surjective with kernel denoted $I_{\mathfrak{p}_i}$, the *inertia group at* $\mathfrak{p}_i$. It turns out that the order of the inertia group is e, the ramification index of $\mathfrak{p}_i$. When $e = 1$, there is a distinguished element of $D_{\mathfrak{p}_i}$, namely the one that maps to the r-power Frobenius in $\mathrm{Gal}(\mathbf{F}_{r'}/\mathbf{F}_r)$. When $e > 1$ we get a distinguished coset of $I_{\mathfrak{p}_i}$ in $D_{\mathfrak{p}_i}$. Changing the choice of $\mathfrak{p}_i$ changes $D_{\mathfrak{p}_i}$, $I_{\mathfrak{p}_i}$ and the distinguished element or coset by conjugation by an element of G. Therefore, we get a well-defined conjugacy class in G depending only on $\mathbf{F}_r$ and P which we denote $\mathrm{Fr}_{\mathbf{F}_r, P}$. Similarly, we write $D_{\mathbf{F}_r, P}$ and $I_{\mathbf{F}_r, P}$ for the conjugacy classes of subgroups of G defined as above. It is not hard to check that $\mathrm{Fr}_{\mathbf{F}_{r^n}, P} = \mathrm{Fr}^n_{\mathbf{F}_r, P}$.

One also associates decomposition and inertia subgroups and a Frobenius element to a prime $\mathfrak{p}$ of $\mathcal{C}$ as follows: we let $\mathbf{F}_r$ be the residue field at $\mathfrak{p}$ and choose $P \in |\mathfrak{p}|$ and then set $D_{\mathfrak{p}} = D_{\mathbf{F}_r, P}$, $I_{\mathfrak{p}} = I_{\mathbf{F}_r, P}$, and $\mathrm{Fr}_{\mathfrak{p}} = \mathrm{Fr}_{\mathbf{F}_r, P}$. The resulting conjugacy classes are well-defined independently of the choice of P. This Frobenius is more analogous to the Frobenius element considered over number fields.

Example: Let $\mathcal{C} \to \mathcal{C}' = \mathbf{P}^1$ be the morphism considered in Example (2) in Section 1.4 and again in Example (2) in Section 1.8. This is a Galois covering with group $G = \{\pm 1\}$. If $a \in \mathbf{F}_r$ is such that $a^3 - 1 \neq 0$, and $P \in \mathbf{P}^1$ is the point $[1 : a]$, then the Frobenius class Fr_P is 1 if $a^3 - 1$ is a square in $\mathbf{F}_r$ and is -1 if it is not a square. If $\mathfrak{p}$ is an $\mathbf{F}_q$-rational prime divisor of $\mathbf{P}^1$, then $\mathrm{Fr}_{\mathfrak{p}}$ is 1 if $\mathfrak{p}$ splits and is -1 if $\mathfrak{p}$ is inert.

The definitions of decomposition and inertia subgroups and Frobenius elements extend to infinite Galois extensions in exactly the same way as in the number field context.

1.10 Cebotarev equidistribution

The classical Cebotarev density theorem says roughly that Frobenius elements are equidistributed in the Galois group of a Galois extension of number fields. To discuss a function field analogue, we keep the notations of the last section so that F/F' is a geometric Galois extension of function fields over $\mathbf{F}_q$, with corresponding morphism of curves $\mathcal{C} \to \mathcal{C}'$ defined over $\mathbf{F}_q$. We consider the distribution of Frobenius conjugacy classes $\mathrm{Fr}_{\mathbf{F}_r, P}$ as P varies over $\mathbf{F}_r$-rational points of $\mathcal{C}'$ for large r.

One analogue of the Cebotarev density theorem for function fields says that the Frobenius classes become equidistributed as r tends to infinity. More precisely, if $C \subset G$ is a conjugacy class, then

$$\lim_{r \to \infty} \frac{|\{P \in \mathcal{C}'(\mathbf{F}_r) \mid \mathrm{Fr}_{\mathbf{F}_r, P} \in C\}|}{|\{P \in \mathcal{C}'(\mathbf{F}_r)\}|} = \frac{|C|}{|G|}$$

where r tends to infinity through powers of q. A useful way to rephrase this is to consider conjugation invariant functions f on G. It make sense to evaluate such a function on a Frobenius conjugacy class and we have

$$\lim_{r \to \infty} \left| \frac{1}{|\mathcal{C}'(\mathbf{F}_r)|} \sum_{P \in \mathcal{C}'(\mathbf{F}_r)} f(\mathrm{Fr}_{\mathbf{F}_r, P}) - \frac{1}{|G|} \sum_{g \in G} f(g) \right| = 0$$

There is a more precise statement about the rate of convergence: given data as above, there exists a constant depending only on F/F' and f such that for all powers r of q,

$$\left| \frac{1}{|\mathcal{C}'(\mathbf{F}_r)|} \sum_{P \in \mathcal{C}'(\mathbf{F}_r)} f(\mathrm{Fr}_{\mathbf{F}_r, P}) - \frac{1}{|G|} \sum_{g \in G} f(g) \right| \leq C r^{-1/2}.$$

The constant C can be made quite explicit in terms of the representation theory of G and the expansion of f in terms of characters. See [KS99b, 9.7.11-13] for details.

As a very simple example of what this means in down-to-earth terms, we return to Example (2) of Section 1.8. In that context, Cebotarev equidistribution says that for large r, for about $1/2$ of the elements $a \in \mathbf{F}_r$, $a^3 - 1$ is a square and for about $1/2$ of the a, it is not a square.

2 ζ-functions and L-functions

In this section we define ζ- and L-functions, give some examples, and discuss the spectral interpretation. Warning: we use a non-standard, radically simplified notation for certain cohomology groups. See Section 4 for references with a more complete treatment.

2.1   The ζ-function of a curve

Let F be a function field with field of constants $\mathbf{F}_q$. Let $\mathcal{C}$ be the corresponding curve and denote by $\mathcal{C}^0$ the set of $\mathbf{F}_q$-rational prime divisors of $\mathcal{C}$. We define the zeta-function of $\mathcal{C}$ in analogy with the Riemann zeta-function:

$$\zeta(\mathcal{C}, s) = \prod_{\mathfrak{p} \in \mathcal{C}^0} (1 - N\mathfrak{p}^{-s})^{-1}$$

where $N\mathfrak{p} = q^{\deg \mathfrak{p}}$ is the number of elements in the residue field at $\mathfrak{p}$. (This function depends not just on the curve $\mathcal{C}$ but also on the constant field $\mathbf{F}_q$ and when we want to make this dependence explicit, we write $\zeta(\mathcal{C}/\mathbf{F}_q, s)$.)

If C_m denotes the number of primes in $\mathcal{C}^0$ of degree m and N_n denotes the number of points of $\mathcal{C}$ defined over $\mathbf{F}_{q^n}$, then we have

$$N_n = \sum_{m \mid n} m C_m.$$

Rearranging formally, we find that

$$\zeta(\mathcal{C}, s) = \exp\left(\sum_{n=1}^{\infty} \frac{N_n}{n} q^{-ns} \right)$$

which makes the diophantine interest of ζ quite visible.

The product defining $\zeta(\mathcal{C}, s)$ and the rearranged sum converge absolutely in the region $\mathrm{Re}\, s > 1$. Using the Riemann-Roch theorem, one can show that $\zeta(\mathcal{C}, s)$ extends to a meromorphic function on all of $\mathbf{C}$, with simple poles at $s = 1$ and $s = 0$ and holomorphic elsewhere, and that it satisfies a functional equation relating s and $1 - s$. (There are no Γ-factors because the product defining ζ is over all places of F.) More precisely,

$$q^{-s(1-g)}\zeta(\mathcal{C}, s) = q^{(s-1)(1-g)}\zeta(\mathcal{C}, 1 - s)$$

where g is the genus of $\mathcal{C}$.

Here are some examples: If F is the rational function field with constant field $\mathbf{F}_q$, so that $\mathcal{C} = \mathbf{P}^1$, then $N_n = q^n + 1$ and so

$$\zeta(\mathcal{C}, s) = \frac{1}{(1 - q^{-s})(1 - q^{1-s})}.$$

Let $\mathcal{C}$ be the curve with affine equation $y^2 = x^3 - x$ over $\mathbf{F}_p$ where $p \equiv 3 \pmod 4$ and $p > 3$. Using the fact that -1 is not a square modulo p, it is easy to check that the number of points on E over $\mathbf{F}_p$ is $p + 1$ and more generally, if f is odd, the number of points on E with coordinates in $\mathbf{F}_{p^f}$ is $p^f + 1$. (One considers pairs $x = a$ and $x = -a$, excluding $x = 0$ and ∞. Since -1 is not a square in $\mathbf{F}_q$, $x^3 - x$ is a square for exactly one of $x = a$ or $x = -a$; when it is a square there are two values of y with $y^2 = x^3 - x$ and none when it is not. Thus the number of solutions with finite non-zero a is $q - 1$ and the total number of solutions is $q + 1$.) A somewhat more elaborate argument using exponential sums allows one to show that for even f, the number of solutions over $\mathbf{F}_{p^f}$ is $p^f + 1 - 2(-p)^{f/2}$. (See Koblitz [Kob93, II.2] or Ireland and Rosen [IR90, Chap. 18] for a nice exposition of this argument.) Using the expression for ζ in terms of the N_n, we conclude that

$$\zeta(\mathcal{C}/\mathbf{F}_p, s) = \frac{(1 - \sqrt{-p}\,p^{-s})(1 + \sqrt{-p}\,p^{-s})}{(1 - p^{-s})(1 - p^{1-s})} = \frac{1 + p^{1-2s}}{(1 - p^{-s})(1 - p^{1-s})}.$$

As a third example we assume that $p > 2$ and $q = p^f \equiv 1 \pmod 3$ and consider the curve $\mathcal{C}$ with affine equation $y^3 = x^4 - x^2$, or rather the smooth, projective curve obtained from this one by desingularization. (This curve is singular at $(x, y) = (0, 0)$, but there is exactly one point over this one in the smooth curve, so for the purposes of counting points we may ignore this.) This curve has genus $g = 2$.

Let $\lambda : \mathbf{F}_q^{\times} \to \mathbf{C}^{\times}$ be a character of order exactly 6 and for $a = 1, 2, 4, 5$ define

$$J_a = \sum_{\substack{x \in \mathbf{F}_q \\ x \neq 0, 1}} \lambda^a(x(1 - x)).$$

It is not hard to check that $|J_i| = q^{1/2}$ and $J_5 = \overline{J}_1$, $J_4 = \overline{J}_2$. Using arguments similar to those in Koblitz or Ireland and Rosen, one verifies that the number of points on $\mathcal{C}$ over $\mathbf{F}_{q^f}$ is $q^f + 1 - \sum_{a \in \{1,2,4,5\}} J_a^f$. This implies that

$$\zeta(\mathcal{C}/\mathbf{F}_q, s) = \frac{\prod_{a \in \{1,2,4,5\}} (1 - J_a q^{-s})}{(1 - q^{-s})(1 - q^{1-s})}.$$

In general, if $\mathcal{C}$ has genus g then $\zeta(\mathcal{C}, s)$ has the form

$$\frac{P(q^{-s})}{(1 - q^{-s})(1 - q^{1-s})}$$

where P is a polynomial of degree $2g$ with integer coefficients and constant term 1. Writing $P(T) = \prod_{i=1}^{2g}(1 - \alpha_i T)$, the functional equation for ζ is equivalent to the fact that the set of inverse roots α_i is invariant under $\alpha_i \mapsto q/\alpha_i$. Moreover, ζ satisfies an analogue of the Riemann hypothesis: all of the inverse roots α_i have absolute value $q^{-1/2}$ and so the zeros of ζ lie on the line $\Re(s) = 1/2$. These results were proven in general by Weil in [Wei48].

More generally, one can define a zeta function for any variety defined over a finite field via a product or exponentiated sum as above. If X is smooth and complete of dimension d, then one knows that $\zeta(X, s)$ is a rational function in q^{-s} of a very special form. More precisely,

$$\zeta(X, s) = \frac{P_1(q^{-s})P_3(q^{-s}) \cdots P_{2d-1}(q^{-s})}{P_0(q^{-s})P_2(q^{-s}) \cdots P_{2d}(q^{-s})}$$

where each P_i is a polynomial with integer coefficients all of whose inverse roots have complex absolute value $q^{i/2}$ (an analogue of the Riemann hypothesis). Moreover, if the inverse roots of P_i are $\alpha_1, \ldots, \alpha_k$, then the inverse roots of P_{2d-i} are $q^d/\alpha_1, \ldots, q^d/\alpha_k$ and so $\zeta(X, s)$ extends to a meromorphic function in the plane and satisfies a functional equation for $s \to d - s$. These properties of the ζ-function were conjectured by Weil in [Wei49] and proved in full generality by Deligne in 1974.

2.2   Spectral interpretation of ζ-functions

Already at the time he made his famous conjectures, Weil envisioned a cohomological explanation for the conjectured properties of the zeta function. This was provided in important cases by Weil and later in full generality by Grothendieck, Deligne, and collaborators.

We fix an auxiliary prime ℓ not equal to the characteristic of $\mathbf{F}_q$. Attached to a curve $\mathcal{C}$ over a finite field $\mathbf{F}_q$ are finite-dimensional $\mathbf{Q}_\ell$-vector spaces $H^0(\mathcal{C})$, $H^1(\mathcal{C})$ and $H^2(\mathcal{C})$ each equipped with an action of $\mathrm{Gal}(\overline{\mathbf{F}}_p/\mathbf{F}_q)$. The ζ-function of $\mathcal{C}$ then has an interpretation in terms of the spectrum of the q-power Frobenius Fr_q, which is a generator of $\mathrm{Gal}(\overline{\mathbf{F}}_p/\mathbf{F}_q)$, namely

$$\zeta(\mathcal{C}, s) = \frac{P_1(q^{-s})}{P_0(q^{-s})P_2(q^{-s})}$$

where

$$P_i(T) = \det\left(1 - T\,\mathrm{Fr}_q\,|H^i(\mathcal{C})\right).$$

(It turns out that the eigenvalues of Fr_q are algebraic numbers, so that we may interpret them as complex numbers. In fact the coefficients of the reversed characteristic polynomials appearing here are integers, so there is no dependence on an embeddings of $\overline{\mathbf{Q}}$ into $\overline{\mathbf{Q}}_\ell$ and $\mathbf{C}$.)

It turns out that $H^0(\mathcal{C})$ is one-dimensional with trivial action of Fr_q, $H^2(\mathcal{C})$ is one-dimensional with Fr_q acting by multiplication by q and $H^1(\mathcal{C})$ is $2g$-dimensional, where g is the genus of g. This shows that $\zeta(\mathcal{C}, s)$ is a rational function in q^{-s} of the form mentioned in the last section.

The functional equation is a manifestation of a Poincaré duality: there are pairings $H^i(\mathcal{C}) \times H^{2-i}(\mathcal{C}) \to H^2(\mathcal{C})$ compatible with the actions of Fr_q and this shows that the eigenvalues of Fr_q on H^i are q divided by the eigenvalues of Fr_q on H^{2-i}, which is the content of the functional equation.

The Riemann hypothesis, namely that the zeros of $\zeta(\mathcal{C}, s)$ lie on the line $\Re(s) = 1/2$, is equivalent to the statement that the eigenvalues of Fr_q on $H^1(\mathcal{C})$ have complex absolute value $q^{1/2}$.

All of the above generalizes to smooth proper varieties of any dimension over $\mathbf{F}_q$. For an X of dimension d, there are finite-dimensional $\mathbf{Q}_\ell$-vector spaces $H^0(X), \ldots, H^{2d}(X)$ with an action of Fr_q; $H^0(X)$ is one-dimensional with trivial Fr_q action and $H^{2d}(X)$ is one-dimensional with Fr_q acting by multiplication by q^d. There is a Poincaré duality pairing $H^i(X) \times H^{2d-i}(X) \to H^{2d}(X)$

which is non-degenerate and compatible with the Frobenius actions. Finally, the eigenvalues of Fr_q on $H^i(X)$ are algebraic integers with absolute value $q^{i/2}$ in every complex embedding.

2.3 Examples of L-functions

Just as in the number field case, we can define L-functions associated to representations of the absolute Galois group of a function field. Before giving the general definitions, we consider three examples.

First, let F be a quadratic extension of $\mathbf{F}_q(t)$, corresponding to a branched cover $\mathcal{C} \to \mathbf{P}^1$ of degree 2. Since $F/\mathbf{F}_q(t)$ is a Galois extension with group $\{\pm 1\}$, we get a quadratic character

$$\chi : \mathrm{Gal}(\overline{\mathbf{F}_q(t)}/\mathbf{F}_q(t)) \to \mathrm{Gal}(F/\mathbf{F}_q(t)) \to \{\pm 1\}.$$

Let us define the L-function of χ as

$$L(\chi, s) = \prod_{\mathfrak{p} \in (\mathbf{P}^1)^0} (1 - \chi(\mathfrak{p}) N\mathfrak{p}^{-s})^{-1}$$

where for unramified $\mathfrak{p}$, $\chi(\mathfrak{p}) = \chi(\mathrm{Fr}_{\mathfrak{p}})$ is 1 if $\mathfrak{p}$ splits in F and -1 if $\mathfrak{p}$ is inert; we set $\chi(\mathfrak{p}) = 0$ if $\mathfrak{p}$ is ramified in F. An elementary (Euler-factor by Euler-factor) computation shows that

$$\zeta(\mathcal{C}, s) = \zeta(\mathbf{P}^1, s) L(\chi, s).$$

On the other hand,

$$\zeta(\mathcal{C}, s) = \frac{P(q^{-s})}{(1 - q^{-s})(1 - q^{1-s})}$$

and

$$\zeta(\mathbf{P}^1, s) = \frac{1}{(1 - q^{-s})(1 - q^{1-s})}$$

and so

$$L(\chi, s) = P(q^{-s}).$$

The functional equation for ζ is equivalent to

$$q^{gs} L(\chi, s) = q^{g(1-s)} L(\chi, 1 - s).$$

This applies in particular to the curve $y^2 = x^3 - x$ considered above: we view it as a degree two cover of the t-line by $(x, y) \mapsto t = x$. It follows that

$$L(\chi, s) = (1 - \sqrt{-p}\,p^{-s})(1 + \sqrt{-p}\,p^{-s}) = 1 + p^{1-2s}.$$

For a second class of examples, consider a Galois extension $F/\mathbf{F}_q(t)$ with Galois group $\mathbf{Z}/d\mathbf{Z}$, corresponding to a degree d cyclic covering of curves $\mathcal{C} \to$

128   *D. Ulmer*

$\mathbf{P}^1$. Let $\chi : \mathrm{Gal}(\overline{\mathbf{F}_q(t)}/\mathbf{F}_q(t)) \to \mathrm{Gal}(F/\mathbf{F}_q(t)) \to \mu_d \subset \overline{\mathbf{Q}}^\times$ be a complex valued character of order exactly d and for $i = 1, \ldots, d-1$ define

$$L(\chi^i, s) = \prod_{\mathfrak{p} \in (\mathbf{P}^1)^0} (1 - \chi^i(\mathfrak{p}) N\mathfrak{p}^{-s})^{-1}$$

where $\chi(\mathfrak{p}) = \chi(\mathrm{Fr}_{\mathfrak{p}})$ for unramified $\mathfrak{p}$ and $\chi(\mathfrak{p}) = 0$ if $\mathfrak{p}$ is ramified in F. Again an elementary calculation shows that

$$\zeta(\mathcal{C}, s) = \zeta(\mathbf{P}^1, s) L(\chi, s) L(\chi^2, s) \cdots L(\chi^{d-1}, s).$$

It turns out that each $L(\chi^i, s)$ for $i = 1, \ldots, d-1$ is a polynomial in q^{-s} and their product is the numerator $P(q^{-s})$ of $\zeta(\mathcal{C}, s)$.

For $d > 2$ a new phenomenon becomes apparent: the functional equation links two distinct L-functions. More precisely, we have

$$q^{N_i s/2} L(\chi^i, s) = \epsilon q^{N_i(1-s)/2} L(\chi^{-i}, 1-s)$$

where $N_i = N_{-i}$ is the degree of $L(\chi^i, s)$ as a polynomial in q^{-s} and ϵ is a complex number of absolute value 1. This will be important later when we discuss symmetry types.

As a specific example of this type, we consider the curve $\mathcal{C}$ defined by $y^3 = x^4 - x^2$, discussed above, viewed as a Galois cover of $\mathbf{P}^1$ of degree 3 via $(x, y) \mapsto t = x$. For a suitable choice of character $\chi : \mathrm{Gal}(F/\mathbf{F}_q(t)) \to \mu_3$, we have $L(\chi, s) = (1 - J_1 q^{-s})(1 - J_4 q^{-s})$ and $L(\chi^2, s) = (1 - J_2 q^{-s})(1 - J_5 q^{-s})$.

A third, more elaborate, class of examples comes from elliptic curves. Let E be an elliptic curve defined over F. This could be given, for example, by a Weierstrass equation

$$y^2 + a_1 xy + a_3 y = x^3 + a_2 x^2 + a_4 x + a_6$$

where the a_i are in F. If E has good reduction at a place $\mathfrak{p}$ of F, we define a local Euler factor by

$$L_{\mathfrak{p}}(E, s) = (1 - a_{\mathfrak{p}} q_{\mathfrak{p}}^{-s} + q_{\mathfrak{p}}^{1-2s})$$

where $q_{\mathfrak{p}}$ is the cardinality of the residue field at $\mathfrak{p}$ and $q_{\mathfrak{p}} - a_{\mathfrak{p}} + 1$ is the number of points on the reduction of E at $\mathfrak{p}$. If E has bad reduction at $\mathfrak{p}$, we define a local factor by

$$L_{\mathfrak{p}}(E, s) = \begin{cases} 1 - q_{\mathfrak{p}}^{-s} & \text{if } E \text{ has split multiplicative reduction at } \mathfrak{p} \\ 1 + q_{\mathfrak{p}}^{-s} & \text{if } E \text{ has non-split multiplicative reduction at } \mathfrak{p} \\ 1 & \text{if } E \text{ has additive reduction at } \mathfrak{p}. \end{cases}$$

Then we define the global (Hasse-Weil) L-function of E as

$$L(E, s) = \prod_{\mathfrak{p}} L_{\mathfrak{p}}(E, s)^{-1}.$$

This *L*-function turns out to be a rational function in q^{-s} and it satisfies a functional equation for $s \to 2 - s$. More precisely, if E is not isomorphic to an elliptic curve defined over $\mathbf{F}_q$, then $L(E, s)$ is a polynomial in q^{-s} whose degree is determined by the genus of the curve corresponding to F and the places of bad reduction of E. In this case,

$$L(E, s) = \prod_{i=1}^{N} (1 - \alpha_i q^{-s})$$

where the set of inverse roots α_i is invariant under $\alpha_i \mapsto q^2 / \alpha_i$ and each of them has complex absolute value q. In particular, the zeros of $L(E, s)$ lie on the line $\Re(s) = 1$.

2.4 *L*-functions attached to Galois representations

As in the number field case, over function fields there are two general classes of *L*-functions, automorphic *L*-functions attached to automorphic representations (generalizing Dirichlet characters, Hecke characters, etc.) and "motivic" *L*-functions attached to representations of Galois groups, and a Langlands philosophy which very roughly speaking says that the latter are the same as the former. In the function field setting there is a quite satisfactory understanding of the analytic properties of motivic *L*-functions which we sketch in this and the following section.

As usual, let $F = \mathbf{F}_q(\mathcal{C})$ be the function field of a curve over $\mathbf{F}_q$. We fix a prime ℓ and write E for a finite extension of $\mathbf{Q}_\ell$ which we may expand as necessary in the course of the discussion. The basic input data is a representation

$$\rho : \mathrm{Gal}(\overline{F}/F) \to \mathrm{GL}_n(E)$$

which is continuous (for the Krull topology on $\mathrm{Gal}(\overline{F}/F)$ and the ℓ-adic topology on $\mathrm{GL}_n(E)$) and unramified outside a finite set of places of F. The latter means that for all but finitely many primes $\mathfrak{p}$, $\rho(I_\mathfrak{p}) = \{1\}$ where $I_\mathfrak{p}$ is the inertia subgroup at $\mathfrak{p}$. We assume that ρ is absolutely irreducible, i.e., is reducible even after extending scalars to $\overline{E}$. We also assume that ρ has a weight $w \in \mathbf{Z}$, which means that for every unramified prime $\mathfrak{p}$, all of the eigenvalues of $\rho(\mathrm{Fr}_\mathfrak{p})$ are algebraic integers and have absolute value $q^{w/2}$ in every complex embedding.

Given ρ, we define an *L*-function by

$$L(\rho, s) = \prod_{\mathfrak{p}} \det\left(1 - \rho(\mathrm{Fr}_\mathfrak{p}) N\mathfrak{p}^{-s} \,\middle|\, (E^n)^{I_\mathfrak{p}}\right)^{-1}.$$

Here $N\mathfrak{p}$ is the cardinality of the residue field at $\mathfrak{p}$, $I_\mathfrak{p}$ is the inertia group at $\mathfrak{p}$, and $(E^n)^{I_\mathfrak{p}}$ denotes the subspace of E^n where $I_\mathfrak{p}$ acts (via ρ) trivially; for almost all $\mathfrak{p}$ this will just be E^n itself. On the space of invariants $(E^n)^{I_\mathfrak{p}}$ there is a well-defined action of the Frobenius elements $\mathrm{Fr}_{\mathfrak{p}}$ and the local factors above are the reciprocals of the reversed characteristic polynomials of the action of $N\mathfrak{p}^{-s}$ times $\rho(\mathrm{Fr}_{\mathfrak{p}})$.

Easy estimates show that the product defining $L(\rho, s)$ converges absolutely in the region $\Re(s) > 1 + w/2$, uniformly on compact subsets, and so defines a holomorphic function there. As we will see in the next section, $L(\rho, s)$ has a meromorphic continuation to all of $\mathbf{C}$ which is entire if and only if ρ restricted to $\mathrm{Gal}(\overline{F}/\overline{\mathbf{F}}_p F)$ contains no copies of the trivial representation. In general, $L(\rho, s)$ satisfies a functional equation

$$L(\rho, s) = \epsilon(\rho, s) L(\rho^{\vee}, 1 - s)$$

where $\rho^{\vee}$ is the dual representation and $\epsilon(\rho, s)$ is an entire function with $\epsilon(\rho, 1/2)$ a complex number of absolute value 1.

The attentive reader may be distressed by the apparent mixture of ℓ-adic and complex numbers in the definition of $L(\rho, s)$. To make things precise, we fix embeddings $\overline{\mathbf{Q}} \hookrightarrow \overline{\mathbf{Q}}_{\ell}$ and $\overline{\mathbf{Q}} \hookrightarrow \mathbf{C}$; since we assumed that the eigenvalues of $\rho(\mathrm{Fr}_{\mathfrak{p}})$ are algebraic numbers we may use the embeddings to regard the coefficients of the reversed characteristic polynomials as complex numbers.

The examples of the previous section can be fit into this general framework as follows. If K/F is a finite Galois extension and $\chi : \mathrm{Gal}(K/F) \to \mu_d \subset E = \mathbf{Q}_{\ell}(\mu_d)$ is a character, then composing with the natural projection $\mathrm{Gal}(\overline{F}/F) \to \mathrm{Gal}(K/F)$ gives a one-dimensional, absolutely irreducible ℓ-adic representation satisfying our hypotheses. It has weight $w = 0$.

The elliptic curve example is somewhat more elaborate. In this case, we consider the ℓ-adic Tate module of E over F, namely $\varprojlim_m E(\overline{F})[\ell^m]$ which is isomorphic to $\mathbf{Z}_{\ell}^2$. There is an action of $\mathrm{Gal}(\overline{F}/F)$ on this Tate module and as ρ we take the dual of this representation. At a prime $\mathfrak{p}$ where E has good reduction, general ℓ-adic results show that the reversed characteristic polynomial of $\mathrm{Fr}_{\mathfrak{p}}$ is just the reversed characteristic polynomial of the $N\mathfrak{p}$-power Frobenius on the group $H^1(E \ (\mathrm{mod} \ \mathfrak{p}))$ mentioned in the discussion of zeta functions. In particular, the coefficients of the local zeta function are given in terms of the number of points on the reduction of E at $\mathfrak{p}$ by the recipe mentioned in the previous section. Something similar, albeit more involved, happens at the places of bad reduction.

2.5   Spectral interpretation of L-functions

There is a spectral interpretation of L-functions which is quite parallel to that of ζ-functions—the key is to think of a representation ρ as providing coefficients for a cohomology theory. Of course we cannot explain the details here, but the idea is this: given ρ, we have cohomology groups $H^i(\mathcal{C}, \rho)$ ($i = 0, 1, 2$) which are finite-dimensional E-vector spaces with an action of $\mathrm{Gal}(\overline{\mathbf{F}}_p/\mathbf{F}_q)$. (For experts, we are taking the lisse sheaf on an open subset of $\mathcal{C}$ associated to ρ, forming its middle extension on $\mathcal{C}$, and taking cohomology on $\mathcal{C} \times \mathrm{Spec}\ \overline{\mathbf{F}}_p$.)

Then

$$L(\rho, s) = \frac{P_1(q^{-s})}{P_0(q^{-s}) P_2(q^{-s})}$$

where

$$P_i(T) = \det\left(1 - T\operatorname{Fr}_q | H^i(\mathcal{C}, \rho)\right).$$

If ρ has weight w then the eigenvalues of Fr_q on $H^i(\mathcal{C}, \rho)$ are algebraic integers with absolute value $q^{(i+w)/2}$ in every complex embedding. Poincaré duality takes the form

$$H^i(\mathcal{C}, \rho) \times H^{2-i}(\mathcal{C}, \rho^\vee) \to H^2(\mathcal{C}, \rho \otimes \rho^\vee) \to H^2(\mathcal{C}).$$

When ρ restricted to $\operatorname{Gal}(\overline{F}/\mathbf{F}_p F)$ has no trivial factors, then $H^0(\mathcal{C}, \rho)$ and $H^2(\mathcal{C}, \rho)$ vanish and so the L-function is a polynomial in q^{-s} whose degree is just the dimension of $H^1(\mathcal{C}, \rho)$. This dimension can be calculated in terms of the dimension and ramification properties of ρ and the genus of $\mathcal{C}$.

2.6 Symmetries

For many interesting representations ρ, there is additional structure coming from the fact that the space where ρ acts admits a Galois-equivariant pairing (at least up to a twist). More precisely, suppose given an absolutely irreducible $\rho : \operatorname{Gal}(\overline{F}/F) \to \operatorname{GL}_n(E)$. Naively we might ask for a pairing

$$\langle \cdot, \cdot \rangle : E^n \times E^n \to E$$

such that $\langle \rho(g)v, \rho(g)v' \rangle = \langle v, v' \rangle$ for all $g \in \operatorname{Gal}(\overline{F}/F)$, but this is not possible when the weight of ρ is non-zero. Instead we ask that

$$\langle \rho(g)v, \rho(g)v' \rangle = \chi_\ell(g)^w \langle v, v' \rangle$$

where $\chi_\ell(g)$ gives the action of g on ℓ-power roots of unity: $\zeta_{\ell^n}^g = \zeta_{\ell^n}^{\chi_\ell(g)}$ for all $\zeta_{\ell^n} \in \mu_{\ell^n}$. When a non-zero (and thus non-degenerate) such pairing exists, we say that ρ is self-dual of weight w. Moreover, the pairing must be either symmetric ($\langle v, v' \rangle = \langle v', v \rangle$) or skew symmetric ($\langle v, v' \rangle = -\langle v', v \rangle$); we say that ρ is orthogonally self-dual or symplectically self-dual respectively.

For example, a finite order character $\chi : \operatorname{Gal}(\overline{F}/F) \to \mu_d$ is self-dual if and only if it is of order 2, in which case it is orthogonally self-dual of weight 0. The representation of $\operatorname{Gal}(\overline{F}/F)$ on the dual of the Tate module of an elliptic curve over F is symplectically self-dual of weight 1.

When ρ self-dual, then so is $H^1(\mathcal{C}, \rho)$, but with the opposite sign and weight $w + 1$. In other words, when ρ is orthogonally (resp. symplectically) self-dual, then there is a skew-symmetric (resp. symmetric) pairing on $H^1(\mathcal{C}, \rho)$ which satisfies $\langle \operatorname{Fr}_q v, \operatorname{Fr}_q v' \rangle = q^{w+1} \langle v, v' \rangle$.

Extending E if necessary, we may choose a basis of $H^1(\mathcal{C}, \rho)$ in which the matrix of the form is the standard one times q^{w+1} and then the matrix of Frobenius in this basis will be $q^{(w+1)/2}$ times an orthogonal or symplectic matrix. Thus extra structure on ρ puts severe restrictions on the action of Frobenius.

At the level of L-functions, these restrictions are reflected in the functional equations: when ρ is symplectically self-dual, the sign in the functional equation is ± 1 (so that the sign sometimes forces vanishing at the central point) whereas when ρ is orthogonally self-dual, the sign in the functional equation is always $+1$ (so that the order of zero at the central point is even).

Note that when ρ is not self-dual, then the Frobenius matrix is *a priori* $q^{(w+1)/2}$ times a general matrix in GL and the functional equation relates two different L-functions and so cannot force zeros at the central point.

3   Families of L-functions

In this section, we come to the *raison d'être* of the article, namely an explanation of how families of L-functions over function fields give rise to well-distributed collections of matrices in classical groups. Rather than attempting to make precise general definitions, we consider several examples which we hope will make the key points clear.

3.1   Arithmetic and geometric families

Let us fix a finite field $\mathbf{F}_q$ and consider all quadratic extensions of the rational function field $\mathbf{F}_q(t)$, or equivalently, all quadratic characters

$$\chi : \mathrm{Gal}(\overline{\mathbf{F}_q(t)}/\mathbf{F}_q(t)) \to \{\pm 1\}.$$

We exclude as trivial the unique character χ factoring through $\mathrm{Gal}(\overline{\mathbf{F}}_p/\mathbf{F}_q)$ which corresponds to the extension $\mathbf{F}_{q^2}(t)$. We want to make statistical statements about the L-functions $L(\chi, s)$ and to do so, the most natural way to partially order them is by the genus of the corresponding field F or what amounts to the same thing, the degree of the conductor of χ.

To keep things as simple as possible, we assume that the characteristic p of $\mathbf{F}_q$ is > 2. In this case, the conductor of χ can be thought of as the set of $\mathfrak{p}$ where χ is ramified and the degree of the conductor of χ is just the sum of the degrees of the places $\mathfrak{p}$ in the conductor. The connection with the genus is given by the Riemann-Hurwitz formula: $g = (\deg(\mathrm{Cond}(\chi)) - 2)/2$.

There are finitely many χ with conductor $\leq N$ (the number is of the order q^N as $N \to \infty$) and so we may consider some quantity associated to $L(\chi, s)$, such as the height of its lowest zero or the spacings between zeros, average over those χ of conductor $\leq N$, and then take a limit as $N \to \infty$. This set-up is entirely analogous to the situation over $\mathbf{Q}$ or a number field and we call this family and ones like it *arithmetic*. It seems likely that many of the results known in the number field situation (e.g., on moments) could be treated in this situation as well, by analogous methods. More ambitiously, Katz and Sarnak

[KS99a] have made several conjectures on arithmetic families which are open and which currently seem just as inaccessible as their number field analogues.

Considerably more can be done in the function field situation if we change the problem slightly. Namely, let us give ourselves the freedom to vary the constant field $\mathbf{F}_q$ as well: We consider quadratic extensions of $\mathbf{F}_{q^n}(t)$ or equivalently quadratic characters $\chi : \mathrm{Gal}(\overline{\mathbf{F}_{q^n}(t)}/\mathbf{F}_{q^n}(t)) \to \{\pm 1\}$, again excluding the character corresponding to $\mathbf{F}_{q^{2n}}(t)$. The number of such characters with conductor of degree $\leq N$ is of the order q^{nN}. We form the average over this set of some quantity associated to $L(\chi, s)$ and then take a limit as $n \to \infty$. This already gives interesting statements, but we may also take a second limit as $N \to \infty$. The advantage of first passing to the limit in n is that we get an infinite collection of L-functions *parameterized by a single algebraic variety*. For this reason we call such families *geometric*.

Let us explain how this parameterization comes about, still assuming for simplicity that $p > 2$. In this case, any quadratic extension F of $\mathbf{F}_{q^n}(t)$ can be obtained by adjoining the square root of a polynomial $f \in \mathbf{F}_{q^n}[t]$. If f is square free the degree of the conductor of χ is essentially the degree of f. (More precisely, it is $\deg(f)$ if $\deg(f)$ is even and $\deg(f) + 1$ if $\deg(f)$ is odd.) For simplicity we restrict to monic polynomials f; the set of monic polynomials of degree N is naturally an affine space of dimension N (using the coefficients of the polynomial as coordinates) and the set of square-free monic polynomials is a Zariski open subset $X \subset \mathbf{A}^N$. Thus we have a natural bijection between certain quadratic characters of conductor N of $\mathrm{Gal}(\overline{\mathbf{F}_{q^n}(t)}/\mathbf{F}_{q^n}(t))$ and $X(\mathbf{F}_{q^n})$, the points of X with coordinates in $\mathbf{F}_{q^n}$. We write χ_f for the character associated to $f \in X(\mathbf{F}_{q^n})$. This geometric structure allows one to bring the powerful tools of arithmetical algebraic geometry to bear, with decisive results.

3.2 Variation of L-functions

We continue with the example of L-functions attached to quadratic characters over $\mathbf{F}_{q^n}(t)$. As we explained in Section 2, $L(\chi_f, s)$ is the numerator of the zeta-function of the hyperelliptic curve $\mathcal{C} \to \mathbf{P}^1$ corresponding to the quadratic extension $F = \mathbf{F}_{q^n}(\sqrt{f})/\mathbf{F}_{q^n}(t)$ cut out by χ_f and it can be computed as the characteristic polynomial of Frobenius on a cohomology group. In particular, there is a symplectic matrix $A_f \in \mathrm{Sp}_{2g}(\mathbf{Q}_\ell)$, well-defined up to conjugacy, such that $L(\chi_f, s) = \det(1 - q^{n(1/2-s)}A_f)$. Thus we have a map from $X(\mathbf{F}_{q^n})$ to conjugacy classes of symplectic matrices.

(The reader uncomfortable with cohomology may proceed as follows: for each point in $f \in X(\mathbf{F}_{q^n})$ we may form the corresponding L-function $L(\chi_f, s) = \prod(1 - \alpha_i q^{n(1/2-s)})$. The α_i are algebraic integers with absolute value 1 in any complex embedding and the collection of them is invariant under $\alpha_i \mapsto \alpha_i^{-1}$. There is thus a well-defined conjugacy class of symplectic matrices A_f so that the α_i are the eigenvalues of A_f. Of course the preceding sentence is equally true with "symplectic" replaced by "orthogonal" or "unitary"; the virtue of the cohomological approach is that it explains why symplectic matrices are the natural choice.)

The first main result is that in a suitable sense, these conjugacy classes become equidistributed as $n \to \infty$. To make this more precise, we use complex matrices and the compact unitary symplectic group USp_{2g}. Namely, we use the fixed embeddings $\overline{\mathbf{Q}} \hookrightarrow \mathbf{C}$ and $\overline{\mathbf{Q}} \hookrightarrow \overline{\mathbf{Q}}_\ell$ to view ℓ-adic matrices as complex matrices. The Weyl unitarian trick and the Peter-Weyl theorem imply that the conjugacy class of A_f in $\mathrm{Sp}_{2g}(\mathbf{C})$ meets the maximal compact subgroup USp_{2g} in a unique USp_{2g}-conjugacy class. We write θ_f for any element of this class. The statement of equidistribution is that as $n \to \infty$, these classes become equidistributed with respect to Haar measure. More precisely, for any continuous, conjugation invariant function h on USp_{2g}, we have

$$\int_{\mathrm{USp}_{2g}} h \, d\mu_{Haar} = \lim_{n \to \infty} \frac{1}{|X(\mathbf{F}_{q^n})|} \sum_{f \in X(\mathbf{F}_{q^n})} h(\theta_f).$$

There is a more precise statement giving the rate of convergence:

$$\left| \int_{\mathrm{USp}_{2g}} h \, d\mu_{Haar} - \frac{1}{|X(\mathbf{F}_{q^n})|} \sum_{f \in X(\mathbf{F}_{q^n})} h(\theta_f) \right| < C q^{-n/2}$$

where C is a constant depending only on X and h.

3.3   Other families

We consider two other examples of geometric families giving rise to general matrices and orthogonal matrices.

First we consider families of cubic L-series. More precisely, fix an integer d and consider the set of monic polynomials in x of degree d with coefficients in extensions of the finite field $\mathbf{F}_q$ where $q \equiv 1 \pmod 3$. The set of all such is naturally the affine space of dimension d, with coordinates given by the coefficients:

$$f = x^d + a_1 x^{d-1} + \cdots + a_{d-1} x + a_d \quad \leftrightarrow \quad (a_1, \ldots, a_d) \in \mathbf{A}^d(\mathbf{F}_{q^n}).$$

We let $X \subset \mathbf{A}^d$ be the Zariski open subset corresponding to polynomials with distinct roots, so that X is obtained from $\mathbf{A}^d$ by removing the zero set of the discriminant, a polynomial in $a_1, \ldots, a_d$. For each extension $\mathbf{F}_{q^n}$ of $\mathbf{F}_q$ and each $f \in X(\mathbf{F}_{q^n})$, the curve with affine equation $y^3 = f(x)$ is a cubic Galois covering of $\mathbf{P}^1$ corresponding to a cubic Galois extension of function fields $F/\mathbf{F}_{q^n}(t)$. There are two non-trivial characters of $\mathrm{Gal}(F/\mathbf{F}_{q^n}(t))$, which we denoted by χ_f and χ_f^{-1}. (We will not explain the details here, but there is a consistent way to choose which is χ_f and which is $\chi_{f^{-1}}$.) The character χ_f gives rise to an L-function $L(\chi_f, s)$ and, via the cohomological machinery discussed in the previous section, to a well-defined conjugacy class of matrices

A_f in $\mathrm{GL}_N(\mathbf{Q}_\ell)$ where $N = d - 2$ and, for convenience, $\ell \equiv 1 \pmod 3$. As we noted in Section 2.3, for cubic characters Poincaré duality and the functional equation link two distinct groups or L-functions and so there is no geometric reason for the Frobenius matrices to lie in a small group and in fact they do not. By results of Katz and the general machinery sketched below, for all sufficiently large d, the Frobenius conjugacy classes are equidistributed in an algebraic group containing the algebraic group SL_N over $\mathbf{Q}_\ell$ with finite index. As before, one makes this precise by using embeddings and Lie theory to deduce for each $f \in X(\mathbf{F}_{q^n})$ a well-defined conjugacy class θ_f in a compact Lie group G with $\mathrm{SU}_N \subset G \subset \mathrm{U}_N$ such that

$$L(\chi_f, s) = \det\left(1 - q^{n(1/2-s)}\theta_f\right) = \prod_{i=1}^{N}(1 - \alpha_i q^{n(1/2-s)})$$

where the α_i are the eigenvalues of θ_f. The equidistribution statement is then that

$$\left| \int_G h \, d\mu_{Haar} - \frac{1}{|X(\mathbf{F}_{q^n})|} \sum_{f \in X(\mathbf{F}_{q^n})} h(\theta_f) \right| < C q^{-n/2}$$

for any continuous, conjugation-invariant function h on G.

For an example of an orthogonal family, we consider the family of quadratic twists of an elliptic curve. More precisely, assume that $p > 3$ and fix an elliptic curve E over $\mathbf{F}_q(t)$ defined by a Weierstrass equation

$$y^2 = x^3 + ax + b$$

with $a, b \in \mathbf{F}_q(t)$. We assume that the j-invariant of E is not in $\mathbf{F}_q$. Fix a degree d. For each monic square-free polynomial $f \in \mathbf{F}_{q^n}[x]$, we may form the quadratic twist E_f of E, with equation

$$fy^2 = x^3 + ax + b \tag{3.1}$$

and its L-function $L(E_f, s)$. If we assume that the zeros of f are disjoint from the points where E has bad reduction, then the degree of $L(E_f, s)$ as a polynomial in q^n is $N = 2d + c$ where c is a constant depending only on E. Let $X \subset \mathbf{A}^d$ be the Zariski open set whose points over $\mathbf{F}_{q^n}$ are the monic, square-free polynomials $f \in \mathbf{F}_q[x]$ with zeros disjoint from the primes dividing the discriminant of E. The cohomological machinery gives us, for each $f \in X(\mathbf{F}_{q^n})$, an orthogonal matrix $A_f \in \mathrm{O}_N(\mathbf{Q}_\ell)$, well-defined up to conjugacy, such that

$$L(E_f, s) = \det\left(1 - q^{n(1-s)}A_f\right).$$

As before, using the embeddings and Lie theory we deduce a conjugacy class θ_f in the compact group $\mathrm{O}_N(\mathbf{R})$. Under further hypotheses on E which we do not discuss one may conclude that in fact $\theta_f \in \mathrm{SO}_N(\mathbf{R})$. (We make these hypotheses only to simplify the equidistribution statement below.) Results of

Katz and Deligne then say that the classes θ_f are equidistributed in the sense that

$$\left| \int_{\mathrm{SO}_N(\mathbf{R})} h \, d\mu_{Haar} - \frac{1}{|X(\mathbf{F}_{q^n})|} \sum_{f \in X(\mathbf{F}_{q^n})} h(\theta_f) \right| < Cq^{-n/2}$$

for any continuous, conjugation-invariant function h on $\mathrm{SO}_N(\mathbf{R})$.

3.4 Idea of proofs

We give a very brief sketch of the main ideas behind the proofs of the equidistribution statements above.

The first ingredient is monodromy. Let X be the variety parameterizing the family under study. Then we have the fundamental group $\pi_1(X)$, which is a quotient of the absolute Galois group of the function field of X over $\mathbf{F}_q$ and which gives automorphisms ("deck transformations") of unramified covers of X. There is a subgroup $\pi_1^{geom}(X) \subset \pi_1(X)$ such that

$$\pi_1(X)/\pi_1^{geom}(X) \cong \mathrm{Gal}(\overline{\mathbf{F}}_p/\mathbf{F}_q).$$

The cohomological machinery gives rise to a representation $\rho : \pi_1(X) \to \mathrm{GL}_N(E)$ (here E is some finite extension of $\mathbf{Q}_\ell$) such that for each point $f \in X(\mathbf{F}_{q^n})$ with Frobenius conjugacy class $\mathrm{Fr}_f \in \pi_1(X)$, we have $\rho(\mathrm{Fr}_f) \in \mathrm{GL}_N(E)$ which is the conjugacy class associated to the L-function named by f. Attached to ρ are two monodromy groups $G^{geom} \subset G^{arith}$. These are defined as the Zariski closures of the images of ρ on $\pi_1^{geom}(X)$ and $\pi_1(X)$ respectively. A basic result of Deligne says that G^{geom} is a semi-simple algebraic group over E. When there is extra structure (i.e., a pairing), then we have an a priori containment $G^{arith} \subset \mathrm{Sp}$ or O. In favorable cases one can establish by geometric methods a lower bound Sp or O or $\mathrm{SL} \subset G^{geom}$ and therefore equalities $G^{geom} = G^{arith} = \mathrm{Sp}$ or O or SL. (Here we are glossing over several technicalities regarding the difference between G^{geom} and G^{arith} and between O and SO.) Part of Katz-Sarnak [KS99b, Chaps. 10-11], most of Katz [Kat02], and several other works of Katz are devoted to these kinds of calculations.

The second main ingredient is a very general equidistribution result of Deligne that says that whatever the arithmetic monodromy group is, the Frobenius classes are equidistributed in it. More precisely, forming classes θ_f in a compact Lie group G associated to G^{arith} and $f \in X(\mathbf{F}_{q^n})$, we have

$$\left| \int_G h \, d\mu_{Haar} - \frac{1}{|X(\mathbf{F}_{q^n})|} \sum_{f \in X(\mathbf{F}_{q^n})} h(\theta_f) \right| < Cq^{-n/2}$$

for all continuous, conjugation-invariant functions h on G. This equidistribution result was proven as a consequence of the Weil conjectures [Del80] and is explained in Katz-Sarnak [KS99b, Chap. 9].

3.5   Large N limits

Another part of the story, the part related to classical random matrix theory, relates to statistical measures of eigenvalues in the large N limit. More precisely, given an $N \times N$ unitary matrix with eigenvalues $e^{2\pi i \phi_j}$ with $0 \leq \phi_1 \leq \cdots \leq \phi_N < 1$ one forms a point measure on $\mathbf{R}$ with mass $1/N$ at each of the normalized spacings $N(\phi_2 - \phi_1), N(\phi_3 - \phi_2), \ldots, N(\phi_N - \phi_{N-1}), N(1 + \phi_1 - \phi_N)$. Averaging this measure over U_N (with respect to Haar measure) yields a measure on $\mathbf{R}$ and it turns out that one may take the limit as $N \to \infty$ and arrive at a measure on $\mathbf{R}$ which is absolutely continuous with respect to Lebesgue measure and has a real analytic density function. Similar results hold for other families of classical groups and it turns out that the measure obtained is the same for the symplectic groups Sp_{2N} and the orthogonal groups O_{2N} and O_{2N+1} (where in the latter case one ignores the forced eigenvalue 1).

Katz and Sarnak also consider other statistical measures of eigenvalues, for example the placement of the eigenvalue closest to 1. In this case there is again a scaling limit as $N \to \infty$ but now the resulting measure on $\mathbf{R}$ depends on the family of classical groups considered. For example, the density function for the symplectic family vanishes at 0, indicating that eigenvalues of symplectic matrices are "repelled" from 1, whereas this is not the case for the unitary and orthogonal families.

These results are purely Lie-theoretic and do not involve any algebraic geometry. We will not attempt to give any details here, but simply refer to Katz-Sarnak [KS99b].

For an example of the application of this in the function field context, we consider families X_g as in Section 3.2 parameterizing quadratic characters χ corresponding to curves $\mathcal{C} \to \mathbf{P}^1$ of genus g. Combining equidistribution results with theorems on large N limits, one sees that integrals with respect to the large N limit measure may be computed using Frobenius matrices. More precisely, suppose that ν_1 is the measure on $\mathbf{R}$ associated to the suitably normalized location of the eigenvalue nearest 1 for symplectic matrices. Then we have

$$\int_{\mathbf{R}} h \, d\nu_1 = \lim_{g \to \infty} \lim_{r \to \infty} \frac{1}{|X_g(\mathbf{F}_r)|} \sum_{f \in X_g(\mathbf{F}_r)} h(\phi_1(\theta_f))$$

for all continuous, compactly supported functions h on $\mathbf{R}$, where θ_f is the symplectic matrix associated to f, $\phi_1(\theta_f)$ is the normalized angle of its eigenvalue closest to 1, and r tends to ∞ through powers of q.

The only point we want to make here is that Katz and Sarnak conjecture that results like this should be true without taking the limit over large finite fields. In other words, one should have

$$\int_{\mathbf{R}} h \, d\nu_1 = \lim_{g \to \infty} \frac{1}{|X_g(\mathbf{F}_q)|} \sum_{f \in X_g(\mathbf{F}_q)} h(\phi_1(\theta_f))$$

This conjecture looks quite deep and will probably require new ideas going beyond the cohomological formalism.

3.6   Applications

We briefly mention three applications to arithmetic of the ideas around function fields and random matrices.

The first application is to guessing the symmetry type of a family of L-functions over a number field. The idea, roughly speaking, is to find a function field analogue of the given family and inspect the cohomology groups computing the L-functions to see whether there is extra symmetry present. If so, the symmetry group should be O, SO, or Sp; if not then it should contain SL. For example, if one looks at the family of quadratic Dirichlet characters over $\mathbf{Q}$, the function field analog is the family of quadratic characters considered in Section 3.1 and so one expects symplectic symmetries. Of course the symplectic group itself is nowhere in sight in the number field context, but one does find computationally that the statistics of low lying zeros obey the distributions associated with symplectic groups. See Katz-Sarnak [KS99a] for more on this and other examples.

The second application is to an analogue of the Goldfeld conjecture. Roughly speaking, this conjecture asserts that in the family of quadratic twists of an elliptic curve over $\mathbf{Q}$, 50% of the curves should have rank 0 and 50% should have rank 1. The most direct function field analogue would concern twists E_f of a given elliptic curve, as in Equation 3.1 above, where $f \in \mathbf{F}_q[x]$ and it would assert that

$$\lim_{d \to \infty} \frac{|\{f \in \mathbf{F}_q[x]|\deg(f) \le d, \, ..., \text{ and } \operatorname{Rank} E_f(\mathbf{F}_q(t)) = 0\}|}{|\{f \in \mathbf{F}_q[x]|\deg(f) \le d, \text{ and } ...\}|} = \frac{1}{2}$$

where "..." stands for conditions on f, namely that f be square free and have zeros disjoint from the points where E has bad reduction. Similarly for rank 1. There are also conjectures where $\operatorname{Rank} E_f(\mathbf{F}_q(t))$ is replaced by $\operatorname{ord}_{s=1} L(E_f, s)$. These conjectures are completely open, although there are some recent nice examples of Chris Hall [Hal04]. But one can do more by allowing ground field extensions. More precisely, Katz proves in [Kat02] that for large d,

$$\lim_{n \to \infty} \frac{|\{f \in \mathbf{F}_{q^n}[x]|\deg(f) \le d, \, ..., \text{ and } \operatorname{ord}_{s=1} L(E_f, s) = 0\}|}{|\{f \in \mathbf{F}_{q^n}[x]|\deg(f) \le d, \text{ and } ...\}|} = \frac{1}{2}$$

under the assumption that E has at least one place of multiplicative reduction. (This hypothesis is needed to ensure that the monodromy group is the full orthogonal group O, rather than SO.) Similar results hold for analytic rank 1 and, with suitable modifications, for cases when the monodromy group is SO. One can deduce results for algebraic ranks by using the inequality $\operatorname{Rank} E_f(\mathbf{F}_{q^n}(t)) \le \operatorname{ord}_{s=1} L(E_f, s)$ which is known in the function field case.

The connection between equidistribution and these results is that with respect to Haar measure, 1/2 of the matrices in the orthogonal group have eigenvalue 1 with multiplicity 1 and 1/2 have eigenvalue 1 with multiplicity 0. Thus when the matrices computing the L-functions $L(E_f, s)$ are equidistributed in O, then we expect a simple zero at $s = 1$ for about 1/2 of the f

and no zero for about $1/2$ of the f. See the introduction of [Kat02] for a lucid discussion of these results and the more general context, including cases where the monodromy is SO.

The third application is to non-vanishing results for twists. Given a function field F over $\mathbf{F}_q$, a Galois representation ρ of $\mathrm{Gal}(\overline{F}/F)$, and an integer $d > 1$, one expects to be able to find infinitely many characters $\chi : \mathrm{Gal}(\overline{F}/F) \to \mu_d$ of order d such that $L(\rho \otimes \chi, s)$ does not vanish at some given point $s = s_0$, for example the center of the functional equation. There are few general results in this direction, but if we modify the problem in the usual way then one can prove quite general theorems. Namely, one considers characters χ of $\mathrm{Gal}(\overline{F}/\mathbf{F}_{q^n}F)$ for varying n and with restrictions on the ramification of χ (for example, that the degree of the conductor of χ be less than some D and the ramification of χ be prime to the ramification of ρ). Then under mild hypotheses, one finds the existence of infinitely many characters χ (indeed a set of positive density in a suitable sense) with $L(\rho \otimes \chi, s)$ non-vanishing at a given point s_0. The precise statements involve both non-vanishing and simple vanishing because there may be vanishing forced by functional equations. The connection with equidistribution is that in any of the classical groups O, Sp, or SL, the set of matrices with a given number as eigenvalue has Haar measure zero (except of course for orthogonal matrices and eigenvalues ± 1, which are related to forced zeros). See [Ulm05] for this and more general non-vanishing results.

4 Further reading

In this section we give a personal and perhaps idiosyncratic overview of some of the literature covering the technology implicit in this article.

For a treatment of number theory in function fields very much parallel to classical algebraic number theory and requiring essentially no algebraic geometry, I recommend [Ros02].

For the basic theory of curves over an algebraically closed ground field, a standard reference in use for generations now is [Ful89]. This gives a student-friendly introduction, with all necessary algebraic background and complete details, of the basic theory of curves over an algebraically closed field. Weil's "Foundations" [Wei62] gives a complete and functional theory for algebraic geometry over arbitrary base fields, but it is quite difficult to read and the language has fallen into disuse—the much more powerful and flexible language of schemes is completely dominant. Various books on diophantine geometry and elliptic curves give short accounts, often incomplete or not entirely accurate, of algebraic geometry over general fields. For careful and complete expositions of the theory of curves over general fields, including the ζ-function and the Riemann hypothesis, two popular references are [Gol03] and [Sti93].

For the basics of general, higher dimensional algebraic geometry, there is no better reference than the first part of [Sha77]. This book gives a masterful exposition of the main themes and goals of the field with excellent taste. Part II of this work, on schemes and complex manifolds, is interesting but not sufficiently detailed to be of use as a primary reference.

One can get an excellent idea of some of the analogies between curves over finite fields and rings of integers in number fields, analogies which motivate many of the ideas in modern arithmetical algebraic geometry, from [Lor96]. Studying this work would be a good first step toward schemes, giving the student a valuable stock of examples and tools.

For an introduction to schemes from many points of view, in particular that of number theory, the best reference by far is a long typescript by Mumford and Lang which was meant to be a successor to "The Red Book" (Springer Lecture Notes 1358) but which was never finished. These notes have excellent discussions of arithmetic schemes, Galois theory of schemes, the various flavors of Frobenius, flatness, issues of inseparability and imperfection, as well as a very down to earth introduction to coherent cohomology. (Some energetic young person would do the community a great service by cleaning up and TeXing these notes.) Some of this material was adapted by Eisenbud and Harris [EH00], including a nice discussion of the functor of points and moduli, but there is much more in the Mumford-Lang notes.

Another excellent and complete reference for the scheme-theoretic tools needed for arithmetical algebraic geometry is [Liu02] which has the virtue of truly being a textbook, with a systematic presentation and lots of exercises.

To my knowledge there is no simple route into the jungle of étale cohomology. Katz's article [Kat94] in the Motives volume gives a clear and succinct statement of the basics, and Iwaniec and Kowalski [IK04, 11.11] give a short introduction to some basic notions with applications to exponential sums. To go deeper, I recommend $SGA4\frac{1}{2}$ [SGA4-1/2] for the main ideas and Milne's masterful text [Mil80], supplemented by the notes on his site (`http://jmilne.org`), for a systematic study.

For wonderful examples of this technology in action I suggest [KS99b] and the papers of Katz referred to there, including [Kat02] (which is the final version of the entry [K-BTBM] in the bibliography of [KS99b]).

Finally, for an in depth introduction to connections between random matrix theory and number theory, I recommend [MHS05], the proceedings of a Newton Institute school on the subject.

References

[SGA4-1/2] P. Deligne. *Cohomologie étale*. Séminaire de Géométrie Algébrique du Bois- Marie SGA $4\frac{1}{2}$. Avec la collaboration de J. F. Boutot, A. Grothendieck, L. Illusie et J. L. Verdier. Lecture Notes in Mathematics, Vol. 569. Springer-Verlag, Berlin-New York, 1977.

[Del80] P. Deligne. La conjecture de Weil. II. *Inst. Hautes Études Sci. Publ. Math.*, (52):137– 252, 1980.

[EH00] D. Eisenbud and J. Harris. *The geometry of schemes*, volume 197 of *Graduate Texts in Mathematics*. Springer-Verlag, New York, 2000.

[Ful89] W. Fulton. *Algebraic curves*. Advanced Book Classics. Addison-Wesley Publishing Company Advanced Book Program, Redwood City, CA, 1989.

[Gol03] D. M. Goldschmidt. *Algebraic functions and projective curves*, volume 215 of *Graduate Texts in Mathematics*. Springer-Verlag, New York, 2003.

[Hal04] C. Hall. L-functions of twisted Legendre curves. Journal of Number Theory, Vol. 119, No. 1, 2006, 128–147.

[IR90] K. Ireland and M. Rosen. *A classical introduction to modern number theory*, volume 84 of *Graduate Texts in Mathematics*. Springer-Verlag, New York, 1990.

[IK04] H. Iwaniec and E. Kowalski. *Analytic number theory*, volume 53 of *American Mathematical Society Colloquium Publications*. American Mathematical Society, Providence, RI, 2004.

[Kat94] N. M. Katz. Review of l-adic cohomology. In *Motives (Seattle, WA, 1991)*, volume 55 of *Proc. Sympos. Pure Math.*, pages 21– 30. Amer. Math. Soc., Providence, RI, 1994.

[Kat02] N. M. Katz. *Twisted L-functions and monodromy*, volume 150 of *Annals of Mathematics Studies*. Princeton University Press, Princeton, NJ, 2002.

[KS99b] N. M. Katz and P. Sarnak. *Random matrices, Frobenius eigenvalues, and monodromy*, volume 45 of *American Mathematical Society Colloquium Publications*. American Mathematical Society, Providence, RI, 1999.

[KS99a] N. M. Katz and P. Sarnak. Zeroes of zeta functions and symmetry. *Bull. Amer. Math. Soc. (N.S.)*, 36:1– 26, 1999.

[Kob93] N. Koblitz. *Introduction to elliptic curves and modular forms*, volume 97 of *Graduate Texts in Mathematics*. Springer-Verlag, New York, 1993.

[Liu02] Q. Liu. *Algebraic geometry and arithmetic curves*, volume 6 of *Oxford Graduate Texts in Mathematics*. Oxford University Press, Oxford, 2002.

[Lor96] D. Lorenzini. *An invitation to arithmetic geometry*, volume 9 of *Graduate Studies in Mathematics*. American Mathematical Society, Providence, RI, 1996.

[MHS05] F. Mezzadri and N. C. Snaith, eds. *Recent perspectives in random matrix theory and number theory*. Volume 322 of the *London Mathematical Society Lecture Note Series*. Cambridge University Press, Cambridge, 2005.

[Mil80] J. S. Milne. *Étale cohomology*, volume 33 of *Princeton Mathematical Series*. Princeton University Press, Princeton, N.J., 1980.

[Ros02] M. Rosen. *Number theory in function fields*, volume 210 of *Graduate Texts in Mathematics*. Springer-Verlag, New York, 2002.

[Sha77] I. R. Shafarevich. *Basic algebraic geometry*. Springer-Verlag, Berlin, 1977.

[Sti93] H. Stichtenoth. *Algebraic function fields and codes*. Universitext. Springer-Verlag, Berlin, 1993.

[Ulm05] D. L. Ulmer. Geometric non-vanishing. *Invent. Math.*, 159:133– 186, 2005.

[Wei48] A. Weil. *Sur les courbes algébriques et les variétés qui s'en déduisent*. Actualités Sci. Ind., no. 1041 = Publ. Inst. Math. Univ. Strasbourg **7** (1945). Hermann et Cie., Paris, 1948.

[Wei49] A. Weil. Numbers of solutions of equations in finite fields. *Bull. Amer. Math. Soc.*, 55:497– 508, 1949.

[Wei62] A. Weil. *Foundations of algebraic geometry*. American Mathematical Society, Providence, R.I., 1962. MR0144898 (26 #2439).

[Wei95] A. Weil. *Basic number theory*. Classics in Mathematics. Springer-Verlag, Berlin, 1995.

Department of Mathematics
University of Arizona
Tucson, AZ
85721 USA

Some applications of symmetric functions theory in random matrix theory

Alex Gamburd *

Dedicated to Persi Diaconis on the occasion of his 60th birthday

1   Introduction

This paper gives a brief and informal introduction to some applications of symmetric functions theory in random matrix theory. It is based on the lecture given as part of the program "Random matrix approaches in number theory" held at the Newton Institute from February to July of 2004; I would like to take this opportunity to thank the organizers for making this program such a wonderful and memorable one.

We begin, in section 2, by presenting a self-contained and fairly complete proof of the Weyl Integration Formula [37], the basic tool for averaging over $U(n)$, following closely Weyl's original derivation. In the course of the proof we also provide a simple qualitative explanation for "quadratic repulsion" of the eigenvalues of unitary matrices. As a straightforward application of the Weyl Integration Formula we derive a simple but striking result of Rains [31] on high powers of unitary matrices.

Following the organizers' brief, in the exposition below I assume no prior knowledge of symmetric function theory. A crucial role in this theory is played by Schur functions and each section provides a glimpse of a different facet of their many-sided nature. In section 3 we introduce Schur functions and prove that they describe irreducible characters of the unitary group, following closely the derivation of Weyl in his classical book [37]. In section 4 we prove asymptotic normality of traces of powers of random unitary matrices as a consequence of Schur-Weyl duality, following Diaconis and Shahshahani [15], who pioneered the symmetric functions theory approach. In section 5, using an alternative definition of Schur functions via Jacobi-Trudi identity, we derive the formula of moments of characteristic polynomials; section 7 is devoted to exploiting some of the consequences of the combinatorial definition of Schur functions.

*The author was supported in part by senior visiting fellowship at the Isaac Newton Institute for Mathematical Sciences and by by NSF grants DMS-0111298 and DMS-0501245.

With the exception of section 6, where we sketch a derivation of the formula for ratios of averages of characteristic polynomials for symplectic group, we restrict our attention to the unitary group and refer to the original papers for derivations pertaining to other compact groups. We also restrict ourself to those aspects of random matrix theory which featured in "Random matrix approaches in number theory" and do not address at all many spectacular applications of symmetric functions and random matrices to problems involving random permutations, referring the reader to beautiful surveys [1, 30] and references therein. The books by Macdonald [27] and Stanley [35] provide a comprehensive introduction to symmetric functions theory; the book by Bump [3] is an excellent reference for representation theory of Lie groups.

2 Weyl Integration Formula

2.1 Unitary Group

Let $U(n)$ denote the group of unitary matrices, that is $n \times n$ complex matrices M such that $\overline{M}^t M = I$; equivalently these are matrices M such that $\langle Mu, Mv \rangle = \langle u, v \rangle$ for all $u, v \in \mathbb{C}$, where

$$\langle u, v \rangle = \sum_{j=1}^{n} u_i \overline{v_j}$$

is the standard hermitian inner product on $\mathbb{C}^n$. The group $G = U(n)$ is a compact connected Lie group with the Lie algebra $\mathfrak{u}(n)$ consisting of $n \times n$ skew-hermitian matrices, that is complex matrices X satisfying

$$\overline{X}^t + X = 0. \tag{2.1}$$

Condition (2.1) arises by differentiating with respect to t the defining condition of $U(n)$

$$\langle e^{tX} u, e^{tX} v \rangle = \langle u, v \rangle \qquad \text{for any } u, v \in \mathbb{C}^n,$$

and then evaluating at $t = 0$ to get

$$\langle Xu, v \rangle + \langle u, Xv \rangle = 0;$$

consequently

$$\langle (\overline{X}^t + X)u, v \rangle = 0 \qquad \text{for any } u, v \in \mathbb{C}^n,$$

which implies (2.1).

Following the standard convention we denote by

$$\mathrm{ad}_B(C) = [B, C] = BC - CB \tag{2.2}$$

and for $g \in U(n)$

$$\mathrm{Ad}g(B) = gBg^{-1}. \tag{2.3}$$

We have a fundamental relation:

$$\operatorname{Ad} \exp t B = \exp(t \operatorname{ad}_B). \tag{2.4}$$

The subgroup of unitary diagonal matrices A (Cartan subgroup) is abelian and isomorphic to $\mathbb{T}^N$, the N-fold power of $\mathbb{T}$, the unit circle in $\mathbb{C}$. An element $a \in A$ is of the form

$$a = \operatorname{diag}(e^{i\theta_1}, \ldots, e^{i\theta_n}).$$

By elementary linear algebra every unitary matrix is conjugate to a diagonal matrix, that is for every $g \in U(n)$ we can write

$$g = hah^{-1} \tag{2.5}$$

for $h \in U(n)$ and $a \in A$. A function f on G is a class function if it is constant on conjugacy classes, that is $f(h^{-1}gh) = f(g)$. Denoting the eigenvalues of g by $x_j = e^{i\theta_j}$, for a class fucntion f we have

$$f(g) = f(a) = f(\operatorname{diag}(e^{i\theta_1}, \ldots, e^{i\theta_n})),$$

where a is an element in A conjugate to g.

2.2   Weyl Integration Formula

Theorem 1 (Weyl Integration Formula [37]). *The expression for normalized Haar measure on $U(n)$ in terms of eigenangles θ_j is given by*

$$\frac{1}{(2\pi)^n} \frac{1}{n!} \prod_{1 \leqslant k < l \leqslant n} |e^{i\theta_k} - e^{i\theta_l}|^2 \, d\theta_1 \ldots d\theta_n. \tag{2.6}$$

The expectation of a class function $\varphi(M) = \varphi(e^{i\theta_1}, \ldots, e^{i\theta_n})$ with respect to Haar measure dM,

$$\mathbb{E}_{U(n)} \varphi(M) = \int_{U(n)} \varphi(M) dM$$

is given by:

$$\frac{1}{(2\pi)^n} \frac{1}{n!} \int_0^{2\pi} \cdots \int_0^{2\pi} \varphi(e^{i\theta_1}, \ldots, e^{i\theta_n}) \prod_{1 \leqslant k < l \leqslant n} |e^{i\theta_k} - e^{i\theta_l}|^2 \, d\theta_1 \ldots d\theta_n. \tag{2.7}$$

Proof   The map

$$\rho : A \times U(n) \to U(n), \qquad \rho(a, g) = gag^{-1} \tag{2.8}$$

is surjective; further it is clear that

$$\rho(a, gb) = \rho(a, g) \quad \text{for} \quad b \in A. \tag{2.9}$$

Hence the map ρ factors to

$$\widetilde{\rho} : A \times (U(n)/A) \to U(n). \tag{2.10}$$

The factored map $\widetilde{\rho}$ is generically finite to one. On the open dense set of matrices with n distinct eigenvalues it is $n!$ to one covering map: two diagonal matrices define the same conjugacy class in $U(n)$ if and only if one can be turned into the other by permuting the diagonal entries.

In replacing h in equation (2.5) by hb as indicated in equation (2.9) the matrix g will stay unaltered if and only if b commutes with $a = \mathrm{diag}(x_1, \ldots, x_n)$. If all the eigenvalues x_i are distinct, b has to be diagonal. However, the situation is different for singular matrices, that is matrices for which two eigenvalues coincide. Rewriting the required commutativity relation $ab = ba$ in the form

$$x_i b_{ik} = b_{ik} x_k \qquad \text{or} \qquad b_{ik}(x_i - x_k) = 0$$

we see that if, for example, $x_1 = x_2$ the matrix b is allowed to be of the form

$$\begin{pmatrix} b_{11} & b_{12} & & & \\ b_{21} & b_{22} & & & \\ & & b_3 & & \\ & & & \ddots & \\ & & & & b_n \end{pmatrix}.$$

We thus see that the singular matrices form a manifold of dimension, not of one, as one might expect, but of *three dimensions less*. Now one can think of $\widetilde{\rho}$ as defining the "polar coordinates" in $U(n)$. Consequently singular elements are like the "center of polar coordinates in three-dimensional space". The formula for the volume element in three-space in terms of polar coordinates contains the factor r^2 which vanishes in second order at the origin. For the same reason, the density in the Weyl integration formula must vanish in the second order with $e^{i\theta_1} - e^{i\theta_2}$, i. e. it must contain the factor $|e^{i\theta_1} - e^{i\theta_2}|^2$. The same holds for all other pairs $e^{i\theta_j}, e^{i\theta_k}$. This is a qualitative reason for "quadratic repulsion" of eigenvalues in unitary group.

Returning to (2.10), we can use $\widetilde{\rho}$ to lift Haar measure dg on $U(n)$ up to $A \times (U(n)/A)$. That is we can find unique measure $d\mu(a, \breve{g})$ on $A \times (U(n)/A)$ such that the set where $\widetilde{\rho}$ is singular has measure zero and on the set where $\widetilde{\rho}$ is finite to one we have the following formula for a function f on $A \times (U(n)/A)$:

$$\int_{A \times (U(n)/A)} f(a, \breve{g}) d\mu(a, \breve{g}) = \int_{U(n)} \left(\sum_{x \in \widetilde{\rho}^{-1}(g)} f(x) \right) dg. \tag{2.11}$$

The coset space $U(n)/A$ also possesses a left-invariant measure $d\breve{g}$. Since Haar measure on $U(n)$ is also conjugation invariant we see that $d\mu$ must be a product measure of the form

$$d\mu(a, \breve{g}) = d\nu(a) d\breve{g}.$$

Since we are dealing with smooth manifolds and smooth maps, $d\nu$ is absolutely continuous with respect to Haar measure da on A:

$$d\nu(a) = \nu(a)da$$

for an appropriate function $\nu(a)$.

Now computation of $\nu(a)$ boils down to computing the Jacobian of the map $\widetilde{\rho}$; we will show below that for $a = \mathrm{diag}(x_1, \ldots, x_n)$ it is given by

$$\nu(a) = c \prod_{1 \le i < j \le n} |x_i - x_j|^2. \tag{2.12}$$

Next, if in (2.11) we take the function f to be a pull-back from $U(n)$ by $\widetilde{\rho}$:

$$f(a, \breve{g}) = \varphi(\widetilde{\rho}(a, \breve{g})) = \varphi(\breve{g}a\breve{g}^{-1}),$$

for some function φ on $U(n)$; we have in light of the discussion above

$$n! \int_{U(n)} \varphi(g)dg = \int_{A \times (U(n)/A)} \varphi(\breve{g}a\breve{g}^{-1})\nu(a)da d\breve{g}.$$

If further φ is a class function, taking into account (2.12) we obtain

$$\int_{U(n)} \varphi(g)dg = \frac{c}{n!} \int_A \varphi(x_1, \ldots, x_n) \prod_{1 \le i < j \le n} |x_i - x_j|^2 da.$$

Now the constant c is easily computed to be equal to 1 (by taking function φ identically equal to 1) and so we obtain the thought-after equation (2.7).

It remains to prove (2.12), that is to compute the Jacobian of the mapping $\widetilde{\rho}$ given by

$$\widetilde{\rho}(a, \breve{g}) = gag^{-1},$$

where $\breve{g} \in U(n)/A$. To this end we make use of the Lie algebra and the exponential map to parametrize a neighborhood of $a \in A$ and of $\breve{g} \in U(n)/A$. The Lie algebra of A, which we'll denote by $\mathfrak{a}$, consists of diagonal matrices $\mathrm{diag}(ia_1, \ldots, ia_n)$ where a_i are real. The neighborhood of $a \in A$ is parametrized by $\alpha \mapsto e^\alpha$ for $\alpha \in \mathfrak{a}$. A neighborhood of $\breve{g}$ is parametrized by $\beta \mapsto ge^\beta A$, with $\beta \in \mathfrak{a}^\perp$, where $\mathfrak{a}^\perp$ is an orthogonal complement of $\mathfrak{a}$ in $\mathfrak{u}(n)$.

So we have parametrized the neighborhood of $(a, \breve{g})$ by

$$(\alpha, \beta) \longmapsto (ae^\alpha, ge^\beta A)$$

and with this parametrization the map $\widetilde{\rho}$ is given by

$$(\alpha, \beta) \longmapsto ge^\beta ae^\alpha e^{-\beta} g^{-1}.$$

Translating on the left by $a^{-1}g^{-1}$ and on the right by g, and using the fact that the Jacobian of translations is one, we are reduced to computing the Jacobian of the map

$$(\alpha, \beta) \longmapsto a^{-1}e^\beta ae^\alpha e^{-\beta}.$$

Now recalling (2.2), (2.3) and (2.4) we see that the differential of this map is given by $(1 - \mathrm{ad}a|_{\mathfrak{a}^\perp})$, and since the Jacobian is the determinant of the differential, we have that

$$\nu(a) = \det(1 - \mathrm{ad}a|_{\mathfrak{a}^\perp}).$$

Finally, recalling (2.1) we note that concretely $\mathfrak{a}^\perp$ is the space of skew-symmetric $n \times n$ matrices with zeros on the diagonal. Consequently, if $a = \mathrm{diag}(x_1, \ldots, x_n)$ we have that

$$\det(1 - \mathrm{ad}a|_{\mathfrak{a}^\perp}) = \prod_{i \neq j}(1 - x_i x_j^{-1}) = \prod_{1 \leq i < j \leq n} |x_i - x_j|^2,$$

where the last equality follows from $|x_i| = 1$ for all i. This completes the proof of (2.12) and consequently also the proof of Weyl integration formula. $\square$

2.3 High Powers of Unitary Matrices

We now use Weyl integration formula to prove, following Rains [31], the following result.

Proposition 1 (Rains [31]). *If M is Haar distributed on $U(N)$ and n is any integer $n \geq N$, then the eigenvalues of M^n are independent and uniformly distributed on the unit circle S^1.*

Proof Denoting the eigenvalues of M by $e^{i\theta_j}$, let $Y_j = e^{in\theta_j}$. We want to show that $(Y_1, \ldots, Y_N)$ are i.i.d. uniform on S^1. By the method of moments it is enough to show that for any $\mathbf{a} = (a_1, \ldots, a_N)$ and $\mathbf{b} = (b_1, \ldots, b_N)$ with $a_i \in \mathbb{N}$ and $b_i \in \mathbb{N}$ we have

$$\mathbb{E}_{U(N)} \prod_{j=1}^{N} Y_j^{a_j} \prod_{k=1}^{N} \overline{Y_k}^{b_k} = \mathbb{E}_{U(N)} \prod_{j=1}^{N} Y_j^{a_j - b_j} = \delta_{\mathbf{ab}}. \tag{2.13}$$

Now by Weyl integration formula,

$$\mathbb{E}_{U(N)} \prod_{j=1}^{N} Y_j^{a_j - b_j} = \frac{1}{(2\pi)^N} \frac{1}{N!} \int_0^{2\pi} \cdots \int_0^{2\pi} e^{\sum_{j=1}^{N} in\theta_j(a_j - b_j)} \times$$
$$\prod_{1 \leq k < l \leq N} (e^{i\theta_k} - e^{i\theta_l})(e^{-i\theta_k} - e^{-i\theta_l}) d\theta_1 \ldots d\theta_N. \tag{2.14}$$

We now examine the expression in the second line of (2.14). For example, for $N = 2$ it is given by

$$(e^{i\theta_1} - e^{i\theta_2})(e^{-i\theta_1} - e^{-i\theta_2}) = 2 - e^{i\theta_1}e^{-i\theta_2} - e^{-i\theta_1}e^{i\theta_2}.$$

In general,

$$\prod_{1 \leq k < l \leq N} (e^{i\theta_k} - e^{i\theta_l})(e^{-i\theta_k} - e^{-i\theta_l})$$

is a Laurent polynomial in $e^{i\theta_j}$ of degree at most $(N-1)$ in any given $e^{i\theta_j}$. Now it is easy to see that for $n \geq N$ only the constant term can contribute, proving (2.13) and consequently Proposition 1. $\qquad\square$

Proposition 1 implies in particular the following result

Corollary 1. *The distribution of* $\mathrm{Tr}(M^n)$ *for* $n \geq N$ *is exactly the same as* $\sum_{j=1}^{N} X_j$ *where* X_j *are independent random variables uniformly distributed on* S^1.

The case of n small (relative to N) will be dealt with in Theorem 2 below. For the discussion of Corollary 1 in the context of the Riemann zeta function see the paper by Montgomery and Soundararajan [29].

3   Schur functions as characters of the unitary group

The reader is likely to have encountered the elementary symmetric functions

$$e_r(x_1, \ldots, x_N) = \sum_{i_1 < \cdots < i_r} x_{i_1} \ldots x_{i_r} \tag{3.1}$$

and power sum symmetric functions

$$p_r(x_1, \ldots, x_N) = \sum_{i=1}^{N} x_i^r. \tag{3.2}$$

To generalize slightly we introduce partition notation.

A partition λ of a nonnegative integer n is a sequence $(\lambda_1, \ldots, \lambda_r) \in \mathbb{N}^r$ satisfying $\lambda_1 \geq \ldots \geq \lambda_r$ and $\sum \lambda_i = n$. We call $|\lambda| = \sum \lambda_i$ the size of λ. The number of parts of λ is the length of λ, denoted $l(\lambda)$. Write $m_i = m_i(\lambda)$ for the number of parts of λ that are equal to i, so we have $\lambda = \langle 1^{m_1} 2^{m_2} \ldots \rangle$.

The Young diagram of a partition λ is defined as the set of points $(i, j) \in \mathbb{Z}^2$ such that $1 \leq i \leq \lambda_j$; it is often convenient to replace the set of points above by squares. The conjugate partition λ' of λ is defined by the condition that the Young diagram of λ' is the transpose of the Young diagram of λ; equivalently $m_i(\lambda') = \lambda_i - \lambda_{i+1}$.

In the figure we exhibited a partition $\lambda = (5, 5, 3, 2) = \langle 1^0 2^1 3^1 5^2 \rangle$; $\lambda \vdash 15$ and $l(\lambda) = 4$.

Now given a partition $\lambda \vdash N$, we define

$$e_\lambda(x_1, \ldots, x_N) = \prod_{j=1}^{N} e_{\lambda_j}(x_1, \ldots, x_n) \tag{3.3}$$

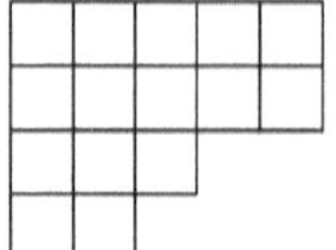

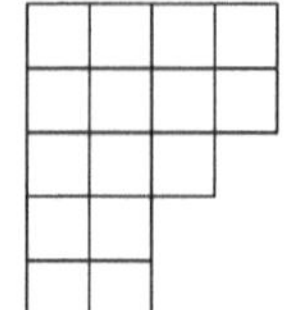

Young diagram of λ Young diagram of λ'

and similarly

$$p_\lambda(x_1,\ldots,x_N) = \prod_{j=1}^{N} p_{\lambda_j}(x_1,\ldots,x_N). \tag{3.4}$$

Both elementary symmetric functions e_λ and power sum symmetric functions p_λ will be important in what follows. However the center stage is occupied by Schur symmetric functions s_λ whose classical definition is as a ratio of two determinants

$$s_\lambda(x_1,\ldots,x_N) = \frac{\det\left(x_i^{\lambda_j+N-j}\right)_{i,j=1}^{N}}{\det\left(x_i^{N-j}\right)_{i,j=1}^{N}}. \tag{3.5}$$

Schur functions arise as irreducible characters of the unitary group and we now give a sketch of the derivation of this fact following the account in Weyl's classical book [37]; we hope this derivation makes it clear how natural the expression (3.5) is.

Let $M \in U(N)$ be a unitary matrix with eigenvalues $e^{i\theta_1},\ldots,e^{i\theta_N}$; let $x_j = e^{i\theta_j}$. We wish to show that the irreducible characters φ of the unitary group $U(N)$ are given by Schur functions s_λ. As a class function φ is a continuous symmetric periodic function of the angles θ_j. Next,

$$\varphi(M) = \varphi(\mathrm{diag}(e^{i\theta_1},\ldots,e^{i\theta_N})),$$

where $\mathrm{diag}(e^{i\theta_1},\ldots,e^{i\theta_N})$ is a diagonal matrix conjugate to M. The subgroup of unitary diagonal matrices A (Cartan subgroup) is abelian and isomorphic to $\mathbb{T}^N$, the N-fold power of $\mathbb{T}$, the unit circle in $\mathbb{C}$. In case $N = 1$ the (irreducible) representations are given by x^m for integer m (recall that we use the notation $x_j = e^{i\theta_j}$). For $N > 1$ the characters are given by finite sums with nonnegative integer coefficients of monomials

$$x^{\mathbf{m}} = x_1^{m_1}\ldots x_N^{m_N}$$

where m_i are integer exponents. To summarize: $\varphi(x_1,\ldots,x_N)$ is a symmetric function of $x_1,\ldots,x_N$ given by $\sum c(\mathbf{m})x^{\mathbf{m}}$, where the coefficients $c(\mathbf{m})$ are nonnegative integers.

Next we rewrite the expression in the Weyl Integration Formula as follows:

$$\prod_{j<k} |x_j - x_k|^2 = \Delta(x)\overline{\Delta(x)},$$

where

$$\Delta = \prod_{j<k}(x_j - x_k) = \det(x_i^{N-j})$$

is a Vandermonde determinant. Now Δ is a skew-symmetric function of $x_1, \ldots, x_N$ and consequently $\varphi\Delta$ is a skew-symmetric function as well. The simplest possible skew-symmetric functions $a_{\mathbf{m}}(x)$ can be obtained by skew-symmetrizing monomials $x^{\mathbf{m}}$:

$$a_{\mathbf{m}}(x_1, \ldots, x_N) = \sum_{w \in S_N} \varepsilon_w w(x^{\mathbf{m}}), \tag{3.6}$$

where

$$w(x^{\mathbf{m}}) = x_1^{m_{w(1)}} \ldots x_N^{m_{w(N)}}$$

and

$$\varepsilon_w = \begin{cases} 1 & \text{if } w \text{ is even} \\ -1 & \text{if } w \text{ is odd.} \end{cases}$$

Since $a_{\mathbf{m}}$ is skew-symmetric, i.e. $w(a_{\mathbf{m}}) = \varepsilon_w(a_{\mathbf{m}})$, we have $a_{\mathbf{m}} = 0$ unless all the m_i are distinct. Hence we can assume that $m_1 > m_2 > \ldots m_N \geq 0$ and so $\mathbf{m} = \lambda + \delta$ where λ is a partition with $l(\lambda) \leq N$ and

$$\delta = \delta_N = (N - 1, N - 2, \ldots, 0).$$

Since $m_j = \lambda_j + N - j$, we get

$$a_{\mathbf{m}} = a_{\lambda+\delta} = \det\left(x_i^{\lambda_j+N-j}\right)_{i,j=1}^N. \tag{3.7}$$

Now since

$$\frac{1}{(2\pi)^N} \int_0^{2\pi} \ldots \int_0^{2\pi} e^{im_1\theta_1} \ldots e^{im_N\theta_N} \overline{e^{in_1\theta_1} \ldots e^{in_N\theta_N}} \, d\theta_1 \ldots d\theta_N$$
$$= \begin{cases} 1 & \text{if } \mathbf{m} = \mathbf{n} \\ 0 & \text{otherwise,} \end{cases}$$

we easily obtain

$$\frac{1}{(2\pi)^N} \int_0^{2\pi} \ldots \int_0^{2\pi} a_{\mathbf{m}}(e^{i\theta_1} \ldots e^{i\theta_N}) \overline{a_{\mathbf{n}}(e^{i\theta_1} \ldots e^{i\theta_N})} \, d\theta_1 \ldots d\theta_N$$
$$= \begin{cases} N! & \text{if } \mathbf{m} = \mathbf{n} \\ 0 & \text{otherwise.} \end{cases}$$

Returning to $\varphi\Delta$, we now write it as a sum of the functions $a_{\mathbf{m}}$, with $\mathbf{m}$ arranged in lexicographic order, $\sum c(\mathbf{m})a_{\mathbf{m}}$; the leading coefficient in this sum is necessarily a positive integer. But since φ is an irreducible character, we have $\mathbb{E}_{U_N}\varphi\overline{\varphi} = 1$, that is

$$\frac{1}{(2\pi)^N}\frac{1}{N!}\int_0^{2\pi}\cdots\int_0^{2\pi}\varphi\Delta(e^{i\theta_1}\ldots e^{i\theta_N})\overline{\varphi\Delta(e^{i\theta_1}\ldots e^{i\theta_N})}d\theta_1\ldots d\theta_N$$
$$= \sum c^2(\mathbf{m}) = 1,$$

which implies that we have only one term $a_{\lambda+\delta}$ in the expansion of φ. This proves that irreducible characters of $U(N)$ are given by Schur function in equation (3.5).

We record the crucial fact derived in this section as follows:

$$\mathbb{E}_{U_N}\left(s_\lambda(M)\overline{s_\mu(M)}\right) = \begin{cases} 1 & \text{if } \lambda = \mu \text{ and } l(\lambda), l(\mu) \leq N \\ 0 & \text{otherwise.} \end{cases} \tag{3.8}$$

4 Frobenius-Schur duality & moments of traces

Frobenius-Schur duality is a relationship between irreducible representations of the unitary group $U(N)$ and the irreducible representations of the symmetric group S_K; see [3] for a modern introduction. Both $U(N)$ and S_K act on the K-fold tensor product $\bigotimes^K \mathbb{C}^N$. The unitary group acts linearly:

$$g(v_1 \otimes v_2 \otimes \ldots v_K) = g(v_1) \otimes g(v_2) \otimes \ldots g(v_K),$$

where $g \in U(N)$ and $v_j \in \mathbb{C}^N$. Symmetric group acts by permuting the factors:

$$w(v_1 \otimes v_2 \otimes \ldots v_K) = v_{w^{-1}(1)} \otimes v_{w^{-1}(2)} \otimes \ldots v_{w^{-1}(K)},$$

where $w \in S_K$ and $v_j \in \mathbb{C}^N$. These actions clearly commute with each other and the crucial fact is that under the joint action of $U(N) \times S_K$ the space $\bigotimes^K \mathbb{C}^N$ decomposes into direct sum

$$\bigotimes^K \mathbb{C}^N = \sum_{\substack{\lambda \vdash K \\ l(\lambda) \leq N}} \rho^\lambda \otimes \sigma^\lambda, \tag{4.1}$$

where ρ^λ is a representation of $U(N)$ and σ^λ representation of S_K labelled by λ.

We denote by χ_μ^λ the value of the irreducible character of the symmetric group S_K associated with λ on the μth conjugacy class. A consequence of Frobenius-Schur duality is the following expression for power sum symmetric functions (3.4) in terms of Schur functions (3.5):

$$p_\mu = \sum_{\lambda \vdash K} \chi_\mu^\lambda s_\lambda. \tag{4.2}$$

We sketch a proof of (4.2). Let $w \in S_K$ belong to conjugacy class μ and let g in $U(N)$ be a matrix with eigenvalues $(x_1, \ldots, x_N)$. We compute the trace of $w \cdot g$ action on $\bigotimes^K \mathbb{C}^N$ in two ways. First, using (4.1) we obtain:

$$\operatorname{tr}(w \cdot g) = \sum_{\substack{\lambda \vdash K \\ l(\lambda) \leq N}} \operatorname{tr}(w \cdot g) \downarrow_{\rho^\lambda \otimes \sigma^\lambda}$$

$$= \sum_{\substack{\lambda \vdash K \\ l(\lambda) \leq N}} \operatorname{tr}(w \downarrow_{\sigma^\lambda}) \operatorname{tr}(g \downarrow_{\rho^\lambda})$$

$$= \sum_{\lambda \vdash K} \chi_\mu^\lambda s_\lambda(x_1, \ldots, x_N).$$

On the other hand, fixing a basis $e_1, \ldots, e_N$ in $\mathbb{C}^N$, and taking for g the diagonal matrix $g = \operatorname{diag}(x_1, \ldots, x_N)$, we have

$$w \cdot g(e_{j_1} \otimes \cdots \otimes e_{j_K}) = e_{j_{w^{-1}(1)}} \otimes \cdots \otimes e_{j_{w^{-1}(K)}} \prod_{m=1}^{K} x_{j_m},$$

and therefore by direct computation we obtain that for a permutation w with conjugacy type μ we have

$$\operatorname{tr}(w \cdot g) = \sum_{\substack{1 \leq j_1, \ldots, j_K \leq N \\ j_m = j_{w^{-1}(m)}}} \prod_{m=1}^{K} x_{j_m} = p_\mu(x_1, \ldots, x_N),$$

establishing (4.2).

We now derive, using (4.2) and (3.8), the asymptotic normality of traces of powers, following the approach of Diaconis and Shahshahani [15], see also [12, 11]. For alternative derivations see Johansson [22] and Soshnikov [32].

Theorem 2. *If M is chosen from Haar measure on U_N, the traces of successive powers have limiting Gaussian distributions: as $N \to \infty$, for any fixed k and Borel sets $B_1, \ldots, B_k$*

$$P(\operatorname{Tr} M \in B_1, \ldots, \operatorname{Tr} M^k \in B_k) \to \prod_{j=1}^{k} P(\sqrt{j}\, Z \in B_j), \qquad (4.3)$$

where Z is standard complex normal.

Theorem 2 follows by method of moments from the following Proposition:

Proposition 2. *Consider* $\mathbf{a} = (a_1, \ldots, a_l)$ *and* $\mathbf{b} = (b_1, \ldots, b_l)$ *with* a_j, b_j *nonnegative natural numbers. Let* $Z_1, \ldots, Z_n$ *be independent standard complex* *normal variables. Then for* $N \geq \max\left(\sum_1^l ja_j, \sum_1^l jb_j\right)$ *we have*

$$\mathbb{E}_{U_N} \prod_{j=1}^{l} (\mathrm{Tr}M^j)^{a_j} \overline{(\mathrm{Tr}M^j)}^{b_j} = \int_{U_N} \prod_{j=1}^{l} (\mathrm{Tr}M^j)^{a_j} \overline{(\mathrm{Tr}M^j)}^{b_j} dM$$

$$= \delta_{\mathbf{ab}} \prod_{j=1}^{l} j^{a_j} a_j! = E\left(\prod_{j=1}^{l} (\sqrt{j}Z_j)^{a_j} (\sqrt{j}\bar{Z}_j)^{b_j}\right). \quad (4.4)$$

To prove Proposition 2 we first note that

$$\mathrm{Tr}M^j = p_j(e^{i\theta_1}, \ldots, e^{i\theta_N}),$$

where p_j are power sum symmetric functions defined in (3.2). Next, if $\mu = \langle 1^{a_1} \ldots l^{a_l} \rangle$ we have

$$\prod_{j=1}^{l} (\mathrm{Tr}M^j)^{a_j} = p_\mu(e^{i\theta_1}, \ldots, e^{i\theta_N}) = p_\mu(M).$$

Now if μ is a partition of K and $\nu = \langle 1^{b_1} \ldots l^{b_l} \rangle$ is a partition of L, using (3.8) and (4.2) we have:

$$\mathbb{E}_{U_N} \prod_{j=1}^{l} (\mathrm{Tr}M^j)^{a_j} \overline{(\mathrm{Tr}M^j)}^{b_j} = \mathbb{E}_{U_N} \left(p_\mu(M)\overline{p_\nu(M)}\right)$$

$$= \mathbb{E}_{U_N} \left(\sum_{\lambda \vdash K} \chi_\mu^\lambda s_\lambda(M) \overline{\sum_{\tau \vdash L} \chi_\tau^\tau s_\tau(M)}\right)$$

$$= \delta_{KL} \sum_{\lambda \vdash K} \chi_\mu^\lambda \chi_\tau^\lambda \delta(\ell(\lambda) \leq N).$$

When $K \leq N$, all partitions of K appear and we can use the second orthogonality relation for characters which we now recall. For any finite group G with irreducible characters χ_i we have

$$\sum_i \chi_i(g)\overline{\chi_i(h)} = \begin{cases} |Z(g)| & \text{if } g \text{ and } h \text{ are conjugate in } G \\ 0 & \text{otherwise} \end{cases}, \quad (4.5)$$

where $Z(g)$ is a centralizer of g. Now since for conjugacy class C_μ given by $\mu = \langle 1^{a_1} \ldots l^{a_l} \rangle$ we have $Z(C_\mu) = \prod j^{a_j} a_j!$ we obtain that

$$\delta_{KL}\delta_{\mu\nu} \prod_{j=1}^{K} j^{a_j} a_j! = \delta_{\mathbf{ab}} \prod j^{a_j} a_j!$$

The last expression equals the joint mixed moments of $\sqrt{j}Z_j$ by an easy calculation, completing the proof of Proposition 2.

5   Moments of characteristic polynomials

For certain functions f an alternative approach to computing the averages $\int_{U(N)} f(M)dM$ over the unitary group can be based on the Heine-Szegö formula [36]:

$$\frac{1}{(2\pi)^N} \int_0^{2\pi} \cdots \int_0^{2\pi} \prod_{j=1}^N f(e^{i\theta_j}) \prod_{1 \leqslant k \leqslant l \leqslant N} |e^{i\theta_k} - e^{i\theta_l}|^2 \, d\theta_1 \ldots d\theta_N = D_N(f).$$

(5.1)

Here $D_N(f)$ is the $N \times N$ Toeplitz determinant with symbol f:

$$D_N(f) = \det\left(\hat{f}(j-k)\right)_{0 \leqslant j,k \leqslant N},$$

(5.2)

where $\hat{f}(r) = \frac{1}{2\pi}\int_0^{2\pi} f(e^{ir\theta}) \, d\theta$.

Johansson [22] gave a proof of Theorem 2 with a very sharp convergence rate by using (5.1) and strong Szegö limit theorem for Toeplitz determinants [21]; on the other hand, as explained in [4], Theorem 2 gives a new proof (and some extensions) of the strong Szegö limit theorem.

In this section we will use Heine-Szegö formula (5.1) together with an alternative definition of Schur functions furnished by Jacobi-Trudi identity:

$$s_\lambda = \det\left(e_{\lambda'_i - i + j}\right)_{i,j=1}^N$$

(5.3)

to derive the expression for the moments of the products of characteristic polynomials of random unitary matrices $P_M(z)$, given by

$$P_M(z) = \det(M - zI) = \prod_{j=1}^N (e^{i\theta_j} - z).$$

(5.4)

We refer the reader to [5] for derivations of the formulae for products and ratios of characteristic polynomials from other classical groups using symmetric functions theory.

It will be more convenient to work with polynomial

$$Q_M(z) = \det(I + Mz),$$

(5.5)

which is related to the characteristic polynomial $P_M(z)$ as follows:

$$Q_M\left(-\frac{1}{z}\right) = \frac{(-1)^N}{z^N} P_M(z).$$

(5.6)

Using Heine-Szegö formula (5.1) we have:

$$\mathbb{E}_{U_N}\left[Q_M(z_1)\ldots Q_M(z_l)\overline{Q_M(z_{l+1})}\ldots\overline{Q_M(z_m)}\right] = \frac{1}{(z_{l+1}\ldots z_m)^N} D_N(f), \quad (5.7)$$

156   A. Gamburd

with

$$f(t) = \frac{1}{t^{m-l}} \prod_{i=1}^{m} (1 + z_i t) = \sum_{r \geqslant l-m} t^r e_{r+m-l}(z_1, \ldots, z_m), \qquad (5.8)$$

where e_r are elementary symmetric functions (3.1). Now the Toeplitz determinant with symbol (5.8) can be computed using the Jacobi-Trudi identity (5.3) and is easily seen to be equal to

$$s_{N^{m-l}}(z_1, \ldots, z_m),$$

where N^{m-l} is a partition consisting of $m - l$ parts equal to N. We have thus obtained the following result, first established in [7]:

Proposition 3. *Notation being as above, we have*

$$\mathbb{E}_{U_N}\left[Q_M(z_1)\ldots Q_M(z_l)\overline{Q_M(z_{l+1})}\ldots\overline{Q_M(z_m)}\right] = \frac{s_{N^{m-l}}(z_1, \ldots, z_m)}{(z_{l+1}\ldots z_m)^N}. \qquad (5.9)$$

In particular, setting all z_j equal to 1, and setting $l = k$ and $m = 2k$, we have, using (5.6) and (5.9):

$$\mathbb{E}_{U_N}|P_M(1)|^{2k} = s_{N^k}(\underbrace{1, \ldots, 1}_{2k}).$$

Now $s_{N^k}(\underbrace{1, \ldots, 1}_{2k})$ is a dimension of an irreducible representation of $U(2k)$ labelled by partition N^k. The value of $s_\lambda(1, \ldots, 1)$ can be computed from (3.5) using l'Hopital rule to give for $\lambda = (\lambda_1, \ldots, \lambda_n)$ the following formula, known as Weyl dimension formula:

$$s_\lambda(1, \ldots, 1) = \frac{\prod_{i<j}(\mu_i - \mu_j)}{\prod_{i<j}(i - j)}, \qquad (5.10)$$

where $\mu_i = \lambda_i + n - i$. It is a nice exercise to deduce that the right-hand side of (5.10) can be expressed in the following equivalent form:

$$s_\lambda(1^n) = \prod_{u \in \lambda} \frac{n + c(u)}{h(u)}, \qquad (5.11)$$

where for a box u in a diagram λ, $h(u)$ is a hook number of u and $c(u)$ is a content number of u, which we now define. Given a diagram λ and a square $u = (i, j) \in \lambda$, the content of λ at u is defined by $c(u) = j - i$. A hook with a vertex u is a set of squares in λ directly to the right or directly below u. We define hook-length (also referred to as hook number) $h(u)$ of λ at u by

$$h(u) = \lambda_i + \lambda_j' - i - j - 1.$$

Equivalently, $h(u)$ is the number of squares directly to the right or directly below u, counting u itself once. For instance, in figure 5.1 we display hook lengths for partition $\lambda = (5, 5, 3, 2)$.

8	7	5	3	2
7	6	4	2	1
4	3	1		
2	1			

Figure 5.1:

Now it is easy to see, that for a partition $\lambda = N^k$ the product of hook numbers is given by

$$\prod_{j=0}^{N-1} \frac{(j+k)!}{j!},$$

whereas the product $\prod_{u\in\lambda}(2k + c(u))$ is given by

$$\prod_{i=1}^{k}\prod_{j=1}^{N}(2k - i + j) = \prod_{j=1}^{N-1} \frac{(j+2k)!}{(j+k)!}.$$

Consequently we have

$$\mathbb{E}_{U_N}|P_M(1)|^{2k} = \prod_{j=0}^{N-1} \frac{j!(j+2k)!}{(j+k)!^2}, \tag{5.12}$$

an expression first obtained (without restriction that k be an integer) by Keating and Snaith in [23] using Selberg's integral. We will give a different expression as well as a combinatorial interpretation of (5.12) in the next section.

6   Averages of ratios of characteristic polynomials

In this section we will sketch a derivation of the formula for averages of ratios of characteristic polynomials in the case of symplectic group, following the approach in [5]. This formula was previously derived by other methods by Conrey, Farmer and Zirnbauer [8] without restriction on the dimension of the group and also by Conrey, Forrester and Snaith [9].

A unitary matrix g is said to be *symplectic* if $gJ^tg = J$ where

$$J = \begin{pmatrix} 0 & I_N \\ -I_N & 0 \end{pmatrix}.$$

A symplectic matrix has determinant equal to 1. The symplectic group $\mathrm{Sp}(2N)$ is the group of $2N \times 2N$ symplectic matrices. The eigenvalues of a symplectic matrix are

$$e^{\pm i\theta_1}, \cdots, e^{\pm i\theta_N}$$

with
$$0 \leqslant \theta_1 \leqslant \theta_2 \leqslant \cdots \leqslant \theta_N \leqslant \pi.$$

The Weyl integration formula [37] for integrating a symmetric function $f(A) = f(\theta_1, \cdots, \theta_N)$ over $\mathrm{Sp}(2N)$ with respect to Haar measure is

$$\mathbb{E}_{\mathrm{Sp}(2N)} f = \int_{\mathrm{Sp}(2N)} f(g) \, dg = \frac{2^{N^2}}{\pi^N N!} \times$$

$$\int_{[0,\pi]^N} f(\theta_1, \cdots, \theta_N) \prod_{1 \leqslant j < k \leqslant N} (\cos \theta_k - \cos \theta_j)^2 \prod_{n=1}^{N} \sin^2 \theta_n \, d\theta_1 \cdots d\theta_N.$$

Denoting the irreducible character of the symplectic group $\mathrm{Sp}(2n)$ labelled by partition λ by $\chi_\lambda^{\mathrm{Sp}_{2n}}$, the Weyl character formula [37] in the case of symplectic group can be written

$$\chi_\lambda^{\mathrm{Sp}_{2n}}(x_1^{\pm 1}, \cdots, x_n^{\pm 1}) = \frac{\det_{1 \leq i,j \leq n}(x_j^{\lambda_i + n - i + 1} - x_j^{-(\lambda_i + n - i + 1)})}{\det_{1 \leqslant i,j \leqslant n}(x_j^{n-i+1} - x_j^{-(n-i+1)})}. \tag{6.1}$$

The determinant in the denominator can be evaluated as by Weyl [37]:

$$\det_{1 \leq i,j \leq n}(x_j^{n-i+1} - x_j^{-(n-i+1)}) = \frac{\prod_{i<j}(x_i - x_j)(x_i x_j - 1) \prod_i (x_i^2 - 1)}{(x_1 \cdots x_n)^n}. \tag{6.2}$$

The following lemma is an easy consequence of Weyl character formula (6.1) and the Laplace expansion:

Lemma 1. *For $\lambda \subseteq \langle N^k \rangle$ let $\tilde{\lambda} = (k - \lambda_N', \cdots, k - \lambda_1')$. Then we have*

$$\prod_{i=1}^{k} \prod_{n=1}^{N} (x_i + x_i^{-1} - t_n - t_n^{-1}) \quad =$$

$$\sum_{\lambda \subseteq N^k} (-1)^{|\tilde{\lambda}|} \chi_\lambda^{\mathrm{Sp}(2k)}(x_1^{\pm 1}, \cdots, x_k^{\pm 1}) \chi_{\tilde{\lambda}}^{\mathrm{Sp}(2N)}(t_1^{\pm 1}, \cdots, t_N^{\pm 1}). \tag{6.3}$$

Proof Using Weyl character formula (6.1) and the Laplace expansion we can rewrite the right-hand side of (6.3) as follows:

$$\det \begin{vmatrix} x_1^{N+k} - x_1^{-(N+k)} & x_1^{N+k-1} - x_1^{-(N+k-1)} & \cdots & x_1^1 - x_1^{-1} \\ \vdots & \vdots & \ddots & \vdots \\ x_k^{N+k} - x_k^{-(N+k)} & x_k^{N+k-1} - x_k^{-(N+k-1)} & \cdots & x_k^1 - x_k^{-1} \\ t_1^{N+k} - t_1^{-(N+k)} & t_1^{N+k-1} - t_1^{-(N+k-1)} & \cdots & t_1^1 - t_1^{-1} \\ \vdots & \vdots & \ddots & \vdots \\ t_N^{N+k} - t_N^{-(N+k)} & t_N^{N+k-1} - t_N^{-(N+k-1)} & \cdots & t_N^1 - t_N^{-1} \end{vmatrix}$$

$$\times \frac{(x_1 \ldots x_k)^k}{\prod_{1 \leqslant i < j \leqslant k}(x_i - x_j)(x_i x_j - 1) \prod_{i=1}^{k}(x_i^2 - 1)}$$

$$\times \frac{(t_1 \ldots t_N)^N}{\prod_{1 \leqslant i < j \leqslant N}(t_i - t_j)(t_i t_j - 1) \prod_{i=1}^{N}(t_i^2 - 1)}. \tag{6.4}$$

Now the determinant in the equation (6.4) can be evaluated using the Weyl denominator formula (6.2) to be equal to

$$\prod_{1\leqslant i<j\leqslant k}(x_i-x_j)(x_ix_j-1)\prod_{i=1}^{k}(x_i^2-1)\prod_{1\leqslant i<j\leqslant N}(t_i-t_j)(t_it_j-1)\prod_{i=1}^{N}(t_i^2-1)\times$$
$$\frac{\prod_{i=1}^{k}\prod_{n=1}^{N}(x_i-t_n)(x_it_n-1)}{(x_1\ldots x_k)^{N+k}(t_1\ldots t_N)^{N+k}}.$$
$$(6.5)$$

Finally combining (6.4) and (6.5) we get that the expression on the right-hand side of (6.3) equals to

$$\frac{\prod_{i=1}^{k}\prod_{n=1}^{N}(x_i-t_n)(x_it_n-1)}{(x_1\ldots x_k)^{N}(t_1\ldots t_N)^{k}}=\prod_{i=1}^{k}\prod_{n=1}^{N}(x_i+x_i^{-1}-t_n-t_n^{-1}),$$

completing the proof. $\qquad\qquad\square$

We will also make use of the following Cauchy-Littlewood identity for $\mathrm{Sp}(2N)$, known to Weyl [37] and Littlewood [26]:

$$\frac{1}{\prod_{n=1}^{N}\prod_{j=1}^{l}(1-y_jt_n)(1-y_jt_n^{-1})}=\qquad(6.6)$$

$$\frac{1}{\prod_{i<j}(1-y_iy_j)}\sum_{\mu}\chi_{\mu}^{\mathrm{Sp}(2N)}(t_1^{\pm1},\cdots,t_N^{\pm1})s_{\mu}(y_1,\cdots,y_l).\qquad(6.7)$$

We are ready to prove the following result.

Theorem 3. *Let y_j be complex numbers with $|y_j|<1$. Suppose $N\geqslant l$. Then we have:*

$$\int_{\mathrm{Sp}(2N)}\frac{\prod_{j=1}^{k}\det(I+x_jg)}{\prod_{i=1}^{l}\det(I-y_ig)}\,dg=$$

$$\sum_{\varepsilon\in\{\pm1\}^{k}}\prod_{j=1}^{k}x_j^{N(1-\varepsilon_j)}\frac{\prod_{i=1}^{k}\prod_{j=1}^{l}(1+x_i^{\varepsilon_i}y_j)}{\prod_{i\leqslant j}(1-x_i^{\varepsilon_i}x_j^{\varepsilon_j})\prod_{1\leqslant i<j\leqslant l}(1-y_iy_j)}.$$

Proof: Lemma 1 implies that

$$\prod_{i=1}^{k}\prod_{n=1}^{N}(x_i+x_i^{-1}+t_n+t_n^{-1})=$$
$$\sum_{\lambda\subseteq N^{k}}\chi_{\lambda}^{\mathrm{Sp}(2k)}(x_1^{\pm1},\cdots,x_k^{\pm1})\chi_{\tilde\lambda}^{\mathrm{Sp}(2N)}(t_1^{\pm1},\cdots,t_N^{\pm1}).\qquad(6.8)$$

Combining (6.8) with Cauchy-Littlewood formula (6.6) we have, with $t_i^{\pm1}$

the eigenvalues of $g \in \mathrm{Sp}(2N)$

$$\frac{\prod_{j=1}^{k} \det(I + x_j g)}{\prod_{i=1}^{l} \det(I - y_i g)} =$$

$$\frac{(x_1 \cdots x_k)^N}{\prod_{i<j}(1 - y_i y_j)} \sum_{\lambda \subseteq N^k} \chi_\lambda^{\mathrm{Sp}(2k)}(x_1^{\pm 1}, \cdots, x_k^{\pm 1}) \chi_{\tilde{\lambda}}^{\mathrm{Sp}(2N)}(t_1^{\pm 1}, \cdots, t_N^{\pm 1})$$

$$\sum_\mu \chi_\mu^{\mathrm{Sp}(2N)}(t_1^{\pm 1}, \cdots, t_N^{\pm 1}) s_\mu(y_1, \cdots, y_l).$$

Since

$$\mathbb{E}_{\mathrm{Sp}(2N)} \chi_\lambda^{\mathrm{Sp}(2N)}(g) \chi_\mu^{\mathrm{Sp}(2N)}(g) = \begin{cases} 1 & \text{if } \lambda = \mu, \ l(\lambda) \leqslant N; \\ 0 & \text{otherwise,} \end{cases}$$

the theorem follows from the following formula which is easily derived from Weyl character formulae using Laplace expansion:

$$\sum_{\lambda \subseteq \langle N^k \rangle} \chi_\lambda^{\mathrm{Sp}(2k)}(x_1^{\pm 1}, \cdots, x_k^{\pm 1}) s_{\tilde{\lambda}}(y_1, \cdots, y_l) =$$

$$\sum_{\varepsilon \in \{\pm 1\}^k} \prod_{j=1}^{k} x_j^{-N\varepsilon_j} \frac{\prod_{i=1}^{k} \prod_{j=1}^{l}(1 + x_i^{\varepsilon_i} y_j)}{\prod_{i \leqslant j}(1 - x_i^{\varepsilon_i} x_j^{\varepsilon_j})}. \tag{6.9}$$

7   Combinatorial definition of Schur functions and some of its consequences

One can give a purely combinatorial definition of Schur functions as follows. A semi-standard Young tableau (SSYT) of shape λ is a filling of the boxes of λ with positive integers such that the rows are weakly increasing and the columns are strictly increasing.

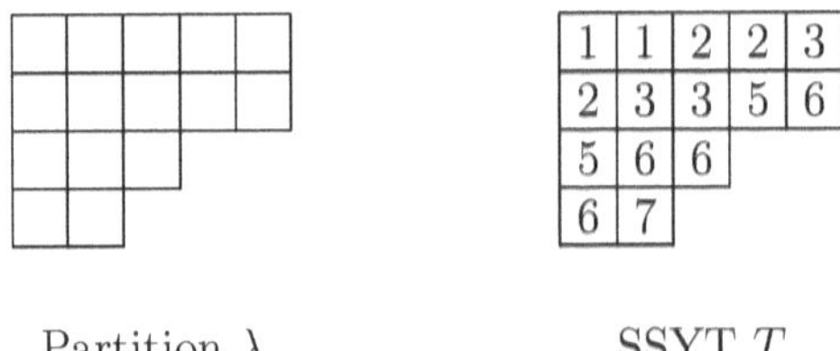

Partition λ SSYT T

Figure 7.1:

In the figure we exhibited a partition $\lambda = (5, 5, 3, 2) = \langle 1^0 2^1 3^1 5^2 \rangle$, and a SSYT T of shape λ (we write $\lambda = \mathrm{sh}(T)$). We say that T has type $\alpha = (\alpha_1, \alpha_2, \ldots)$, denoted $\alpha = \mathrm{type}(T)$, if T has $\alpha_i = \alpha_i(T)$ parts equal to i. Thus, the SSYT in the figure has type $(2, 3, 3, 0, 2, 4, 1)$. For any SSYT T of type α write

$$x^T = x_1^{\alpha_1(T)} x_2^{\alpha_2(T)} \cdots.$$

In our example we have

$$x^T = x_1^2 x_2^3 x_3^3 x_4^0 x_5^2 x_6^4 x_7^1.$$

Let λ be a partition. A purely combinatorial definition of Schur function s_λ in the variables $x = (x_1, x_2, \ldots, x_N)$ is given by the formal power series

$$s_\lambda(x) = \sum_T x^T, \tag{7.1}$$

where the sum is over all SSYT's T of shape λ.

For instance, SSYT of shape $(2, 1)$ with largest part at most three are given by

$$
\begin{array}{ccccccc}
\begin{array}{cc}1&1\\2\end{array} &
\begin{array}{cc}1&2\\2\end{array} &
\begin{array}{cc}1&1\\3\end{array} &
\begin{array}{cc}1&3\\3\end{array} &
\begin{array}{cc}2&2\\3\end{array} &
\begin{array}{cc}2&3\\3\end{array} &
\begin{array}{cc}1&2\\3\end{array} &
\begin{array}{cc}1&3\\2\end{array}
\end{array}
$$

Consequently,

$$s_{21}(x_1, x_2, x_3) = x_1^2 x_2 + x_1 x_2^2 + x_1^2 x_3 + x_1 x_3^2 + x_2^2 x_3 + x_2 x_3^2 + 2x_1 x_2 x_3.$$

The number of SSYT of shape λ and type α is denoted $K_{\lambda\alpha}$, and is called the Kostka number. We have

$$s_\lambda = \sum_\alpha K_{\lambda\alpha} x^\alpha. \tag{7.2}$$

7.1   Relationship to plane partitions

We now use combinatorial definition of Schur functions to describe the relation between moments of characteristic polynomials of random unitary matrices and plane partitions. A plane partition, $\mathcal{P}$, is a finite set of lattice points with positive integer coefficients $\{(i, j, k)\} \subseteq \mathbb{N}^3$ with the property that if $(a, b, c) \in \mathcal{P}$ and $1 \leq i \leq a$, $1 \leq j \leq b$, $1 \leq k \leq c$, then $(i, j, k) \in \mathcal{P}$. A plane partition is symmetric if $(i, j, k) \in \mathcal{P}$ if and only if $(j, i, k) \in \mathcal{P}$. The height of stack (i, j) is the largest value of k for which there exist a point (i, j, k) in the plane partition.

The study of plane partitions was initiated by MacMahon [28] who proved that the generating function for plane partition fitting in the box

$$\mathcal{B}(a, b, c) = \{(i, j, k) | 1 \leq i \leq a, 1 \leq j \leq b, 1 \leq k \leq c\} \tag{7.3}$$

is given by

$$\prod_{i=1}^{a} \prod_{j=1}^{b} \prod_{k=1}^{c} \frac{1 - q^{i+j+k-1}}{1 - q^{i+j+k-2}}; \tag{7.4}$$

in particular, the total number of plane partitions fitting inside $\mathcal{B}(a, b, c)$ is given by

$$\prod_{i=1}^{a}\prod_{j=1}^{b}\prod_{k=1}^{c}\frac{i+j+k-1}{i+j+k-2}.$$

We can express the generating function for plane partitions given in (7.4) in terms of Schur functions as follows. From the combinatorial definition of Schur functions it follows that

$$s_{b^a}(q^{a+c}, q^{a+c+1}, \ldots, q)$$

where b^a denotes a partition with a parts all of which equal to b, is the generating function for plane partitions strictly decreasing down columns with exactly a rows each of length b and with the largest stack of height less than or equal to $a + c$. Removing $a - i + 1$ from row i we bijectively obtain a plane partition which is a subset of $\mathcal{B}(a, b, c)$. Consequently an alternative expression for (7.4) is given by

$$q^{-ba(a+1)/2}s_{b^a}(q^{a+c}, q^{a+c+1}, \ldots, q). \tag{7.5}$$

Now from (5.9) it follows that for $a = c = m$, $b = N$ (7.5) can be expressed as

$$\mathbb{E}_{M \in U(N)}\prod_{j=1}^{m}\det(I_N + \alpha_j M)\det(I_N + \overline{\beta_j M})$$

with $\alpha_i = q^{n+i}$ and $\beta_i = q^i$. In particular, we obtain the following result.

Proposition 4. *The expected value of the 2m-s moment of a characteristic polynomial of random unitary matrix M in $U(N)$ is equal to the total number of plane partitions fitting in $\mathcal{B}(m, m, N)$:*

$$E_{U(N)}|P_M(1)|^{2m} = \prod_{i=1}^{m}\prod_{j=1}^{m}\prod_{k=1}^{N}\frac{i+j+k-1}{i+j+k-2}. \tag{7.6}$$

We refer the reader to [18] for analogous results pertaining to other groups.

7.2   Random matrices and magic squares

We will now use the combinatorial definition of Schur functions to compute the moments of coefficients of characteristic polynomial of unitary matrices. We write

$$P_M(z) = \det(M - zI) = \prod_{j=1}^{N}(e^{i\theta_j} - z) = (-1)^N\sum_{j=0}^{N}\mathrm{Sc}_j(M)z^{N-j}(-1)^j, \tag{7.7}$$

where $\mathrm{Sc}_j(M)$ is the j-th *secular coefficient* of the characteristic polynomial. Note that

$$\mathrm{Sc}_1(M) = \mathrm{Tr}(M), \tag{7.8}$$

and
$$\mathrm{Sc}_N(M) = \det(M). \tag{7.9}$$

The moments of traces were discussed in section 4. Moments of the higher secular coefficients were studied by Haake and collaborators [19, 20] who obtained:
$$\mathbb{E}_{U(N)}\mathrm{Sc}_j(M) = 0, \quad \mathbb{E}_{U(N)}|\mathrm{Sc}_j(M)|^2 = 1; \tag{7.10}$$

and posed the question of computing the higher moments. The answer is given by Proposition 5, which we state below after pausing to give the following definition.

Definition 1. If A is an m by n matrix with nonnegative integer entries and with row and column sums
$$r_i = \sum_{j=1}^{n} a_{ij},$$
$$c_j = \sum_{i=1}^{m} a_{ij};$$

then the the row-sum vector $\mathrm{row}(A)$ and column-sum vector $\mathrm{col}(A)$ are defined by
$$\mathrm{row}(A) = (r_1, \ldots, r_m),$$
$$\mathrm{col}(A) = (c_1, \ldots, c_n).$$

Given two partitions $\mu = (\mu_1, \ldots, \mu_m)$ and $\tilde{\mu} = (\tilde{\mu}_1, \ldots, \tilde{\mu}_n)$ we denote by $N_{\mu\tilde{\mu}}$ the number of nonnegative integer matrices A with $\mathrm{row}(A) = \mu$ and $\mathrm{col}(A) = \tilde{\mu}$.

For example, for $\mu = (2, 1, 1)$ and $\tilde{\mu} = (3, 1)$ we have $N_{\mu\tilde{\mu}} = 3$; and the matrices in question are
$$\begin{bmatrix} 2 & 0 \\ 1 & 0 \\ 0 & 1 \end{bmatrix}, \begin{bmatrix} 2 & 0 \\ 0 & 1 \\ 1 & 0 \end{bmatrix}, \begin{bmatrix} 1 & 1 \\ 1 & 0 \\ 1 & 0 \end{bmatrix}.$$

For $\mu = (2, 2, 1)$ and $\tilde{\mu} = (3, 1, 1)$ we have $N_{\mu\tilde{\mu}} = 8$; and the matrices in question are
$$\begin{bmatrix} 0 & 1 & 1 \\ 2 & 0 & 0 \\ 1 & 0 & 0 \end{bmatrix}, \begin{bmatrix} 1 & 1 & 0 \\ 1 & 0 & 1 \\ 1 & 0 & 0 \end{bmatrix}, \begin{bmatrix} 1 & 0 & 1 \\ 1 & 1 & 0 \\ 1 & 0 & 0 \end{bmatrix}, \begin{bmatrix} 2 & 0 & 0 \\ 0 & 1 & 1 \\ 1 & 0 & 0 \end{bmatrix},$$
$$\begin{bmatrix} 2 & 0 & 0 \\ 1 & 1 & 0 \\ 0 & 0 & 1 \end{bmatrix}, \begin{bmatrix} 2 & 0 & 0 \\ 1 & 0 & 1 \\ 0 & 1 & 0 \end{bmatrix}, \begin{bmatrix} 1 & 1 & 0 \\ 2 & 0 & 0 \\ 0 & 0 & 1 \end{bmatrix}, \begin{bmatrix} 1 & 0 & 1 \\ 2 & 0 & 0 \\ 0 & 1 & 0 \end{bmatrix}.$$

We are ready to state the following result, proved in [13] where analogous results are established for other compact groups; see also [18] for further generalizations.

Proposition 5. *(a) Consider* $\mathbf{a} = (a_1, \ldots, a_l)$ *and* $\mathbf{b} = (b_1, \ldots, b_l)$ *with* a_j, b_j *nonnegative natural numbers. Then for* $N \geqslant \max\left(\sum_1^l ja_j, \sum_1^l jb_j\right)$ *we have*

$$\mathbb{E}_{U_N} \prod_{j=1}^{l} (\mathrm{Sc}_j(M))^{a_j} \overline{(\mathrm{Sc}_j(M))}^{b_j} = N_{\mu\tilde{\mu}}. \tag{7.11}$$

Here μ *and* $\tilde{\mu}$ *are partitions* $\mu = \langle 1^{a_1} \ldots l^{a_l} \rangle$, $\tilde{\mu} = \langle 1^{b_1} \ldots l^{b_l} \rangle$ *and* $N_{\mu\tilde{\mu}}$ *is the number of nonnegative integer matrices* A *with* $\mathrm{row}(A) = \mu$ *and* $\mathrm{col}(A) = \tilde{\mu}$.
 (b) In particular, for $N \geqslant jk$ *we have*

$$E_{U(N)} |\mathrm{Sc}_j(M)|^{2k} = H_k(j), \tag{7.12}$$

where $H_k(j)$ *is the number of* $k \times k$ *nonnegative integer matrices with each row and column summing up to* j *– "magic squares".*

The reader is likely to have encountered objects, which following Ehrhart [17] are refered to as "historical magic squares". These are square matrices of order k, whose entries are nonnegative integers $(1, \ldots, k^2)$ and whose rows and columns sum up to the same number. The oldest such object,

$$\begin{bmatrix} 4 & 9 & 2 \\ 3 & 5 & 7 \\ 8 & 1 & 6 \end{bmatrix} \tag{7.13}$$

first appeared in ancient Chinese literature under the name *Lo Shu* in the third millennium BC and repeatedly reappeared in the cabbalistic and occult literature in the middle ages. Not knowing ancient Chinese, Latin, or Hebrew it is difficult to understand what is "magic" about *Lo Shu*; it is quite easy to understand however why it keeps reappearing: there is (modulo reflections) only one historic magic square of order 3.

Following MacMahon [28] and Stanley [33], what is referred to as magic squares in modern combinatorics are square matrices of order k, whose entries are nonnegative integers and whose rows and columns sum up to the same number j. The number of magic squares of order k with row and column sum j, denoted by $H_k(j)$, is of great interest; see [14] and references therein. The first few values are easily obtained:

$$H_k(1) = k!, \tag{7.14}$$

corresponding to all k by k permutation matrices (this is the k-th moment of the traces; see section 4);

$$H_1(j) = 1, \tag{7.15}$$

corresponding to 1×1 matrix $[j]$.

We also easily obtain $H_2(j) = j + 1$, corresponding to $\begin{bmatrix} i & j-i \\ j-i & i \end{bmatrix}$, but the value of $H_3(j)$ is considerably more involved:

$$H_3(j) = \binom{j+2}{4} + \binom{j+3}{4} + \binom{j+4}{4}. \tag{7.16}$$

This expression was first obtained by Mac Mahon in 1915 [28] and a simple proof was found only a few years ago by M. Bona [2].

The study of $H_k(j)$ was initiated by by Mac Mahon in 1915 [28]; in the early seventies Stanley and Ehrhart [33, 34, 16] proved that $H_k(j)$ is a polynomial in j of degree $(k-1)^2$ satisfying the following relations

$$H_k(-1) = H_k(-2) = \cdots = H_k(-k+1) = 0, \tag{7.17}$$

and

$$H_k(-k-j) = (-1)^{k-1} H_k(j). \tag{7.18}$$

They further showed that the leading coefficient of $H_k(j)$ is the relative volume of $\mathcal{B}_k$ - the k-th Birkhoff polytope, which is the convex hull of permutation matrices:

$$\mathcal{B}_k = \left\{ (x_{ij}) \in \mathbb{R}^{k^2} \;\middle|\; x_{ij} \geqslant 0; \quad \sum_{i=1}^{k} x_{ij} = 1; \quad \sum_{j=1}^{k} x_{ij} = 1 \right\}. \tag{7.19}$$

To prove proposition 5 we first note that

$$\mathrm{Sc}_j(M) = e_j(M), \tag{7.20}$$

where e_j are the elementary symmetric functions defined in (3.1), and that

$$\prod_{j=1}^{l} (\mathrm{Sc}_j(M))^{a_j} \overline{(\mathrm{Sc}_j(M))}^{b_j} = e_\mu(M) e_{\tilde{\mu}}(\overline{M}), \tag{7.21}$$

where μ and $\tilde{\mu}$ are partitions $\mu = \langle 1^{a_1} \ldots l^{a_l} \rangle$, $\tilde{\mu} = \langle 1^{b_1} \ldots l^{b_l} \rangle$ and e_μ, $e_{\tilde{\mu}}$ are elementary symmetric functions defined in (3.3). We express the elementary symmetric functions in terms of Schur functions:

$$e_\mu = \sum_{\lambda} K_{\lambda' \mu} s_\lambda, \tag{7.22}$$

where $K_{\lambda\mu}$ is the Kostka number defined preceding (7.2).

We now integrate over the unitary group and use the fact that the Schur function are irreducible characters expressed in (3.8), to obtain:

$$\int_{U(N)} e_\mu(M) e_{\tilde{\mu}}(\overline{M}) dM = \sum_{\lambda' \vdash |\mu| = |\tilde{\mu}|} K_{\lambda'\mu} K_{\lambda'\tilde{\mu}} = N_{\mu\tilde{\mu}} \tag{7.23}$$

where $N_{\mu\tilde{\mu}}$ is the number of nonnegative integer matrices A with $\mathrm{row}(A) = \mu$ and $\mathrm{col}(A) = \tilde{\mu}$. The last equality in (7.23) is the consequence of the Knuth correspondence [25], establishing a bijection between $\mathbb{N}$ -matrices A of finite support and ordered pairs of (P, Q) of SSYT of the same shape with $\mathrm{type}(P) = \mathrm{col}(A)$ and $\mathrm{type}(Q) = \mathrm{row}(A)$. This completes proof of Proposition 5.

7.3 Pseudomoments of the Riemann zeta-function and pseudomagic squares

From Proposition 5 it follows that if we consider *truncated* characteristic polynomial

$$P_{M,l}(z) = \sum_{j=0}^{l} \mathrm{Sc}_j(M) z^{N-j} (-1)^j, \tag{7.24}$$

we have for $N \geq lk$

$$\mathbb{E}_{U(N)} |P_{M,l}(1)|^{2k} = G_k(l), \tag{7.25}$$

where $G_k(l)$ denotes the number of $k \times k$ nonnegative integer matrices with row and column sums less than or equal to l (referred to as "pseudomagic squares" by Ehrhart [17]):

$$G_k(l) = \mathrm{card} \left\{ (x_{ij}) \in \mathbb{Z}^{k^2} \ \middle| \ x_{ij} \geqslant 0; \ \sum_{i=1}^{k} x_{ij} \leqslant l; \ \sum_{j=1}^{k} x_{ij} \leqslant l \right\}. \tag{7.26}$$

Ehrhart [17] proved that $G_k(l)$ is a polynomial in l of degree k^2 with leading coefficient given by $\gamma_k = \mathrm{vol}(\mathcal{P}_k)$, where $\mathcal{P}_k$ is the convex polytope of substochastic matrices in $\mathbb{R}^{k^2}$ defined by the following inequalities:

$$\mathcal{P}_k = \left\{ (x_{ij}) \in \mathbb{R}^{k^2} \ \middle| \ x_{ij} \geqslant 0; \ \sum_{i=1}^{k} x_{ij} \leqslant 1; \ \sum_{j=1}^{k} x_{ij} \leqslant 1 \right\}. \tag{7.27}$$

For example,

$$G_2(l) = \frac{1}{6}(l+1)(l+2)(l^2+3l+3),$$

and

$$\mathrm{vol}(\mathcal{P}_2) = \frac{1}{6}.$$

A basic conjecture of Keating and Snaith [23] (considerably refined and extended in [6]) asserts that for moments of the Riemann zeta function

$$\zeta(s) = \sum_{n=1}^{\infty} \frac{1}{n^s}$$

we have

$$\lim_{T \to \infty} \frac{1}{T} \int_0^T \left| \zeta\left(\frac{1}{2}+it\right) \right|^{2k} dt \sim a_k g_k \log^{k^2} T, \tag{7.28}$$

where a_k is an arithmetic factor given by

$$a_k = \prod_p \left(1 - \frac{1}{p}\right)^{k^2} \sum_{j=0}^{\infty} \frac{d_k(p^j)^2}{p^j}, \tag{7.29}$$

and g_k is a "geometric" factor given by

$$g_k = \lim_{N\to\infty} \frac{\mathbb{E}_{U(N)}|P_M(1)|^{2k}}{N^{k^2}} = \prod_{j=0}^{k-1} \frac{j!}{(j+n)!}. \tag{7.30}$$

In [10] the following result is proved for moments of the "truncated" Riemann zeta function (that is, for moments of its partial sums):

$$\lim_{T\to\infty} \frac{1}{T} \int_0^T \left| \sum_{n=1}^{X} \frac{1}{n^{\frac{1}{2}+it}} \right|^{2k} dt \sim a_k \gamma_k (\log X)^{k^2}, \tag{7.31}$$

where a_k is an arithmetic factor, given by (7.29), and γ_k is the geometric factor, $\gamma_k = \mathrm{vol}(\mathcal{P}_k)$.

Note that using (7.25) we can rewrite the geometric factor γ_k in a manner similar to the expression for g_k in (7.30) as follows:

$$\gamma_k = \lim_{l\to\infty} \frac{\mathbb{E}_{U(lk)}|P_{M,l}(1)|^{2k}}{l^{k^2}}. \tag{7.32}$$

References

[1] D. Aldous, P. Diaconis, *Longest increasing subsequences: from patience sorting to the Baik-Deift-Johansson theorem*, BAMS, **36**, 1999, 413-432.

[2] M. Bona, *A New Proof of the Formula for the Number of the 3×3 magic squares*, Mathematics Magazine **70**, 1997, 201-203.

[3] D. Bump, *Lie Groups*, Springer-Verlag, NY, 2004.

[4] D. Bump and P. Diaconis, *Toeplitz minors*, Journal of Combinatorial Theory A, **97**, 2002, 252-271.

[5] D. Bump and A. Gamburd, *On the averages of characteristic polynomials from classical groups*, `math-ph/0502043`, to appear in Communications in Mathematical Physics.

[6] J. B. Conrey, D. W. Farmer, J. P. Keating, M. O. Rubinstein, and N. C. Snaith, *Integral moments of L-functions*, Proc. Lond. Math. Soc., **91**(1), 2005, 33-104, `math.NT/0206018`.

[7] J. B. Conrey, D. W. Farmer, J. P. Keating, M. O. Rubinstein, and N. C. Snaith, *Autocorrelation of random matrix polynomials*, Communications in Mathematical Physics, **237**, 2003, 365-395.

[8] B. Conrey, D. Farmer, and M. Zirnbauer, Autocorrelation of ratios of characteristic polynomials, preprint, 2004.

[9] J. B. Conrey, P. Forrester, and N. C. Snaith. *Averages of ratios of characteristic polynomials for compact classical groups.* Int. Math. Res. Notices, 7, 397-431, 2005.

[10] B. Conrey and A. Gamburd, *Pseudomoments of the Riemann zeta-function and pseudomagic squares,* Journal of Number Theory, **117**(2), 2006, 263-278.

[11] P. Diaconis, *Patterns in eigenvaues,* Bulletin of the AMS, **40**, 2003, 155-179.

[12] P. Diaconis and S. Evans,*Linear Functionals of Eigenvalues of Random Matrices.* Transactions Amer. Math. Soc. **353**, 2001, 2615–2633.

[13] P. Diaconis and A. Gamburd, *Random matrices, magic squares and matching polynomials,* Electronic Journal of Combinatorics, **11**(2), 2004, #R2.

[14] P. Diaconis and A. Gangolli, *Rectangular arrays with fixed margins,* in: D. Aldous, P. P. Varaiya, J. Spencer, J. M. Steele (Eds.), Discrete Probability and Algorithms, IMA Volumes Math. Appl., **72**, 1995, 15-41.

[15] P.Diaconis and M.Shahshahani, *On the eigenvalues of random matrices,* J. Appl. Probab., **31A**, 1994, 49-62.

[16] E. Ehrhart, *Sur les carrés magiques,* C. R. Acad. Sci. Paris, **227 A**, 1973, 575-577.

[17] E. Ehrhart, *Polynômes arithmétiques et Méthode des Polyèdres en combinatoire,* Birkhäuser, 1977.

[18] P. Forrester and A. Gamburd, *Counting formulas associated with some random matrix averages,* preprint.

[19] F. Haake, *Quantum signatures of chaos,* Springer, 2000.

[20] F. Haake, M. Kus, H. -J. Sommers, H. Schomerus, K. Zyckowski, *Secular determinants of random unitary matrices,* J. Phys. A.: Math. Gen. **29**, 1996, 3641-3658.

[21] K. Johansson, *On Szegö's Asymptotic Formula for Toeplitz Determinants and Generalizations.* Bull. Sc. Math., **112**, 1988, 257–304.

[22] K. Johansson, *On Random Matrices from the Compact Classical Groups* Ann. Math. **145**, 1997, 519–545.

[23] J. P. Keating and N. C. Snaith, *Random matrix theory and* $\zeta(\frac{1}{2}+it)$, Commun. Math. Phys., **214**, 2000, 57-89.

[24] J. P. Keating and N. C. Snaith, *Random matrix theory and L-functions at* $s = \frac{1}{2}$, Commun. Math. Phys., **214**, 2000, 91-110.

[25] D. Knuth, *Permutations, matrices, and generalized Young tableaux,* Pacific J. Math., **34**, 1970, 709-727.

[26] D. E. Littlewood. *The Theory of Group Characters and Matrix Representations of Groups.* Oxford University Press, New York, 1940.

[27] I. G. Macdonald, *Symmetric functions and Hall polynomials*, Second edition, Oxford University Press, New York, 1995.

[28] P. A. MacMahon, *Combinatory Analysis*, Cambridge University Press, 1915.

[29] H. Montgomery and K. Soundararajan, *Primes in short intervals*, Commun. Math. Phys., **2**52(1-3), 2004, 589-617, `math.NT/0409258`.

[30] A. Okounkov, *Symmetric functions and random partitions*, in Symmetric Functions 2001: Surveys of Developments and Perspectives, edited by S. Fomin, Kluwer, 2002.

[31] E. Rains, *High powers of random elements of compact Lie groups*, Probab Theory Relat. Fields, **107**, 1997, 219-241.

[32] A. Soshnikov *The Central Limit Theorem for Local Linear Statistics in Classical Compact Groups and Related Combinatorial Identities*. Ann. Probab., **28**, 2000, 1353–1370.

[33] R. Stanley, *Linear homogeneous diophantine equations and magic labelings of graphs*, Duke Math. J., **40**, 1973, 607-632.

[34] R. Stanley, *Combinatorial reciprocity theorems*, Adv. in Math., **14**, 1974, 194-253.

[35] R. Stanley. *Enumerative Combinatorics, Vol. 2*. Cambridge University Press, 1999.

[36] G. Szegö, *Orthogonal Polynomials*, AMS, 1967.

[37] H. Weyl, *Classical Groups*, Princeton University Press, Princeton, 1946.

School of Mathemtaics,
Institute for Advanced Study,
Princeton, NJ 08540

Department of Mathematics,
University of California,
Santa Cruz

agamburd@math.ias.edu

The distribution of ranks in families of quadratic twists of elliptic curves

A. Silverberg

This paper gives a very brief survey, in the form of a table, of some results and conjectures about densities of ranks of elliptic curves over $\mathbb{Q}$ in families of quadratic twists. The table summarizes some of the knowledge to date on the density of quadratic twists of rank r or $\geq r$, for some small values of r. We also give a commentary on the table. For a more extensive survey of ranks of elliptic curves over $\mathbb{Q}$, see [RS02]. The author thanks Roger Heath-Brown for helpful comments.

We first fix notation. If E is an elliptic curve of the form $y^2 = f(x)$, let $E^{(d)}$ denote $dy^2 = f(x)$, the quadratic twist of E by d. If E is an elliptic curve over $\mathbb{Q}$, it suffices to consider d that are squarefree integers. Let

$$N_*(X) = \#\{\text{squarefree } d \in \mathbb{Z} : |d| \leq X, \text{rank}(E^{(d)}(\mathbb{Q})) \text{ is } *\},$$

where $*$ can be "2", "odd", "≥ 3", etc. The table below gives a summary of some of the known results (and conjectures) to date on the rate of growth of $N_r(X)$ and $N_{\geq r}(X)$.

It is well-known (see Theorem 333 of [HW79]) that

$$N_{\geq 0}(X) = \#\{\text{squarefree } d \in \mathbb{Z} : |d| \leq X\}$$
$$\sim 2X \prod_p (1 - \frac{1}{p^2}) = \frac{2}{\zeta(2)}X = \frac{12}{\pi^2}X.$$

The first part of the Birch and Swinnerton-Dyer Conjecture [BSD63/5] says that the rank of an elliptic curve E over $\mathbb{Q}$ should be equal to the analytic rank (i.e., the order of vanishing at $s = 1$ of the L-function of E over $\mathbb{Q}$). In particular, the Birch and Swinnerton-Dyer Conjecture implies the Parity Conjecture, which says that the rank has the same parity as the analytic rank. The parity of the analytic rank can be read off from the sign in the functional equation for the L-function. Using the way that the sign varies as one twists the curve, one can show that the Parity Conjecture implies that, as $|d|$ grows, the ranks of the quadratic twists by squarefree $d \in \mathbb{Z}$ of a fixed elliptic curve over $\mathbb{Q}$ are even half the time and odd half the time. (See the N_{odd}, N_{even} table entry.)

In 1979, Goldfeld [G79] conjectured that for every fixed elliptic curve, the average rank of its quadratic twists is $\frac{1}{2}$.

[1] Supported by NSA grant MDA904-03-1-0033.

Goldfeld Conjecture. *If E is an elliptic curve over $\mathbb{Q}$, then*

$$\lim_{X \to \infty} \frac{\sum_{\text{squarefree } d \in \mathbb{Z}, |d| \leq X} \operatorname{rank}(E^{(d)}(\mathbb{Q}))}{\#\{\text{squarefree } d \in \mathbb{Z} : |d| \leq X\}} = \frac{1}{2}.$$

If one assumes both the Parity Conjecture and Goldfeld's Conjecture, then for every fixed elliptic curve, the ranks of the quadratic twists should be zero half the time and one half the time. (See the last N_0 and N_1 table entries.)

In the table, "w/PC" means that the result is conditional on the Parity Conjecture, and "w/PC & GC" means that the result is conditional on both the Parity Conjecture and Goldfeld's Conjecture. Further, "w/RMTC" means this is a conjecture made in [CKRS02] (see Conjecture 1 and (7) of [CKRS02]), which is based on Random Matrix Theory. Unless otherwise stated, the table entries hold for all elliptic curves over $\mathbb{Q}$.

All "$\gg$" and "$\ll$" entries in the table, and in the discussion below, should be read as "there is a positive constant, depending on E but not on X, such that for all sufficiently large X, we have ...".

$N_{\geq 0}(X)$	$\sim \frac{12}{\pi^2} X$	
$N_{\text{odd}}(X)$ $N_{\text{even}}(X)$	$\sim \frac{6}{\pi^2} X$	w/PC
$N_0(X)$	$\gg X/\log X$ $\gg X$ $\sim \frac{6}{\pi^2} X$	[OS98] for some E [HB94, J98, V98, W99, Y03] w/PC & GC
$N_1(X)$	$\gg X^{1-\epsilon}$ $\gg X$ $\sim \frac{6}{\pi^2} X$	[PP97] for some E [V98] w/PC & GC
$N_{\geq 1}(X)$	$\geq \frac{6}{\pi^2} X$ $\sim \frac{6}{\pi^2} X$	w/PC w/PC & GC
$N_{\geq 2}(X)$	$\gg X^{\frac{1}{7}}/\log X$ $\gg X^{\frac{1}{3}}$ $\gg X^{\frac{1}{2}}$ $\gg X^{\frac{3}{4}-\epsilon}, \ll X^{\frac{3}{4}+\epsilon}$	$j(E) \neq 0, 1728$ [ST95] for some E [ST95, RS01] w/PC [ST95, GM91] w/RMTC [CKRS02]
$N_{\geq 3}(X)$	$\gg X^{\frac{1}{6}}$ $\gg X^{\frac{1}{3}}$	for some E [ST95, RS01] for some E w/PC [ST95, RS01]
$N_{\geq 4}(X)$	$\to \infty$ $\gg X^{\frac{1}{6}}$	for some E [Me98, RS04] for some E w/PC [RS01, RS04]
$N_{\geq 5}(X)$	$\to \infty$	for some E w/PC [Me98, RS04]

Ono and Skinner [OS98], using results of Waldspurger and of Friedberg and Hoffstein, show that $N_0(X) \gg X/\log X$ for all elliptic curves. It was known earlier that $N_0(X) \gg X/\log X$ for certain elliptic curves (see for example [R74] for $y^2 = x^3 - x$).

Work of Monsky [Mo90], Birch [B69, B70], and Heegner [H52] shows that certain elliptic curves E have $\text{rank}(E^{(p)}(\mathbb{Q})) \geq 1$ for all primes p in certain congruence classes (for example, for $y^2 = x^3 - x$ and all primes $p \equiv 5$ or 7 (mod 8)), and thus $N_{\geq 1}(X) \gg X/\log X$ (in fact, $N_1(X) \gg X/\log X$).

For the elliptic curve $y^2 = x^3 - x$, Heath-Brown (see Theorem 2 of [HB94]) shows that $N_0(X) > (.279)6X/\pi^2$, and, subject to the Parity Conjecture, $N_1(X) > (.559)6X/\pi^2$. Results along these lines for other elliptic curves can be found in [J98, V98, W99, V99, O01, Y03]. Vatsal [V98] shows that $N_1(X) \gg X$ and $N_0(X) \gg X$ for the curve $y^2 + y = x^3 + x^2 + x$, unconditionally. Ono [O01] shows that if E has no rational point of order 2, then there is some $\alpha(E)$ with $0 < \alpha(E) < 1$ such that $N_0(X) \gg X/(\log X)^{1-\alpha(E)}$.

The methods for finding lower bounds for $N_{\geq r}(X)$ when $r \geq 2$ involve finding twists $E^{(g(T))}$ of E over $\mathbb{Q}(T)$ of rank $\geq r$, and specializing T. By Theorem C of [Si83], for all but finitely many $t \in \mathbb{Q}$ one has $\text{rank}(E^{(g(t))}) \geq r$. Using sieve theory, one can find a lower bound on the number of squarefree integers d such that $|d| \leq X$ and $d = g(t)z^2$ for some $t, z \in \mathbb{Q}$ (so $\text{rank}(E^{(d)}(\mathbb{Q})) = \text{rank}(E^{(g(t))}(\mathbb{Q}))$); see [GM91, ST95]. Mestre [Me92, Me98, Me00] has techniques for finding elliptic curves of "large" rank over $\mathbb{Q}(T)$ (and over $\mathbb{Q}$). See also work of Howe, Leprévost, and Poonen [HLP00]. Gouvêa and Mazur [GM91] showed that, assuming the Parity Conjecture, $N_{\geq 2,\text{even}}(X) \gg X^{1/2-\varepsilon}$. Stewart and Top [ST95], without assuming the Parity Conjecture, showed that if $j(E) \neq 0$ or 1728, then $N_{\geq 2}(X) \gg X^{1/7}/(\log X)^2$. For elliptic curves of the form

$$y^2 = ax^3 + bx^2 + bx + a$$

with $a, b \in \mathbb{Q}$ and $a(3a - b)(a + b) \neq 0$, they showed $N_{\geq 2}(X) \gg X^{1/3}$. For

$$y^2 = x(x - 1)(x - (\frac{b^2 + 1}{2b})^2)$$

with $b \in \mathbb{Q} - \{0, 1, -1\}$, they showed $N_{\geq 3}(X) \gg X^{1/6}$ (based on an idea of Schoen [Sc90] who they say also obtained it independently).

Building on the work of Mestre, Gouvêa-Mazur, and Stewart-Top, Rubin and Silverberg [RS01, RS04] show that $N_{\geq 2}(X) \gg X^{\frac{1}{3}}$ holds, for example, for every elliptic curve $y^2 = x^3 + ax + b$ such that the cubic has a non-zero rational root. Further, $N_{\geq 3}(X) \gg X^{\frac{1}{3}}$ holds (subject to the Parity Conjecture), for example, for every elliptic curve E of the form $y^2 = x^3 + ax + b$ such that the cubic has three real roots (equivalently, the discriminant $\Delta(E) > 0$) and either

(a) the largest or the smallest root is rational, or

(b) E has a rational subgroup of order 3.

In particular, it holds for all elliptic curves over $\mathbb{Q}$ for which all the 2-torsion is rational.

Conjecture 1 of [CKRS02] predicts that $N_{\geq 2}(X) \sim CX^{3/4} \log^m(X)$ for some C and m. This prediction is based on random matrix theory. The workshop announcement [CFMS04] states that random matrix theory also predicts that $N_3(X) \sim CX^{1/4} \log^m(X)$, which is inconsistent with the $N_{\geq 3}(X) \gg X^{\frac{1}{3}}$ results stated above[1].

In [Me98], Mestre stated that if E is an elliptic curve over $\mathbb{Q}$ with torsion subgroup isomorphic to $\mathbb{Z}/8\mathbb{Z} \times \mathbb{Z}/2\mathbb{Z}$, then E has infinitely many (non-isomorphic) quadratic twists with rank at least 4 over $\mathbb{Q}$. Theorems 3.2, 3.6, and 5.1 and Corollary 5.2 of [RS04] give other "$N_{\geq r}(X) \to \infty$" results for $r = 4$ and, assuming the Parity Conjecture, for $r = 5$.

All the $N_{\geq r}(X)$ entries in the table apply for example to the elliptic curves

$$y^2 = x(x + n)(x - n - 1) \tag{1.1}$$

with $n = 1$, 8, 16, 21, 56, 65, 85, 96, 161, 176, 208, 261, 341, 408, 456, 533, and 560 (see Corollary 5.2 of [RS04]). All except the $N_{\geq 5}(X)$ entry are known to hold for the elliptic curves

$$y^2 = x(x - 1)(x + \frac{a^2 - 1}{a^2 + 2}) \tag{1.2}$$

with $a \in \mathbb{Q} - \{0, \pm 1\}$ (see Theorem 5.5 of [RS01] and Theorem 3.2 of [RS04]), and all of those entries except the $N_{\geq 4}(X) \to \infty$ entry are known to hold even when $a = 0$. We remark that (1.2) is the quadratic twist by $\frac{a^2+2}{3}$ of (1.1) with $n = \frac{a^2-1}{3}$, and (1.2) for $a = 0$ is the quadratic twist by 2 of the curve $y^2 = x^3 - x$.

References

[B69]   B. J. Birch, *Diophantine analysis and modular functions*, in Algebraic Geometry (Internat. Colloq., Tata Inst. Fund. Res., Bombay, 1968), 35–42, Oxford Univ. Press, London, 1969.

[B70]   B. J. Birch, *Elliptic curves and modular functions*, in Symposia Mathematica, Vol. IV (INDAM, Rome, 1968/69), 27–32 Academic Press, London, 1970.

[BSD63/5]   B. J. Birch, H. P. F. Swinnerton-Dyer, *Notes on elliptic curves. I, II*, J. Reine Angew. Math. **212** (1963), 7–25; **218** (1965), 79–108.

---

[1] **Note added in proof:** The $X^{1/4}$ conjecture for $N_3(X)$ mentioned in the workshop announcement has been revised; see *Discretisation for odd quadratic twists* by Conrey, Rubinstein, Snaith, and Watkins in this volume for the current conjectures.

[CFMS04]　J. B. Conrey, D. Farmer, F. Mezzadri, N. C. Snaith, workshop announcement: *Special Week on Ranks of Elliptic Curves and Random Matrix Theory, 9–13 February, 2004*,
http://www.newton.cam.ac.uk/programmes/RMA/rmaw01.html .

[CKRS02]　J. B. Conrey, J. P. Keating, M. O. Rubinstein, N. C. Snaith, *On the frequency of vanishing of quadratic twists of modular L-functions*, in Number theory for the millennium, I (Urbana, IL, 2000), A. K. Peters, Natick, MA, 2002, 301–315.

[G79]　D. Goldfeld, *Conjectures on elliptic curves over quadratic fields*, in Number theory, Carbondale 1979 (Proc. Southern Illinois Conf., Southern Illinois Univ., Carbondale, Ill., 1979), M. B. Nathanson, ed., Lect. Notes in Math. **751**, Springer, Berlin, 1979, 108–118.

[GM91]　F. Gouvêa, B. Mazur, *The square-free sieve and the rank of elliptic curves*, J. Amer. Math. Soc. **4** (1991), 1–23.

[HW79]　G. H. Hardy, E. M. Wright, An introduction to the theory of numbers, fifth edition, The Clarendon Press, Oxford University Press, New York, 1979.

[HB94]　D. R. Heath-Brown, *The size of Selmer groups for the congruent number problem. II*, Invent. Math. **118** (1994), 331–370.

[H52]　K. Heegner, *Diophantische Analysis und Modulfunktionen*, Math. Z. **56** (1952), 227–253.

[HLP00]　E. Howe, F. Leprévost, B. Poonen, *Large torsion subgroups of split Jacobians of curves of genus two or three*, Forum Math. **12** (2000), 315–364.

[J98]　K. James, *L-series with nonzero central critical value*, J. Amer. Math. Soc. **11** (1998), 635–641.

[Me92]　J-F. Mestre, *Rang de courbes elliptiques d'invariant donné*, C. R. Acad. Sci. Paris **314** (1992), 919–922.

[Me98]　J-F. Mestre, *Rang de certaines familles de courbes elliptiques d'invariant donné*, C. R. Acad. Sci. Paris **327** (1998), 763–764.

[Me00]　J-F. Mestre, *Ranks of twists of elliptic curves*, lecture at MSRI, September 11, 2000.

[Mo90]　P. Monsky, *Mock Heegner points and congruent numbers*, Math. Z. **204** (1990), 45–67.

[O01]　K. Ono, *Nonvanishing of quadratic twists of modular L-functions and applications to elliptic curves*, J. Reine Angew. Math. **533** (2001), 81–97.

[OS98] K. Ono, C. Skinner, *Non-vanishing of quadratic twists of modular L-functions*, Invent. Math. **134** (1998), 651–660.

[PP97] A. Perelli, J. Pomykała, *Averages of twisted elliptic L-functions*, Acta Arith. **80** (1997), 149–163.

[R74] M. J. Razar, *The non-vanishing of $L(1)$ for certain elliptic curves with no first descents*, Amer. J. Math. **96** (1974), 104–126.

[RS01] K. Rubin, A. Silverberg, *Rank frequencies for quadratic twists of elliptic curves*, Exper. Math. **10** (2001), 559–569.

[RS02] K. Rubin, A. Silverberg, *Ranks of elliptic curves*, Bull. Amer. Math. Soc. **39** (2002), 455–474.

[RS04] K. Rubin, A. Silverberg, *Twists of elliptic curves of rank at least four*, in this volume.

[Sc90] C. Schoen, *Bounds for rational points on twists of constant hyperelliptic curves*, J. Reine Angew. Math. **411** (1990), 196–204.

[Si83] J. H. Silverman, *Heights and the specialization map for families of abelian varieties*, J. Reine Angew. Math. **342** (1983), 197–211.

[ST95] C. L. Stewart, J. Top, *On ranks of twists of elliptic curves and power-free values of binary forms*, J. Amer. Math. Soc. **8** (1995), 943–973.

[V98] V. Vatsal, *Rank-one twists of a certain elliptic curve*, Math. Ann. **311** (1998), 791–794.

[V99] V. Vatsal, *Canonical periods and congruence formulae*, Duke Math. J. **98** (1999), 397–419.

[W99] S. Wong, *Elliptic curves and class number divisibility*, Internat. Math. Res. Notices **1999** (1999), no. 12, 661–672.

[Y03] G. Yu, *Rank 0 quadratic twists of a family of elliptic curves*, Compositio Math. **135** (2003), 331–356.

Department of Mathematics,
University of California at Irvine,
Irvine, CA 92697
USA

Twists of elliptic curves of rank at least four

K. Rubin and A. Silverberg

Abstract

We give infinite families of elliptic curves over $\mathbb{Q}$ such that each curve has infinitely many non-isomorphic quadratic twists of rank at least 4. Assuming the Parity Conjecture, we also give elliptic curves over $\mathbb{Q}$ with infinitely many non-isomorphic quadratic twists of odd rank at least 5.

1 Introduction

Mestre [Me92] showed that every elliptic curve over $\mathbb{Q}$ has infinitely many (non-isomorphic) quadratic twists of rank at least 2 over $\mathbb{Q}$, and he gave [Me98, Me00] several infinite families of elliptic curves over $\mathbb{Q}$ with infinitely many (non-isomorphic) quadratic twists of rank at least 3. Further, he stated ([Me98]) that if E is an elliptic curve over $\mathbb{Q}$ with torsion subgroup isomorphic to $\mathbb{Z}/8\mathbb{Z} \times \mathbb{Z}/2\mathbb{Z}$, then there are infinitely many (non-isomorphic) quadratic twists of E with rank at least 4 over $\mathbb{Q}$.

In this paper (Theorems 3.2 and 3.6) we give additional infinite families of elliptic curves over $\mathbb{Q}$ with infinitely many (non-isomorphic) quadratic twists of rank at least 4. The family of elliptic curves in Theorem 3.2 is parametrized by the projective line. The family of elliptic curves in Theorem 3.6 is parametrized by an elliptic curve of rank one. In both cases, the twists are parametrized by an elliptic curve of rank at least one.

In addition, we find elliptic curves over $\mathbb{Q}$ that, assuming the Parity Conjecture, have infinitely many (non-isomorphic) quadratic twists of odd rank at least 5 (see Theorem 5.1 and Corollary 5.2). The proof relies on work of Rohrlich [R93].

In Theorem 5.6 of [RS01] we gave an infinite family of elliptic curves over $\mathbb{Q}$ for which the number of twists of even rank at least 4 grows at least like $X^{1/6}$, if the Parity Conjecture holds. In Theorem 3.5 below we give a different infinite family for which this holds.

The results are obtained by extending the method of [RS01] (we learned at [Me00] that this was one of the methods used independently and earlier by Mestre to obtain the results announced in [Me98]).

---

[1]Supported by NSF grant DMS-0140378.
[2]Supported by NSA grant MDA904-03-1-0033.

Definition 1.1. *If $E : y^2 = f(x)$ is an elliptic curve, let $E^{(d)}$ denote $dy^2 = f(x)$, the quadratic twist of E by d.*

2   The general method

We first give a more explicit version of Lemma 2.1 of [RS01].

Lemma 2.1. *Suppose that E is an elliptic curve over a field F, that $K_1, \ldots, K_n$ are distinct separable extensions of F of degree at most 2, and that for $i = 1, \ldots, n$, there are points $P_i \in E(K_i)$ of infinite order. Suppose also that if $K_i \neq F$, then $\sigma(P_i) = -P_i$, where σ is the non-trivial element of $\mathrm{Gal}(K_i/F)$. Let K denote the compositum $K_1 \cdots K_n$. Then $\{P_1, \ldots, P_n\}$ is an independent set in $E(K)$.*

Proof Let $G = \mathrm{Gal}(K/F)$. Let $\chi_i : \mathrm{Gal}(K_i/F) \to \{\pm 1\}$ denote the non-trivial character if $K_i \neq F$, and the trivial character if $K_i = F$. Let $e_i = \sum_{\sigma \in G} \chi_i(\sigma)\sigma$. Then for all i and j,

$$e_i(P_j) = \sum_{\sigma \in G} \chi_i(\sigma)(\sigma(P_j)) = \sum_{\sigma \in G} \chi_i(\sigma)(\chi_j(\sigma)P_j)$$

$$= \left(\sum_{\sigma \in G} \chi_i(\sigma)\chi_j(\sigma) \right) P_j = \begin{cases} O & \text{if } i \neq j \\ |G|P_j & \text{if } i = j. \end{cases}$$

Suppose $\sum_j n_j P_j = O$. Then $O = e_i(\sum_j n_j P_j) = |G|n_i P_i$ for every i. Since P_i has infinite order, $n_i = 0$ for every i. $\qquad\square$

Definition 2.2.   (i) *If $k(t) \in \mathbb{Z}[t]$, we say that $k(t)$ is* squarefree *if $k(t)$ is not divisible by the square of any non-constant polynomial in $\mathbb{Z}[t]$.*

  (ii) *Suppose $g(t) \in \mathbb{Q}(t)$. A* squarefree part *of $g(t)$ is a squarefree $k(t) \in \mathbb{Z}[t]$ such that $g(t) = k(t)j(t)^2$ for some $j(t) \in \mathbb{Q}(t)$.*

The following result is a variant of Corollary 2.2 of [RS01].

Proposition 2.3. *Suppose $f(x) \in \mathbb{Q}[x]$ is a separable cubic, and E is the elliptic curve $y^2 = f(x)$. Let $h_1(t) = t$, suppose we have non-constant $h_2(t), \ldots, h_r(t) \in \mathbb{Q}(t)$, let $k_i(t)$ be a squarefree part of $f(h_i(t))/f(t)$, and suppose that $k_1(t), \ldots, k_r(t)$ are distinct modulo $(\mathbb{Q}^*)^2$. Then:*

  (i) *the rank of $E^{(f(t))}\big(\mathbb{Q}\big(t, \sqrt{k_2(t)}, \ldots, \sqrt{k_r(t)}\big)\big)$ is at least r;*

  (ii) *if C is the curve defined by the equations $s_i^2 = k_i(t)$ for $i = 1, \ldots, r$, then for all but at most finitely many rational points $(\tau, \sigma_1, \ldots, \sigma_r) \in C(\mathbb{Q})$, the rank of $E^{(f(\tau))}(\mathbb{Q})$ is at least r.*

Proof Apply Lemma 2.1 to the elliptic curve $E^{(f(t))}$ over the field $F = \mathbb{Q}(t)$, with $K_i = F\big(\sqrt{k_i(t)}\big)$ (so $K_1 = F$). Since the polynomials k_i are squarefree and distinct modulo $(\mathbb{Q}^*)^2$, the fields K_i are distinct. For $i = 1, \ldots, r$, let

$$P_i = \big(h_i(t), \sqrt{f(h_i(t))/f(t)}\big) \in E^{(f(t))}\big(\mathbb{Q}(t, \sqrt{k_i(t)})\big).$$

Note that P_i has infinite order, since its x-coordinate is not constant. Now (i) follows. Part (ii) now follows from Theorem C of [S83]. $\qquad\square$

Retain the setting of Proposition 2.3. Suppose from now on that each h_i is a linear fractional transformation that permutes the roots of f. Then by Proposition 2.9 of [RS01], $k_i(t)$ is linear. More precisely, $k_1(t) = 1$, and if $h_i(t) = \frac{\alpha t + \beta}{t + \delta}$ with $\alpha, \beta, \delta \in \mathbb{Q}$, then $k_i(t) = f(\alpha)(t + \delta)$ and $f(h_i(t))/f(t) = k_i(t)(t + \delta)^{-4}$.

In [RS01] we considered the case where $r \leq 3$. Suppose $r = 3$. Then

$$\mathbb{Q}(C) = \mathbb{Q}\big(t, \sqrt{k_2(t)}, \sqrt{k_3(t)}\big),$$

and the genus of C is zero, where C was defined in Proposition 2.3(ii). Our goal was to choose h_2 and h_3 so that the corresponding curve C has a rational point (and therefore has infinitely many rational points). We considered pairs of the five non-trivial linear fractional transformations that permute the roots of f, until we found h_2 and h_3 for which we could find a rational point on the corresponding curve C. We used this to parametrize the rational points on C, i.e., we found an explicit $u \in \mathbb{Q}\big(t, \sqrt{k_2(t)}, \sqrt{k_3(t)}\big)$ so that

$$\mathbb{Q}(C) = \mathbb{Q}\big(t, \sqrt{k_2(t)}, \sqrt{k_3(t)}\big) = \mathbb{Q}(u).$$

We then computed t as a function of u, i.e., $t = t(u) \in \mathbb{Q}(u)$. The map $u \mapsto \big(t(u), \sqrt{k_2(t(u))}, \sqrt{k_3(t(u))}\big)$ defines an isomorphism from $\mathbb{P}^1(\mathbb{Q})$ onto $C(\mathbb{Q})$. By Proposition 2.3(ii), for all but finitely many $u \in \mathbb{Q}$, the rank of $E^{(f(t(u)))}(\mathbb{Q})$ is at least 3.

In this paper, we consider the case $r = 4$. Then the genus of C is one. We will start with a pair h_2, h_3 as above, and, among the remaining three candidates for h_4, look for one for which we can see enough rational points on the corresponding curve C to ensure that C is an elliptic curve of positive rank. We have

$$\mathbb{Q}(C) = \mathbb{Q}\big(t, \sqrt{k_2(t)}, \sqrt{k_3(t)}, \sqrt{k_4(t)}\big) = \mathbb{Q}(u, v)$$

with $v^2 = k_4(t(u))$. A rational point on the elliptic curve C corresponds to a pair $u_0, v_0 \in \mathbb{Q}$ such that $v_0^2 = k_4(t(u_0))$. By Proposition 2.3(ii), for all but finitely many such (u_0, v_0), the rank of $E^{(f(t(u_0)))}(\mathbb{Q})$ is at least 4.

3   Rank ≥ 4

From now on we consider elliptic curves of the form

$$y^2 = x(x - 1)(x - \lambda)$$

where $\lambda \in \mathbb{Q} - \{0, 1\}$.

Definition 3.1. *We fix a numbering of the linear fractional transformations $h_i(t)$ in $\mathbb{Q}(t)$ that permute the set $\{0, 1, \lambda\}$, along with corresponding squarefree parts $k_i(t)$:*

$$h_1(t) = t, \qquad\qquad k_1(t) = 1,$$

$$h_2(t) = \frac{t - \lambda}{(2 - \lambda)t - 1}, \qquad k_2(t) = (1 - \lambda)((\lambda - 2)t + 1),$$

$$h_3(t) = \frac{\lambda^2(t - 1)}{(\lambda^2 - \lambda + 1)t - \lambda}, \quad k_3(t) = \lambda(1 - \lambda)((\lambda^2 - \lambda + 1)t - \lambda),$$

$$h_4(t) = \frac{\lambda t}{(\lambda + 1)t - \lambda}, \qquad k_4(t) = \lambda((\lambda + 1)t - \lambda),$$

$$h_5(t) = \frac{\lambda^2(t - 1)}{t(2\lambda - 1) - \lambda^2}, \qquad k_5(t) = \lambda(\lambda - 1)((1 - 2\lambda)t + \lambda^2),$$

$$h_6(t) = \frac{\lambda(2 - \lambda)}{(\lambda^2 - \lambda + 1)t - \lambda^2}, \quad k_6(t) = \lambda((\lambda - 1)((\lambda^2 - \lambda + 1)t - \lambda^2).$$

Theorem 3.2. *Suppose $a \in \mathbb{Q} - \{0, 1, -1\}$. Let $\eta = a^2$, let*

$$f_\eta(x) = x(x - 1)\left(x - \frac{1 - \eta}{\eta + 2}\right),$$

and let E_η be $y^2 = f_\eta(x)$. Let C_η be the curve

$$v^2 = (\eta + 1)^2 u^4 + 4\eta(2\eta^2 + 3\eta + 1)u^3 +$$
$$2(7\eta^4 + 7\eta^3 + 2\eta^2 + \eta + 1)u^2 + 4(2\eta^5 + \eta^4 - 2\eta^2 - \eta)u + (\eta^3 - 1)^2,$$

and let

$$t_\eta(u) = \frac{2(1 - \eta)T_\eta(u)}{3((\eta + 1)u^2 + 1 - \eta^3)^2}$$

where

$$T_\eta(u) = (\eta + 1)^2 u^4 + 2\eta(2\eta^2 + 3\eta + 1)u^3 +$$
$$2(3\eta^4 + 3\eta^3 + \eta^2 + \eta + 1)u^2 + 2\eta(\eta^3 - 1)(2\eta + 1)u + \eta^6 - 2\eta^3 + 1.$$

Then:

(i) *E_η and C_η are elliptic curves over $\mathbb{Q}$;*

(ii) $\mathrm{rank}(C_\eta(\mathbb{Q})) \geq 1$;

(iii) *for all but possibly finitely many $(u, v) \in C_\eta(\mathbb{Q})$, the quadratic twist of E_η by $f_\eta \circ t_\eta(u)$ has rank at least 4 over $\mathbb{Q}$;*

(iv) *there are infinitely many non-isomorphic quadratic twists of E_η of rank at least 4 over $\mathbb{Q}$.*

Proof We proved Theorem 4.2(a) of [RS01] by noticing that when $\tau = \frac{2\lambda}{\lambda+1}$, then

$$k_3(\tau)/k_2(\tau) = \lambda^2 \quad \text{and} \quad k_2(\tau) = \frac{(\lambda-1)^2(-2\lambda+1)}{\lambda+1}.$$

We wanted $k_2(\tau)$ and $k_3(\tau)$ to be squares. Note that $\frac{-2\lambda+1}{\lambda+1} = a^2$ if and only if $\lambda = \frac{1-a^2}{2+a^2}$, and when these hold then $k_2(\frac{2\lambda}{\lambda+1})$ and $k_3(\frac{2\lambda}{\lambda+1})$ are both squares, and $(\frac{2\lambda}{\lambda+1}, (\lambda-1)a, \lambda(\lambda-1)a) \in C_{a^2} = C_\eta$. Further, we found that

$$\mathbb{Q}\big(t, \sqrt{k_2(t)}, \sqrt{k_3(t)}\big) = \mathbb{Q}(u)$$

with $t = t_\eta(u)$ as in the statement of this theorem.

The curve C_η in the statement of this theorem is $v^2 = k_4(t_\eta(u))$. We observed that $(0, \eta^3 - 1) \in C_\eta(\mathbb{Q})$. We have

$$\mathbb{Q}(C_\eta) = \mathbb{Q}\left(u, \sqrt{k_4(t_\eta(u))}\right) = \mathbb{Q}\big(t, \sqrt{k_2(t)}, \sqrt{k_3(t)}, \sqrt{k_4(t)}\big).$$

By Proposition 2.3(i) (or Corollary 2.2 of [RS01] with $g_i(t) = k_i(t)f_\eta(t)$), the rank of $E_\eta^{(f_\eta \circ t_\eta(u))}(\mathbb{Q}(C_\eta))$ is at least 4. By Proposition 2.3(ii), the rank of $E_\eta^{(f_\eta \circ t_\eta(u))}(\mathbb{Q})$ is at least 4 for all but finitely many $(u, v) \in C_\eta(\mathbb{Q})$. More explicitly, for $i = 1, \ldots, 4$, write

$$f_\eta \circ h_i(t) = f_\eta(t) \cdot k_i(t) \cdot j_i(t)^2$$

with $j_i(t) \in \mathbb{Q}(t)$. Then the points

$$\left(h_i \circ t_\eta(u), j_i \circ t_\eta(u)\sqrt{k_i \circ t_\eta(u)}\right) \in E_\eta^{(f_\eta(t_\eta(u)))}(\mathbb{Q}(u, v))$$

are

$$(t_\eta(u), 1),$$

$$\left(h_1 \circ t_\eta(u), \left(\frac{-(\eta+1)u^2 + \eta^3 - 1}{a((\eta+1)u^2 + 2(\eta^2 - 1)u + \eta^3 - 1)}\right)^3\right),$$

$$\left(h_2 \circ t_\eta(u), \left(\frac{-(\eta+1)u^2 + \eta^3 - 1}{a((\eta+1)u^2 + 2(\eta^2 + \eta + 1)u + \eta^3 - 1)}\right)^3\right),$$

$$\left(h_3 \circ t_\eta(u), \left(\frac{-(\eta+1)u^2 + \eta^3 - 1}{v}\right)^3\right).$$

They give four independent points in $E_\eta^{(f_\eta \circ t_\eta(u))}(\mathbb{Q}(C_\eta))$, by Lemma 2.1 above. (The fact that the first three are independent in $E_\eta^{(f_\eta \circ t_\eta(u))}(\mathbb{Q}(u))$ was essentially shown in the proof of Theorem 4.2 of [RS01].)

We next write down a (generalized) Weierstrass model for C_η. Let B_η be the elliptic curve $y^2 = (x - \alpha)(x - \beta)(x - \gamma)$ where

$$\alpha = -2(\eta^2 - 1)(\eta^2 + \eta + 1),$$
$$\beta = -2(\eta^2 - 1)(3\eta^2 + \eta - 1),$$
$$\gamma = -2(\eta^2 + \eta + 1)(3\eta^2 + 2\eta + 1).$$

There is a birational isomorphism from C_η to B_η that takes $(0, \eta^3 - 1) \in C_\eta(\mathbb{Q})$ to the identity element and takes the point

$$P_a := (-(a + 1)(\eta + a + 1), -(a + 1)(\eta + a + 1)(\eta + 2)(a\eta - 2\eta - 1))$$

in $C_\eta(\mathbb{Q})$ (with $a^2 = \eta$) to

$$Q_a := (2(\eta^3 - 1), 8a\eta(\eta + 2)(\eta^3 - 1)) \in B_\eta(\mathbb{Q}).$$

We used PARI/GP and Mathematica to check that for $1 \leq n \leq 10$ and $n = 12$, the denominator of the x-coordinate of nQ_a has no nonzero rational roots. Thus by Mazur's Theorem [Ma77], Q_a has infinite order for every $a \in \mathbb{Q} - \{0, 1, -1\}$, giving (ii). (In fact, $\mathbb{Z} \times \mathbb{Z}/4\mathbb{Z} \times \mathbb{Z}/2\mathbb{Z} \subseteq B_\eta(\mathbb{Q})$, since

$$(2(\eta^2 - 1)(\eta^2 + \eta + 1), 8\eta(2\eta + 1)(\eta^2 - 1)(\eta^2 + \eta + 1))$$

is a point of order four in $B_\eta(\mathbb{Q})$.)

Suppose $\eta \in \mathbb{Q} - \{0, 1\}$ is the square of a rational number. We checked that the degree 12 polynomial $f_\eta \circ t_\eta(u)$ is then always separable, so for each squarefree $d \in \mathbb{Z}$, the hyperelliptic curve $f_\eta \circ t_\eta(u) = dz^2$ has genus 5, and thus has only finitely many rational solutions (u, z). In other words, for each such η and d, the set of $u \in \mathbb{Q}$ such that $f_\eta \circ t_\eta(u)$ and d differ by a rational square is finite. Thus, since $C_\eta(\mathbb{Q})$ is infinite, for each w there are infinitely many non-isomorphic quadratic twists of E_η of rank at least 4 over $\mathbb{Q}$, proving (iv). $\qquad\square$

Corollary 3.3. *There are infinitely many $j \in \mathbb{Q}$ such that every elliptic curve E over $\mathbb{Q}$ with $j(E) = j$ has infinitely many quadratic twists of rank at least 4 over $\mathbb{Q}$.*

Proof Apply Theorem 3.2(iv) with $j = j\left(E_\eta^{(f_\eta \circ t_\eta(u))}\right)$. $\qquad\square$

Corollary 3.3 also follows from results stated in [Me98].

Remark 3.4. Among the E_η in Theorem 3.2 are infinitely many elliptic curves that are not twists of curves isogenous to elliptic curves with torsion subgroup $\mathbb{Z}/8\mathbb{Z} \times \mathbb{Z}/2\mathbb{Z}$, and thus give many new examples not given in [Me98]. For example, if E_η has good reduction at $p = 3$ or 5 or 7 (for example, if $a \in \mathbb{Z}$ and a is divisible by 3 or 5 or 7), then E_η has no quadratic twist $E_\eta^{(d)}$ isogenous over $\mathbb{Q}$ to an elliptic curve A with torsion subgroup $\mathbb{Z}/8\mathbb{Z} \times \mathbb{Z}/2\mathbb{Z}$, as can be seen as follows. If A has good reduction at p, then the Weil bound gives $\#A(\mathbb{F}_p) \leq 1 + p + 2\sqrt{p} < 16$, a contradiction (since the 2-torsion injects under reduction modulo primes of good reduction). Therefore A and $E_\eta^{(d)}$ have bad reduction at p, so p ramifies in $K = \mathbb{Q}(\sqrt{d})$. If $\mathcal{P}$ is the prime of K above p, then A has bad reduction at $\mathcal{P}$, since otherwise

$$16 \leq \#A(\mathcal{O}_K/\mathcal{P}) \leq 1 + N(\mathcal{P}) + 2\sqrt{N(\mathcal{P})} = 1 + p + 2\sqrt{p} < 16.$$

Thus $E_\eta^{(d)}$ has bad reduction at $\mathcal{P}$, contradicting the fact that $E_\eta^{(d)}$ is isomorphic over K to E_η which has good reduction at $\mathcal{P}$.

We next show that if we assume the Parity Conjecture, we can obtain a stronger conclusion than that of Theorem 3.2 for a larger class of elliptic curves.

If E is an elliptic curve over $\mathbb{Q}$, let

$$N_*(X) = \#\{\text{squarefree } d \in \mathbb{Z} : |d| \leq X, \operatorname{rank}(E^{(d)}(\mathbb{Q})) \text{ is } *\}.$$

In [RS01] we showed that for $y^2 = x(x-1)(x-\frac{1-a^2}{a^2+2})$ with $a \in \mathbb{Q}-\{0,1,-1\}$, we have $N_3(X) \gg X^{1/6}$ (for $X \gg 1$). We also showed, subject to the Parity Conjecture, that for every elliptic curve with all its two-torsion rational and a rational cyclic subgroup of order four, $N_{\geq 4, \text{even}}(X) \gg X^{1/6}$ (for $X \gg 1$).

Theorem 3.5. *Let E be $y^2 = x(x-1)(x-\frac{1-a^2}{a^2+2})$ where $a \in \mathbb{Q} - \{0,1,-1\}$ (as in Theorem 3.2). Suppose that the Parity Conjecture holds for all quadratic twists of E. If $|a| > 1$, then $N_{\geq 4, \text{even}}(X) \gg X^{1/6}$ for $X \gg 1$.*

Proof Suppose t_η is the function defined in Theorem 3.2 above (with $\eta = a^2$). In Theorem 4.2(a) of [RS01] we showed that there is a degree 12 polynomial $g(u) \in \mathbb{Q}[u]$ that differs from $f \circ t_\eta(u)$ by a square, is a product of 3 quartics, and satisfies $\operatorname{rank}(E^{(g(u))}(\mathbb{Q}(u)) \geq 3$. One can show that for every $a \in \mathbb{Q}-\{0,1,-1\}$ with $|a| > 1$, $g(u)$ has at least one real root. The result now follows from Corollary 5.2 of [RS01]. $\qquad\square$

Theorem 3.6. *Let A be the elliptic curve $y^2 = 4x^4 - 2x^2 - 1$. For every $a \in \mathbb{Q}^*$, let*

$$f_a(x) = x(x-1)(x+2a^2),$$

and let E_a be the elliptic curve $y^2 = f_a(x)$. Let C_a be the genus one curve $v^2 = (4a^2+1)^2(4a^4-2a^2-1)u^4 + 4a(4a^2+1)(4a^4+2a^2+1)u^3 - 2(16a^8 + 4a^6 + 10a^4 + 3a^2 - 1)u^2 + a(a^2+1)(4a^4+2a^2+1)u + (a^2+1)^2(4a^4-2a^2-1)$, and let

$$t_a(u) = \frac{T_a(u)}{2((4a^2+1)u^2 + a^2 + 1)^2}$$

where $T_a(u) = -(2a^2-1)(4a^2+1)^2u^4 - 4a(4a^2+1)(2a^2+1)u^3 + 2(4a^4+3a^2+1)(2a^2-1)u^2 + 4a(a^2+1)(2a^2+1)u - (a^2+1)^2(2a^2-1)$. Then:

(i) $A(\mathbb{Q}) \cong \mathbb{Z} \oplus \mathbb{Z}/2\mathbb{Z}$;

(ii) *if $(a, b) \in A(\mathbb{Q})$, then C_a is an elliptic curve over $\mathbb{Q}$ and*

$$\mathrm{rank}(C_a(\mathbb{Q})) \geq 1;$$

(iii) *for all but possibly finitely many $(u, v) \in C_a(\mathbb{Q})$, the quadratic twist of E_a by $f_a \circ t_a(u)$ has rank at least 4 over $\mathbb{Q}$;*

(iv) *if $(a, b) \in A(\mathbb{Q})$, then there are infinitely many non-isomorphic quadratic twists of E_a of rank at least 4 over $\mathbb{Q}$.*

Proof   Part (i) is easy. The rest of the proof proceeds in the same way as that of Theorem 3.2, where now we use the functions h_1, h_2, h_5, and h_3 given at the beginning of this section. Since the curve C_a is $v^2 = k_3(t_a(u))$, we have (see also Theorem 4.1 of [RS01])

$$\mathbb{Q}(C_a) = \mathbb{Q}\left(u, \sqrt{k_3(t_a(u))}\right) = \mathbb{Q}\left(t, \sqrt{k_2(t)}, \sqrt{k_5(t)}, \sqrt{k_3(t)}\right).$$

Let B_a be $y^2 = (x - \alpha)(x - \beta)(x - \gamma)$ where

$$\begin{aligned}
\alpha &= 2(4a^4 - 2a^2 - 1)(4a^2 + 1)(a^2 + 1), \\
\beta &= 2(4a^6 + 2a^4 + 5a^2 + 1)(4a^2 + 1), \\
\gamma &= -2(16a^6 + 4a^4 - 2a^2 + 1)(a^2 + 1).
\end{aligned}$$

It is easy to check that B_a is an elliptic curve whenever $a \in \mathbb{Q}^*$. Suppose that $(a, b) \in A(\mathbb{Q})$. Then there is a birational isomorphism from C_a to B_a that takes the rational point $(0, (a^2 + 1)b)$ to the identity element and takes the point $(a, (4a^4 + 2a^2 + 1)b) \in C_a(\mathbb{Q})$ to the point

$$Q_{(a,b)} := \left(\frac{g(a)(4a^6 - 2a^4 - a^2 - 2)}{a^2}, \frac{-4g(a)(4a^4 + 2a^2 + 1)b}{a^3} \right) \in B_a(\mathbb{Q}),$$

where $g(a) = 2(a^2 + 1)(4a^2 + 1)$. For $a \notin \{0, 1, -1\}$, we used PARI/GP and Mathematica to check that $nQ_{(a,b)} \neq O$ for $1 \leq n \leq 10$ and $n = 12$. Thus by Mazur's Theorem [Ma77], $\mathrm{rank}(B_a(\mathbb{Q})) \geq 1$. Further, the rank of $B_1(\mathbb{Q})$ $(= B_{-1}(\mathbb{Q}))$ is one. We now have (ii). The points

$$(t_a(u), 1),$$

$$\left(h_2 \circ t_a(u), \left(\frac{2a((4a^2 + 1)u^2 + a^2 + 1)}{-(4a^2 + 1)u^2 + 2a(4a^2 + 1)u + a^2 + 1} \right)^3 \right),$$

$$\left(h_5 \circ t_a(u), \left(\frac{(4a^2 + 1)u^2 + a^2 + 1}{a(4a^2 + 1)u^2 + 2(a^2 + 1)u - a(a^2 + 1)} \right)^3 \right),$$

$$\left(h_3 \circ t_a(u), \left(\frac{2a((4a^2 + 1)u^2 + a^2 + 1)}{v} \right)^3 \right)$$

give four independent points in $E_a^{(f_a \circ t_a(u))}(\mathbb{Q}(C_a))$. $\square$

4 Root numbers

Definition 4.1. *If E is an elliptic curve over $\mathbb{Q}$, let N_E denote the conductor of E, let w_E denote the global root number, i.e., the sign in the functional equation for $L(E,s)$, and let $w_{E,p}$ denote the local root number at a prime $p \le \infty$. Write $w_E(d)$ for $w_{E^{(d)}}$ and write $w_{E,p}(d)$ for $w_{E^{(d)},p}$.*

Definition 4.2. *If $\alpha \in \mathbb{Q}^*$ and $n \in \mathbb{Z}^+$, then:*

(i) *$\alpha \equiv 1 \bmod^{\times} n$ means that $\alpha - 1 \in n\mathbb{Z}_\ell$ for all primes $\ell \mid n$;*

(ii) *$\alpha \equiv 1 \bmod^{\times} n\infty$ means that $\alpha \equiv 1 \bmod^{\times} n$ and $\alpha > 0$.*

Lemma 4.3. *Suppose E is an elliptic curve over $\mathbb{Q}$, $d, d' \in \mathbb{Q}^*$, and there exists $\beta \in \mathbb{Q}^*$ such that $\beta^2 d/d' \equiv 1 \bmod^{\times} 8N_E\infty$. Then $w_E(d) = w_E(d')$.*

Proof Taking the squarefree parts of d and d', we can reduce to the case where d and d' are squarefree integers.

If $p < \infty$ and $p \nmid dN_E$, then $E^{(d)}$ has good reduction over $\mathbb{Q}_p$, so $w_{E,p}(d) = 1$ (see Proposition 2(iv) of [R93]). Similarly for d'. Thus,

$$w_E(d) = \prod_{p \le \infty} w_{E,p}(d) = \prod_{p \mid dN_E\infty} w_{E,p}(d). \tag{4.1}$$

If d/d' is a square in $\mathbb{Q}_p^*$, then $E^{(d)}$ and $E^{(d')}$ are isomorphic over $\mathbb{Q}_p$, so $w_{E,p}(d) = w_{E,p}(d')$ for all $p \le \infty$. In particular, since $d/d' > 0$, it follows that $w_{E,\infty}(d) = w_{E,\infty}(d')$. If $p \mid 2N_E$, then d/d' is a square in $\mathbb{Q}_p^*$ (since $\beta^2 d/d' \equiv 1 \bmod^{\times} 8N_E$), so $w_{E,p}(d) = w_{E,p}(d')$. If $p \mid 2N_E$, then p divides d if and only if p divides d' (since $2\mathrm{ord}_p(\beta) + \mathrm{ord}_p(d) = \mathrm{ord}_p(d')$, and d and d' are squarefree). Thus,

$$\frac{\prod_{p \mid dN_E\infty} w_{E,p}(d)}{\prod_{p \mid dN_E\infty} w_{E,p}(d')} = \frac{\prod_{p \mid d,\, p \nmid 2N_E} w_{E,p}(d)}{\prod_{p \mid d',\, p \nmid 2N_E} w_{E,p}(d')}. \tag{4.2}$$

Suppose $p \nmid N_E$, so E has good reduction at p. Since E and $E^{(d)}$ are isomorphic over $\mathbb{Q}_p(\sqrt{d})$, $E^{(d)}$ has good reduction over $\mathbb{Q}_p(\sqrt{d})$. If $p \mid d$, then $\mathbb{Q}_p(\sqrt{d})$ is the smallest extension of $\mathbb{Q}_p$ over which $E^{(d)}$ has good reduction (and similarly for d'). By (iii) and (v) of Proposition 2 of [R93] with $e = 2$, we have

$$w_{E,p}(d) = \left(\frac{-1}{p}\right) \tag{4.3}$$

if $p \mid d$ and $p \nmid 2N_E$, where $\left(\frac{-1}{m}\right)$ is the Jacobi symbol.

By (4.1), (4.2), and (4.3), we have

$$\frac{w_E(d)}{w_E(d')} = \frac{\prod_{p \mid d,\, p \nmid 2N_E} \left(\frac{-1}{p}\right)}{\prod_{p \mid d',\, p \nmid 2N_E} \left(\frac{-1}{p}\right)} = \frac{\left(\frac{-1}{f}\right)}{\left(\frac{-1}{f'}\right)},$$

where $f = d/\gcd(d, 2N_E)$ and $f' = d'/\gcd(d', 2N_E)$. Note that $f/f' = d/d'$. Then $\beta^2 f/f' \equiv 1 \bmod^{\times} 4$, so $f \equiv f' \pmod 4$, so $\left(\frac{-1}{f}\right) = \left(\frac{-1}{f'}\right)$. $\qquad\square$

Lemma 4.4. *Suppose E and B are elliptic curves over $\mathbb{Q}$, $B(\mathbb{Q})$ has infinite order, $P \in B(\mathbb{Q})$, r is a rational function in $\mathbb{Q}(B)$, and P is not a zero or pole of r. Then there exist a $Q \in B(\mathbb{Q})$ of infinite order and an open neighborhood U of O in $B(\mathbb{R})$ such that if $k \in \mathbb{Z}$ and $kQ \in U$ then $w_E(r(P+kQ)) = w_E(r(P))$.*

Proof Let
$$V = B(\mathbb{R}) \times \prod_{p \mid 2N_E} B(\mathbb{Q}_p)$$
and let $g(z) = r(P + z)/r(P) \in \mathbb{Q}(B)$. Then $g(O) = 1$, and g induces a function
$$g : V - \{\text{poles of } g\} \to \mathbb{R} \times \prod_{p \mid 2N_E} \mathbb{Q}_p,$$
which is continuous at O. Let $B_n(\mathbb{Q}_p)$ denote the subset of $B(\mathbb{Q}_p)$ of points that, in a minimal Weierstrass model for B, have

$$\mathrm{ord}_p(x\text{-coordinate}) \leq -2n$$

(see Exercise 7.4 on p. 187 of [S86]). Then the $B_n(\mathbb{Q}_p)$'s form a basis for the open sets around O in $B(\mathbb{Q}_p)$, and are subgroups of finite index in $B(\mathbb{Q}_p)$. Since g is continuous at O, there is an open neighborhood U of O in $B(\mathbb{R})$ and for every $p \mid 2N_E$ there is an $n_p \in \mathbb{Z}^{\geq 0}$ such that

$$g\Big(U \times \prod_{p \mid 2N_E} B_{n_p}(\mathbb{Q}_p)\Big) \subseteq \mathbb{R}^+ \times \prod_{p \mid 2N_E} (1 + 8N_E\mathbb{Z}_p).$$

Let $k_p = [B(\mathbb{Q}_p) : B_{n_p}(\mathbb{Q}_p)]$ and let $Q_0 \in B(\mathbb{Q})$ be a point of infinite order. Let $Q = (\mathrm{lcm}_{p \mid 2N_E}\{k_p\})Q_0 \in B(\mathbb{Q})$. Then Q has infinite order, and $Q \in B_{n_p}(\mathbb{Q}_p)$ for all $p \mid 2N_E$. Now apply Lemma 4.3 with $d = r(P + kQ)$, $d' = r(P)$, and $\beta = 1$. $\qquad\square$

Lemma 4.5. *Suppose B is an elliptic curve over $\mathbb{Q}$, $Q \in B(\mathbb{Q})$ is a point of infinite order, and U is an open subset of the identity component $B(\mathbb{R})^0$ of $B(\mathbb{R})$. Then $\{k \in \mathbb{Z} : kQ \in U\}$ is infinite.*

Proof Replacing Q by $2Q$, we may assume that $Q \in B(\mathbb{R})^0$. Note that $B(\mathbb{R})^0$ is isomorphic to the unit circle in $\mathbb{C}^*$, so every infinite subgroup is dense. Thus $\{kQ : k \in \mathbb{Z}\}$ is dense in $B(\mathbb{R})^0$, and the lemma follows. $\qquad\square$

5   Rank ≥ 5

Theorem 5.1. *Suppose $a \in \mathbb{Q} - \{0, 1, -1\}$ and $\eta = a^2$. Suppose E_η, f_η, and t_η are as in Theorem 3.2. If $w_{E_\eta}(f_\eta \circ t_\eta(u_1)) = -1$ for some $(u_1, v_1) \in B_\eta(\mathbb{Q})$, and the Parity Conjecture holds for all quadratic twists of E_η, then E_η has infinitely many non-isomorphic quadratic twists of odd rank ≥ 5 over $\mathbb{Q}$.*

Proof Let $P = (u_1, v_1)$, and let $r(z) = f_\eta \circ t_\eta \circ x(z) \in \mathbb{Q}(B_\eta)$, where the function x gives the x-coordinate of a point. By Lemmas 4.4 and 4.5 with $E = E_\eta$ and $B = B_\eta$, there are $Q \in B_\eta(\mathbb{Q})$ and infinitely many $k \in \mathbb{Z}$ such that

$$w_{E_\eta}(r(P + kQ)) = w_{E_\eta}(r(P)) = -1,$$

so by the Parity Conjecture, $E_\eta^{(r(P+kQ))}(\mathbb{Q})$ has odd rank.

For all but finitely many $k \in \mathbb{Z}$, the rank of $E_\eta^{(r(P+kQ))}(\mathbb{Q})$ is at least 4, by Theorem 3.2(iii). Thus for infinitely many k, the rank of $E_\eta^{(r(P+kQ))}(\mathbb{Q})$ is at least 5. As argued in the proof of Theorem 3.2, for each squarefree $d \in \mathbb{Q}^*$, the set of $u \in \mathbb{Q}$ such that $f_\eta \circ t_\eta(u)$ and d differ by a rational square is finite, since the hyperelliptic curve $f_\eta \circ t_\eta(u) = dz^2$ has only finitely many rational solutions (u, z). Thus there are infinitely many non-isomorphic quadratic twists of E_η of odd rank at least 5 over $\mathbb{Q}$. $\qquad\qquad\square$

Corollary 5.2. *Suppose*

$$a \in \{2, 5, 6, 7, 8, 12, 13, 14, 15, 16, 17, 18, 21, 22,$$
$$23, 24, 25, 26, 28, 30, 32, 33, 35, 36, 37, 39, 40, 41\}.$$

If the Parity Conjecture holds for all quadratic twists of

$$E_{a^2} : y^2 = x(x-1)\left(x - \frac{1-a^2}{a^2+2}\right),$$

then E_{a^2} has infinitely many non-isomorphic quadratic twists of odd rank ≥ 5 over $\mathbb{Q}$.

Proof With $\eta = a^2$ and $P_a = (u_0, v_0) \in C_\eta(\mathbb{Q})$ as in the proof of Theorem 3.2, and $P'_\eta = (u_1, v_1) = (1 - \eta, (1 - \eta)(2 + \eta)) \in C_\eta(\mathbb{Q})$, one can check that for each of the above a's, at least one of $w_{E_\eta}(f_\eta \circ t_\eta(u_0))$ and $w_{E_\eta}(f_\eta \circ t_\eta(u_1))$ is -1. The result now follows from Theorem 5.1. $\qquad\qquad\square$

References

[Ma77] B. Mazur, *Modular curves and the Eisenstein ideal*, Inst. Hautes Études Sci. Publ. Math. **47** (1977), 33–186.

[Me92] J-F. Mestre, *Rang de courbes elliptiques d'invariant donné*, C. R. Acad. Sci. Paris **314** (1992), 919–922.

[Me98] J-F. Mestre, *Rang de certaines familles de courbes elliptiques d'invariant donné*, C. R. Acad. Sci. Paris **327** (1998), 763–764.

[Me00] J-F. Mestre, *Ranks of twists of elliptic curves*, lecture at MSRI, September 11, 2000.

[R93]   D. Rohrlich, *Variation of the root number in families of elliptic curves*, Compositio Math. **87** (1993), 119–151.

[RS01] K. Rubin, A. Silverberg, *Rank frequencies for quadratic twists of elliptic curves*, Exper. Math. **10** (2001), 559–569.

[S83]   J. H. Silverman, *Heights and the specialization map for families of abelian varieties*, J. Reine Angew. Math. **342** (1983), 197–211.

[S86]   J. H. Silverman, The arithmetic of elliptic curves, Graduate Texts in Mathematics **106**, Springer-Verlag, New York, 1986.

[ST95] C. L. Stewart, J. Top, *On ranks of twists of elliptic curves and power-free values of binary forms*, J. Amer. Math. Soc. **8** (1995), 943–973.

Department of Mathematics,
University of California at Irvine,
Irvine, CA 92697, USA

The powers of logarithm for quadratic twists

Christophe Delaunay and Mark Watkins

Abstract

We briefly describe how to get the power of logarithm in the asymptotic for the number of vanishings in the family of even quadratic twists of a given elliptic curve. There are four different possibilities, largely dependent on the rational 2-torsion structure of the curve we twist.

1 Introduction

Let E be a rational elliptic curve of conductor N and Δ its discriminant, with E_d its dth quadratic twist. The seminal paper [CKRS] modelled the value-distribution of $L(E_d, 1)$ via random matrix theory and applied a discretisation process to the coefficients of an associated modular form of weight $3/2$. This led to the conjecture that asymptotically there are $c_E X^{3/4}(\log X)^{3/8-1}$ twists by *prime $p < X$* with even functional equation and $L(E_p, 1) = 0$, where the $3/8$ comes from random matrix theory, and the -1 comes from the prime number theorem.

We wish to determine a similar heuristic for the asymptotic for the number of twists by *all* fundamental discriminants $|d| < X$ such that $L(E_d, s)$ has even functional equation and $L(E_d, 1) = 0$. We find that the power of logarithm that we obtain depends on the growth rate of various local Tamagawa numbers of twists of E. Because of this, it is somewhat unfortunate that isogenous curves need not have the same local Tamagawa numbers. This is most particularly a problem when we have a curve with full rational 2-torsion and it is isogenous to one that only has one rational 2-torsion point; in this case, we should work with the curve with full 2-torsion. This makes the statement of our result a bit messy, but we have:

Heuristic 1.1. *Let E be a rational elliptic curve. Then the number of even quadratic twists E_d with $L(E_d, 1) = 0$ and $|d| < X$ is asymptotically $c'_E X^{3/4}(\log X)^{b_E+3/8}$ where $c'_E > 0$ and*

- *$b_E = 1$ when E (or a curve isogenous to it) has full rational 2-torsion,*

- *$b_E = \sqrt{2}/2$ when E has one rational 2-torsion point (and no curve isogenous to E has full 2-torsion),*

- *$b_E = 1/3$ when E has no rational 2-torsion and Δ is square.*

- $b_E = \sqrt{2}/2 - 1/3$ *when E has no rational 2-torsion and Δ is not square.*

The $3/8$ in the exponent comes from random matrix theory, and so we only need concern ourselves with calculting b_E. Also, we do not consider the constant c'_E, as that would greatly complicate the discussion.

2   Discussion

The discretisation for the values of $L(E_d, 1)$ can be re-interpreted as saying that

$$L(E_d, 1) < \Omega(E_d)g(E_d)/T(E_d)^2 \implies L(E_d, 1) = 0$$

where Ω is the real period, g is the product of the Tamagawa factors, T is the order of the torsion subgroup. This comes from the Birch and Swinnerton-Dyer conjecture and the fact that the order of the Shafarevich-Tate group is an integer. From random matrix theory, we expect that there is some constant $c > 0$ such that the probability that $L(E_d, 1) \leq t$ tends to $ct^{1/2}(\log|d|)^{3/8}$ as $t \to 0$. Combining this distribution with the discretisation, we get that (as $|d| \to \infty$)

$$\mathrm{Prob}\big[L(E_d, 1) = 0\big] \sim c\sqrt{\Omega(E_d)g(E_d)/T(E_d)^2}(\log|d|)^{3/8}.$$

This becomes useful upon realising how these quantities vary in twist families. In particular, we have (up to a factor of 2 that we ignore) that $\Omega(E_d) = \Omega(E)/\sqrt{|d|}$ while $T(E_d)$ is constant for $|d|$ sufficiently large. This reduces our problem to a determination of how the Tamagawa product $g(E_d)$ varies; from the above we have that

$$\mathrm{Prob}\big[L(E_d, 1) = 0\big] \approx c'\sqrt{g(E_d)}(\log|d|)^{3/8}/|d|^{1/4},$$

and so the number of twists should be (here the d are fundamental)

$$N(X) \sim \sum_{\substack{|d|<X \\ E_d \text{ even}}} \mathrm{Prob}\big[L(E_d, 1) = 0\big] \approx \sum_{\substack{|d|<X \\ E_d \text{ even}}} c'\sqrt{g(E_d)}(\log|d|)^{3/8}/|d|^{1/4}.$$

and by partial summation we have that

$$N(X) \approx c''X^{3/4}(\log X)^{3/8} \sum_{\substack{|d|<X \\ E_d \text{ even}}} \sqrt{g(E_d)},$$

We now compute the expected average value of $\sqrt{g(E_d)}$ via an analysis of the splitting behaviour of the cubic polynomial associated to E.

3 Tamagawa numbers

For simplicity, we now restrict to twisting by positive fundamental discriminants d with $\gcd(d, N) = 1$ and even sign in the functional equation.[1] We first isolate the contribution to the Tamagawa factor $g(E_d)$ coming from the primes that divide the discriminant of E, and call this $g(E)$. Writing $g_p(E_d)$ for the local Tamagawa number at p for the twist E_d, we have, up to a bounded factor B_d which includes $G(E)$ and other contributions from bad primes, that

$$g(E_d) = B_d \cdot \prod_{p\mid d} g_p(E_d).$$

We shall ignore B_d for the remainder of the discussion, as consideration of it does not change the power of logarithm. Again possibly ignoring a finite set of bad primes, when we twist by d, for primes $p\mid d$ the local Tamagawa number $g_p(E_d)$ at p for E_d is either 1, 2, or 4.[2] If we write E in the form $y^2 = f(x)$, these correspond to the cubic f having 0, 1, or 3 roots modulo p (provided that this model for E is good at p).

We assume that we can use the Chebotarev density theorem to determine the frequency of each splitting behaviour of the cubic f. When E has full 2-torsion, the cubic f splits completely over the rationals, so we have $g_p(E_d) = 4$ for all $p\mid d$. When E has one rational 2-torsion point, the cubic f splits over $\mathbf{Q}$ as a quadratic factor times a linear factor, and the quadratic splits into two linear factors precisely when its discriminant is square mod p; thus asymptotically half the primes $p\mid d$ have $g_p(E_d) = 2$, and the other half yield $g_p(E_d) = 4$. Finally, when f is irreducible over the rationals, we have two cases, depending upon whether[3] Δ is square: when it is square (such as with $x^3 - 3x + 1$), asymptotically $1/3$ of the primes have $g_p(E_d) = 4$ and the other $2/3$ have $g_p(E_d) = 1$; when the discriminant is not square, the local Tamagawa factors are $g_p(E_d) = 1, 2, 4$ in proportions $1/3$, $1/2$, and $1/6$.[4]

---

[1]A rigourous accounting would also separate the d into congruence classes modulo the discriminant (see [D]) but we omit this so as to focus on the main ideas. Indeed, the more pedantic analysis would only modify the constant c'_E and not the power of logarithm in the asymptotic.

[2]We can note that for $p > 2$ we have $g_p(E_d) = g_p(E_{p^*})$ where $p^* = p(-1)^{(p-1)/2}$, which essentially eliminates the dependence on d.

[3]The fact that the elliptic curve discriminant Δ and the discriminant of the cubic differ by a factor of 16 does not affect our analysis.

[4]Our use of the Chebotarev density theorem is not quite legitimate in general. We need to be more careful about our restriction to *even* twists (a condition that is given by congruences modulo N), which can give incompatibility conditions, especially in the case where f is irreducible and has non-square discriminant, as here the splitting condition cannot be given by congruence conditions modulo N.

4   Analytic number theory

The problem of computing the average value of $\sqrt{g(E_d)}$ is now essentially one of analytic number theory; for simplicity,[5] we explain how to compute the average value at positive fundamental discriminants d of the multiplicative function $h(d) = \sqrt{g(E_d)}$, and so wish to compute an asymptotic for

$$F(X) = \sum_{d \leq X} \mu^\star(d)^2 h(d),$$

where the modified Möbius function $(\mu^\star)^2$ is the characteristic function of (positive) fundamental discriminants (this differs from μ^2 only at the prime 2). We analyse $F(X)$ via the behaviour of the logarithm of the Euler product $\prod_p \big(1 + h(p)/p^s\big)$ as $s \to 1^+$. Explicitly, as $s \to 1^+$ we have that (ignoring the modification at the prime 2)

$$\log \prod_p \left(1 + \frac{h(p)}{p^s}\right) \sim \sum_p \frac{h(p)}{p^s} \sim -(t_1 + t_2\sqrt{2} + t_4\sqrt{4}) \log(s-1),$$

where t_k is the probability that h takes on the value $\sqrt{k}$, and the last step is in analogy with the fact that $\sum_p 1/p^s \sim -\log(s-1)$. Via exponentiation we obtain $\prod_p \big(1 + h(p)/p^s\big) \sim c/(s-1)^A$ for some constant $c \neq 0$, where $A = (t_1 + t_2\sqrt{2} + t_4\sqrt{4}) > 0$. An application of the Tauberian theorem (or Perron's formula) then gives us that $F(X) \sim c'X(\log X)^{A-1}$ for some $c' \neq 0$.

Finally, we conclude by computing the value of A in each of the four cases:

- $(t_1, t_2, t_4) = (0, 0, 1)$ and so $A = 2$ for the case of full 2-torsion;

- $(t_1, t_2, t_4) = (0, 1/2, 1/2)$ and so $A = 1 + \sqrt{2}/2$ for the case of one rational 2-torsion point;

- $(t_1, t_2, t_4) = (2/3, 0, 1/3)$ and so $A = 4/3$ when there is no 2-torsion and Δ is square;

- $(t_1, t_2, t_4) = (1/3, 1/2, 1/6)$ and so $A = 2/3 + \sqrt{2}/2$ when there is no 2-torsion and Δ is non-square.

5   Acknowledgments

The second author was partially supported by Engineering and Physical Sciences Research Council (EPSRC) grant GR/T00658/01 (United Kingdom).

---

[5]For computations regarding the restriction to d with $\gcd(d, N) = 1$ and even sign, see [D, §6], especially Theorem 6.3 and Theorem 6.8 with $k = -1/2$; essentially Dirichlet characters mod N can be used to isolate the desired congruence classes.

References

[CKRS]   J. B. Conrey, J. P. Keating, M. O. Rubinstein, N. C. Snaith, *On the frequency of vanishing of quadratic twists of modular L-functions.* In *Number theory for the millennium, I* (Urbana, IL, 2000), edited by M. A. Bennett, B. C. Berndt, N. Boston, H. G. Diamond, A. J. Hildebrand and W. Philipp, A K Peters, Natick, MA (2002), 301–315. Available online at `arxiv.org/math.NT/0012043`

[D]   C. Delaunay, *Moments of the Orders of Tate-Shafarevich Groups.* International Journal of Number Theory, **1** (2005), no. 2, 243–264.

Christophe Delaunay
Institut Camille Jordan
Université Claude Bernard Lyon 1
43, avenue du 11 novembre 1918
69622 Villeurbanne Cedex - France

Mark Watkins
School of Mathematics
University of Bristol
Bristol BS8 1TW
UK

Note on the frequency of vanishing of L-functions of elliptic curves in a family of quadratic twists

Christophe Delaunay

Abstract

In this note, we give an example of an elliptic curve E such that for all prime discriminants $d < 0$ for which the sign of the functional equation of the L-function of the quadratic twist E_d of E by d is $+1$, we have $L(E_d, 1) \neq 0$. Furthermore, using the Birch and Swinnerton-Dyer conjecture, we prove that the Tate-Shafarevitch group of E_d, for all such d, has a trivial 2-part. Our method can be generalized to other examples.

1   Notations and introduction

Let E be an elliptic curve defined over $\mathbb{Q}$ with conductor N and with L-function $L(E, s) = \sum_n a(n) n^{-s}$. From the work of Wiles, Taylor ([Wil], [Tay-Wil]) and Breuil, Conrad, Diamond, Taylor ([Bre et al.]), $L(E, s)$ can be continued to the whole complex plane and satisfies a functional equation:

$$\Lambda(E, s) = \varepsilon(E) \Lambda(E, 2 - s),$$

where $\varepsilon(E) = \pm 1$ is the sign of the functional equation and $\Lambda(E, s)$ is given by:

$$\Lambda(E, s) = \left(\frac{\sqrt{N(E)}}{2\pi} \right)^s \Gamma(s) L(E, s).$$

Let d be a fundamental discriminant, $\left(\frac{d}{\cdot}\right)$ its associated quadratic character and E_d the quadratic twist of E by d. Furthermore, we assume that d is prime to N. Hence the conductor of E_d is Nd^2, and the sign of the functional equation of $L(E_d, s) = \sum_n a(n) \left(\frac{d}{n}\right) n^{-s}$ is

$$\varepsilon(E_d) = \varepsilon(E) \left(\frac{d}{-N(E)} \right).$$

Classical questions are concerned with the distribution of the special values $L(E_d, 1)$ as d runs through discriminants with $\varepsilon(E_d) = 1$. For example, one can ask for the density of those d such that $L(E_d, 1) = 0$ or for the density of those d such that $p \mid |\text{III}(E_d)|$, where p is a fixed prime and $\text{III}(E_d)$ is the Tate-Shafarevitch group[1] of E_d. The Birch and Swinnerton-Dyer conjecture gives a precise link between the value $L(E_d, 1)$ and the order $|\text{III}(E_d)|$ of the Tate-Shafarevitch group.

Using elementary arithmetic on quadratic forms, we prove that if E is the elliptic curve "17a1" in Cremona's table ([Cre]) then for all prime discriminants $d < 0$ with $\varepsilon(E_d) = 1$, we have $L(E_d, 1) \neq 0$ and (using the Birch and Swinnerton-Dyer conjecture) $2 \nmid |\text{III}(E_d)|$. Other examples can be handled with the same method.

2   The example

Throughout this section E denotes the elliptic curve with conductor $N = 17$ defined by:

$$E \; : \; y^2 + xy + y = x^3 - x^2 - x - 14.$$

We consider the quadratic twists E_d of E by discriminants $d < 0$ coprime with N such that $\varepsilon(E_d) = 1$ (i.e. $d \equiv 1, 2, 4, 8, 9, 13, 15, 16 \mod 17$). By a theorem of Waldspurger ([Wal]), the values of $L(E_d, 1)$ are related to the coefficients $c(n)$ of a weight $3/2$ modular form. More precisely, we have:

$$L(E_d, 1) = \frac{\kappa}{\sqrt{|d|}} c(|d|)^2, \tag{2.1}$$

where κ is a constant (here $\kappa \approx 2.74573911$) and where the modular form of weight $3/2$, computed by Tornaria ([Tor]), is given by:

$$\sum_n c(n) q^n = \frac{\theta_1(q) - \theta_2(q)}{2}$$

with:

$$\theta_1(q) \;=\; \sum_{(x,y,z) \in \mathbb{Z}^3} q^{3x^2 + 23y^2 + 23z^2 - 2xy - 2xz - 22yz}$$

$$\theta_2(q) \;=\; \sum_{(x,y,z) \in \mathbb{Z}^3} q^{7x^2 + 11y^2 + 20z^2 - 6xy - 4xz - 8yz}.$$

We have:

Theorem 1. *If $d < 0$ is a prime discriminant with $\varepsilon(E_d) = 1$, the coefficient $c(-d)$ is odd.*

---

[1]The Tate-Shafarevitch group of an elliptic curve E is some "cumbersome" group which, roughly speaking, measures the obstruction of a certain "local-global" principle (see [Sil] for a precise definition). It is conjectured that it is a finite group and, if so, one can prove that its order is a perfect square.

Proof Let d be such a discriminant and $p = -d$. Remark that we have $p \equiv 3$ mod 4 and that $\left(\frac{-17}{p}\right) = 1$. We let:

$$
\begin{aligned}
Q_1(x, y, z) &= 3x^2 + 23y^2 + 23z^2 - 2xy - 2xz - 22yz \\
Q_2(x, y, z) &= 7x^2 + 11y^2 + 20z^2 - 6xy - 4xz - 8yz.
\end{aligned}
$$

We consider the following sets:

$$
\begin{aligned}
R_1 &= \{(x, y, z) \in \mathbb{Z}^3,\ Q_1(x, y, z) = p\} \\
R_2 &= \{(x, y, z) \in \mathbb{Z}^3,\ Q_2(x, y, z) = p\}
\end{aligned}
$$

and we must prove that $|R_1| - |R_2| \equiv 2 \mod 4$. In fact, the two ternary quadratic forms Q_1 and Q_2 are invariant by the involutions $\iota : (x, y, z) \mapsto (-x, -y, -z)$ and $\tau : (x, y, z) \mapsto (x - z, y - z, -z)$. Hence, if $P = (x, y, z) \in R_i$ (for $i = 1, 2$), then $P, \iota(P), \tau(P)$ and $\iota \circ \tau(P)$ also belong to R_i and these 4 points are distinct except if $z = 0$. Thus,

$$
|R_i| \equiv |\{(x, y) \in \mathbb{Z}^2, Q_i(x, y, 0) = p\}| \quad \mod 4.
$$

Hence for $|R_1| \mod 4$, we are led to study the number of solutions of

$$
p = 3x^2 - 2xy + 23y^2 = q_1(x, y)
$$

and for $|R_2| \mod 4$ the number of solutions of

$$
p = 7x^2 - 6xy + 11y^2 = q_2(x, y).
$$

There are 8 classes of primitive quadratic forms with discriminant $\Delta = -2^4 \times 17$ modulo $SL_2(\mathbb{Z})$. As a set of representatives we can choose the 8 following ones:

$$
\begin{aligned}
q_1(x, y) &= 3x^2 - 2xy + 23y^2 & \overline{q_1}(x, y) &= 3x^2 + 2xy + 23y^2 \\
q_2(x, y) &= 7x^2 - 6xy + 11y^2 & \overline{q_2}(x, y) &= 7x^2 + 6xy + 11y^2 \\
q_3(x, y) &= 8x^2 - 4xy + 9y^2 & \overline{q_3}(x, y) &= 8x^2 + 4xy + 9y^2 \\
q_4(x, y) &= 4x^2 + 17y^2 & q_5(x, y) &= x^2 + 68y^2
\end{aligned}
$$

Since we have $\left(\frac{-17}{p}\right) = 1$, the prime p must be represented by one of these forms. Since $p \equiv 3 \mod 4$, the prime p cannot be represented by the forms $q_3, \overline{q_3}, q_4$ and q_5. Hence, we have two possibilities:

- The prime p is represented by q_1 with only 2 solutions (and so it is for $\overline{q_1}$) and p is not represented by q_2 (neither by $\overline{q_2}$).

- The prime p is not represented by q_1 hence it is by q_2 with only 2 solutions (and so it is by $\overline{q_2}$).

In each case, we conclude that $|R_1| - |R_2| \equiv 2 \mod 4$ and so $c(p)$ is odd. $\quad\square$

Corollary 2. *Let $d < 0$ be a prime discriminant such that $\varepsilon(E_d) = 1$, then we have $L(E_d, 1) \neq 0$.*

Proof Indeed, $c(|d|)$ is odd so by equation (2.1) we have $L(E_d, 1) \neq 0$. □

Remarks.

1- By the results of [Kol] and [BFH] or [Mu-Mu] we deduce that the rank of $E_d(\mathbb{Q})$ is 0 and that its Tate-Shafarevich group is finite.

2- It is a classical question to understand the ratio of d such that $L(E_d, 1) = 0$. Using random matrix theory and its link with L-functions, Conrey, Keating, Rubinstein and Snaith ([CKRS], [CKRS2]) have conjectured that if E is an elliptic curve over $\mathbb{Q}$, there exists a constant $c_E \geqslant 0$ such that:

$$\sum_{\substack{p \leqslant T \\ -p \text{ discriminant} \\ \varepsilon(E_{-p})=1 \\ L(E_{-p},1)=0}} 1 \sim c_E T^{3/4} \log(T)^{-5/8}$$

So, corollary 2 implies that the constant c_E can be 0.

Corollary 3 (under the Birch and Swinnerton-Dyer conjecture). *For all prime discriminants $d < 0$ such that $\varepsilon(E_d) = 1$ we have $2 \nmid |\text{III}(E_d)|$.*

Proof For such a discriminant, we already know that $L(E_d, 1) \neq 0$ and, in our example, the Birch and Swinnerton-Dyer conjecture predicts that:

$$|\text{III}(E_d)| = c(|d|)^2$$

Hence, $|\text{III}(E_d)|$ is odd. □

Remarks.

1- This seems to be in contradiction with the heuristic in [Del] which would have suggested a density of about 58% of $|\text{III}(E_d)|$ divisible by 2. Note that for odd primes p, the numerical data, performed by Rubinstein ([Rub]), about the density of $|\text{III}(E_d)|$ divisible by p are in close agreement with the predictions given by the heuristic (except for the p dividing $|E(\mathbb{Q})_{\text{tors}}|$). In fact, as we have seen, the effect of taking only prime discriminants d has a large consequence on the 2-divisibility of $S(E_d)$. This effect seems to disappear if we consider all discriminants $d < 0$ such that $\varepsilon(E_d) = 1$. For example, the density of the fundamental discriminants $-10^8 < d < 0$ such that $2 \mid S(E_d)$ is about 61.3%. We expect that the correct density is the one predicted by the heuristic but that the convergence is slow.

2- Using a 2-descent argument, it can be directly proved that the 2-parts of the Tate-Shafarevich groups $\text{III}(E_d)$ are all trivial and that the rank of E_d are all 0 whenever $d < 0$ runs through fundamental prime discriminants such that $\varepsilon(E_d) = 1$ ([Ant-Bun-Fre, exemple 1]). Hence, our results may also be seen as a check of the 2-part of the Birch and Swinnerton-Dyer conjecture for our family of quadratic twists.

3- Using also a 2-descent argument, one can obtain similar results for the case of "odd" quadratic twists of E by prime discriminants $d < 0$ ([De-Ro]): more precisely, if $d < 0$ is a prime discriminant such that $\varepsilon(E_d) = -1$ then we have $L'(E_d, 1) \neq 0$ and $2 \nmid |\text{III}(E_d)|$ (using a weak version of the Birch and Swinnerton-Dyer conjecture).

3   Generalization

Of course, our method can easily be adapted for many other examples (for instance $E =$"15a1", "21a1", "33a1"...). However, when the conductor N of the elliptic curve E is not prime, then the discriminants $d < 0$ should satisfy some more local conditions at the primes dividing N; indeed, if we want, for example, to apply the Kohnen-Zagier's theorem ([Koh-Zag]) for finding the weight $3/2$ modular form, we must have, for all prime $\ell \mid N$, $\left(\frac{d}{\ell}\right) = \varepsilon_\ell$, where ε_ℓ is the eigenvalue of the Atkin-Lehner operator at ℓ. For instance, if we take $E = 15a1$, we prove, using the same technics as above, that for all prime discriminants $d < 0$ such that $\left(\frac{d}{3}\right) = 1$ and $\left(\frac{d}{5}\right) = -1$ then $L(E_d, 1) \neq 0$ and $S(E_d)$ is not divisible by 2.

References

[Ant-Bun-Fre] J. A. Antoniadis M. Bungert and G. Frey, *Properties of twists of elliptic curves*, J. Reine Angew. Math. **405** (1990), 1–28.

[Bre et al.] C. Breuil, B. Conrad, F. Diamond and R. Taylor, *On the modularity of elliptic curves over $\mathbb{Q}$: wild 3-adic exercises*, J. Amer. Math. Soc. **14** (2001), no. 4, 843–939.

[BFH] D. Bump, S. Friedberg and J. Hoffstein, *Non vanishing theorem for L-functions of modular forms and their derivatives*, Invent. Math. **102** (1990), 543–618.

[CKRS] J. B. Conrey, J. P. Keating, M. O. Rubinstein and N. C. Snaith, *On the frequency of vanishing of quadratic twists of modular L-functions*, Number theory for the millennium, I (Urbana, IL, 2000), 301–315, A K Peters, Natick, MA, 2002.

[CKRS2] J. B. Conrey, J. P. Keating, M. O. Rubinstein and N. C. Snaith, *Random Matrix Theory and the Fourier Coefficients of Half-Integral Weight Forms*, Experimental Mathematics **15** (2006), arXiv:math.nt/0412083.

[Cre] J. Cremona, Elliptic curves data, available on http://www.maths.nott.ac.uk/personal/jec/ftp/data/INDEX.html.

[Del] C. Delaunay, *Heuristics on Tate-Shafarevitch groups of elliptic curves defined over $\mathbb{Q}$*, Exp. Math. **10** (2001), no. 2, 191–196.

[De-Ro] C. Delaunay and X. -F. Roblot, work in progress.

[Koh-Zag] W. Kohnen and D. Zagier, *Values of L-series of modular forms at the center of the critical strip*, Invent. Math. **64** (1981), 173–198.

[Kol] V. A. Kolyvagin, *Finiteness of $E(\mathbb{Q})$ and $III(E/\mathbb{Q})$ for a subclass of Weil curves*, (Russian) Izd. Akad. Nauk. Ser. Mat **52** (1988) no.6, 1154–1180; translation in Math USSR Izd. **33** no. 3 (1989), 473–499.

[Mu-Mu] M. R. Murty and V. K. Murty, *Mean values of derivatives of modular L-series*, Annals of Math. **133** (1991), 447–475.

[Tay-Wil] R. Taylor and A. Wiles, *Ring-theoretic properties of certain Hecke algebras*, Ann. of Math. (2)**141** (1995), no. 3, 553–572.

[Rub] M. Rubinstein, *Numerical data*, available at
www.math.uwaterloo.ca/~mrubinst/L_function/VALUES/DEGREE_2/
ELLIPTIC/QUADRATIC_TWISTS/WEIGHT_THREE_HALVES/

[Sil] J. Silverman, *The arithmetic of elliptic curves*, Graduate texts in Math. **106**, Springer-Verlag, New-York, (1986).

[Tor] G. Tornaría, *Numerical data*, available at
http://www.math.utexas.edu/users/tornaria/cnt/

[Wal] J.-L. Waldspurger, *Sur les coefficients de Fourier des formes modulaires de poids demi-entier*, J. Math. Pures Appl. (9) **60** (1981), pp. 375–484.

[Wil] A. Wiles, *Modular elliptic curves and Fermat's last theorem*, Ann. of Math. (2) **141** (1995), no.3, 443–551.

Institut Camille Jordan
Université Claude Bernard Lyon 1
43, avenue du 11 novembre 1918
69622 Villeurbanne Cedex - France

Discretisation for odd quadratic twists

*J. Brian Conrey, Michael O. Rubinstein, Nina C. Snaith
and Mark Watkins*

Abstract

The discretisation problem for even quadratic twists is almost understood, with the main question now being how the arithmetic Delaunay
heuristic interacts with the analytic random matrix theory prediction.
The situation for odd quadratic twists is much more mysterious, as the
height of a point enters the picture, which does not necessarily take
integral values (as does the order of the Shafarevich-Tate group). We
discuss a couple of models and present data on this question.

1   Introduction

Let $E : y^2 = x^3 + Ax + B$ be a fixed rational elliptic curve, and consider the
sets $S^+(X)$ and $S^-(X)$ of quadratic twists of E that contain respectively the
even[1] and odd twists $E_d : dy^2 = x^3 + Ax + B$ with $|d| < X$ a fundamental
discriminant. For even twists, the Birch–Swinnerton-Dyer conjecture [BSD]
states that

$$L(E_d, 1) = \Omega_d \frac{g_d \cdot \#\text{III}_d}{T_d^2}$$

where Ω_d is the real period, g_d is the global Tamagawa number, III_d is the
Shafarevich-Tate group,[2] and T_d is the order of the torsion subgroup, all of
these quantities being with respect to the quadratic twist E_d. Random matrix
theory applied with orthogonal symmetry [CKRS] predicts that

$$\text{Prob}\big[L(E_d, 1) \leq x\big] \approx x^{1/2}(\log|d|)^{3/8} \quad \text{as } x \to 0, \tag{1.1}$$

where we use the $\approx$ notation to indicate that the quotient of the two sides
tends to an unspecified constant that depends on E. Since $\#\text{III}_d$ is a square
while g_d and T_d are well-understood integers, we get a discretisation from (1.1)
— we expect that $L(E_d, 1) = 0$ if, say, we have that $L(E_d, 1) < g_d\Omega_d/T_d^2$.
Because Ω_d essentially acts like $\approx 1/\sqrt{|d|}$, this gives a rough prediction that

$$\text{Prob}\big[L(E_d, 1) = 0\big] \approx (\log|d|)^C/|d|^{1/4}$$

---

[1] A twist is even if the order of vanishing of its L-function at $s = 1$ (that is, its analytic
rank) is even, which is the same as saying that the sign of its functional equation is $+1$;
similarly for odd twists.

[2] We allow the order to be zero, in which case we suspect a curve of higher rank.

as $|d| \to \infty$, where the constant C is well-understood, largely dependent on the rational 2-torsion structure of E. Finally, these heuristics lead to a conjecture about the number of positive rank twists in $S^+(X)$, namely that there should be about $\approx X^{3/4}(\log X)^C$ of them as $X \to \infty$.

The situation is somewhat different for odd twists; here we have that $L(E_d, 1) = 0$ from the functional equation, and now the BSD conjecture takes into account the regulator R_d:

$$L'(E_d, 1) = \Omega_d \frac{g_d \cdot R_d \# \text{III}_d}{T_d^2}.$$

This regulator is rather mysterious, and, as in the case of regulators and class numbers for real quadratic fields, does not seem totally disjoint from the Shafarevich-Tate group. The heuristic of Delaunay [D] gives some idea of how we might expect $\#\text{III}$ to be distributed, but for the regulator we have only the lower bound of size $c \log |d|$ of Silverman [Si] and a conjectured upper bound[3] of $|d|^{1/2+\varepsilon}$ of Lang [L].

Also, the analogue of (1.1) has a different exponent; we have[4]

$$\text{Prob}\left[L'(E_d, 1) \leq x\right] \approx x^{3/2}(\log |d|)^{3/8} \quad \text{as } x \to 0. \tag{1.2}$$

In analogy with the class number problem[5] we might be so bold as to guess that $R_d \# \text{III}_d$ is always large if nonzero, say as big as $|d|^{1/2-\varepsilon}$. Since Ω_d acts like $\approx 1/\sqrt{|d|}$, this then implies that $L'(E_d, 1) \gg 1/|d|^\varepsilon$. More generally, we might guess that

$$\text{¿} \quad L'(E_d, 1) \gg 1/|d|^\theta \qquad \text{for curves of analytic rank 1} \quad \text{?} \tag{1.3}$$

This, our early suspicion, has not been confirmed by experiment (see Section 4), but we do not yet totally discount the possibility. A more formalised discussion could include the following. For $L(E_d, 1)$, we used a sharp-cutoff, but this might not be so applicable for curves of positive rank, due to the fungibility of the regulator. We can re-write the rank 0 case as follows: first let $D_0(t)$ be the density function such that $\int_0^t D_0(t)\, dt = \text{Prob}\left[L(E_d, 1) \leq t\right]$; then our above discretisation just says that $\text{Prob}\left[L(E_d, 1) = 0\right] = \int_0^\infty D_0(t)\chi(t)\, dt$ where $\chi(t)$ is the characteristic function[6] of the interval $\left[0, g_d\Omega_d/T_d^2\right]$. In the rank 1 case, we similarly have a distribution $D_1(t)$ which integrates to the probability, and so we might guess that

$$\text{Prob}\left[L'(E_d, 1) = 0\right] = \int_0^\infty D_1(t)F(t)\, dt \tag{1.4}$$

---

[3] Assuming BSD and GRH we essentially get Lang's conjecture; in place of GRH, by bounding $L'(E_d, 1)$ via convexity, we get a crude upper bound of $|d|^{1+\varepsilon}$.

[4] The exponent on the logarithm is $-r^2/2 + r/2 + 3/8$, where r is the order of the zero enforced at $s = 1$; see [Sn1] for the general case, and [Sn2] for the case $r = 1$.

[5] Note, however, that our L-values are at the center of the critical strip, while those for the class number problem are at the edge.

[6] Although some have expressed doubt that such a sharp cutoff is correct in the rank 0 case, we have no evidence either way; it is thought that this should only affect the leading constant in the heuristic.

for some[7] "cutoff" function $F(t)$. The condition of (1.3) could have $F(t)$ be the characteristic function of $\left[0, \left(g_d \Omega_d / T_d^2\right) \cdot d^{1/2-\theta}\right]$. For simplicity, we shall present our heuristics in terms of (1.3), though all of our discussion could use the more general formulation of (1.4).

From (1.3), in analogy with the above argument for the rank 0 case (and ignoring logarithmic factors) we obtain that as $|d| \to \infty$ we have

$$\operatorname{Prob}\left[L'(E_d, 1) = 0\right] \approx 1/|d|^{3\theta/2},$$

so that the number of twists of rank greater than 1 should be about $X^{1-3\theta/2}$ as $X \to \infty$. Note that the only provable (assuming BSD) bound is (essentially) that $L'(E_d, 1) \gg 1/\sqrt{|d|}$, which would lead to a prediction of only $X^{1/4}$ odd twists of rank greater than 1. However, for an infinite family of curves E and under the assumption of the Parity Conjecture, Rubin and Silverberg [RS][RS2, §8.2] can prove that there are $\gg X^{1/3}$ twists of rank at least 3.

The above conjecture (1.3) would imply that R_d and III_d are linked in a mysterious way; if we have a generator of small height (so that R_d is small), then this tends to make III_d be larger than general. The constructions of Rubin and Silverberg by their very nature yield points that are of height that is polynomial in $\log|d|$ — indeed, almost any parametrised family will have this feature, as writing down points of larger height is not feasible. These facts together suggest that by taking families with small generators we can generate large values of III. However, this does not work quite so simply in practise — we do get large values of III, but not always (as we will see in Section 4). This is one of the reasons why we might suggest the "statistical" version (1.4) with a more general cutoff function F rather than simply (1.3).

2   A model from Heegner points (largely due to Birch)

Suppose that E has rank zero and $d < 0$ is a fundamental discriminant that is a square modulo $4N$, where N is the conductor of E, and also assume for simplicity that $\gcd(d, 6N) = 1$. By work of Gross and Zagier [GZ], we have a construction for a point P_d on E_d that gives a torsion point precisely when the rank of E_d is greater than 1; indeed, the height λ of the constructed point is proportional to $L'(E_d, 1)$:

$$\lambda(P_d) = \frac{\sqrt{|d|}}{4\Omega_{\mathrm{vol}}} L(E, 1) L'(E_d, 1),$$

---

[7]Such a general cutoff function could be rephrased, for instance, as predicting that (1.3) holds except for a proportionately small number of exceptions. As an example, taking $F(t) \equiv 1$ on $\left[0, 1/\sqrt{d}\right]$ and $F(t) = 1/t\sqrt{d}$ for $\left[1/\sqrt{d}, 1\right]$ and $F(t) \equiv 0$ for $t > 1$, we get $\operatorname{Prob}\left[L'(E_d, 1) = 0\right] \approx 1/d^{1/2}$, and so $\theta = 1/3$ in this statistical sense. Our cluelessness about F stems from our inability to model heights of points.

where here Ω_{vol} is the area of the fundamental parallelogram associated to a minimal model for E. When the rank of E_d is 1, the point P_d has infinite order but is not in general a generator of the free part of the group of rational points; the index of P_d depends on $\#\mathrm{III}_d$, but cancels out in the end.

The construction of the point P_d goes via class-field theory; we get a point U_d over the Hilbert class field via a complex multiplication result largely due to Shimura, and then sum the conjugates to get a point first in the imaginary quadratic field $\mathbf{Q}(\sqrt{d})$ and then in $\mathbf{Q}$ itself. The number of conjugates of U_d in the Hilbert class field is essentially the class number h of $\mathbf{Q}(\sqrt{d})$. These points, all being conjugate, have the same height. To get the height of the resulting point in $\mathbf{Q}$, we model the situation by assuming that we are summing h unit vectors in h-dimensional space; this leads to the prediction that the height is almost surely close to h, which is of size $\sqrt{|d|}$. If we assume the height of U_d is not too small we then get a prediction of $L'(E_d, 1) \gg 1/|d|^\varepsilon$, leading to about $X^{1-\varepsilon}$ twists in $S^-(X)$ which have rank 3 or greater. However, it is not clear why the height of U_d might not be of size $1/|d|^C$ itself, as its coordinates are in a field whose degree is of size $\sqrt{|d|}$.

We can try to test the validity of this model by taking d with $L'(E_d, 1)$ small and then computing the height of the point U_d in the Hilbert class field. However, when the class field has large degree (that is, when the class number is large), it will be difficult to recognise the coordinates of U_d, so we cannot take $|d|$ too large here. We were thus unable to generate enough examples to perform any real test of the model.[8]

3   Alternative ideas

A less profound idea is to assert that the connection between rank 1 and rank 3 twists should be the same as the connection between rank 0 and rank 2 twists, to a first approximation. Heuristics and random matrix theory [CKRS] give $X^{3/4+\varepsilon}$ rank 2 curves amongst even quadratic twists up to X. If we thus guess that there about $X^{3/4}$ twists of rank 3 up to X, via reverse-engineering the argument of two sections previous, this could then be used to determine a value of $\theta = 1/6$.

We note that there are two random matrix models that have been proposed for modeling the zeros of L-functions associated with elliptic curves. The prediction (1.2) of Snaith [Sn1] is extended to higher ranks by looking at a zero-dimensional subset of $SO(\text{even})$ (for even twists) or $SO(\text{odd})$ (for

---

[8]In upcoming work with S. R. Donnelly, the fourth-named author has instead fixed d (of small class number) and varied over E, to attempt to test the model. Also, as this model predicts something (up to guessing the height of U_d) about the distribution for the height of P_d, it thus also indicates something about the $L'(E_d, 1)$ distribution; our use of the Heegner point model to choose the cutoff for (1.3) does not immediately seem more justified than reversing this interaction and using (1.3) to choose the cutoff for the distribution given by the model here.

odd twists) with r eigenvalues conditioned to lie at 1. This model predicts $\text{Prob}[L^{(r)}(E_d, 1) \leqslant x] \approx x^{r+1/2}(\log x)^{-r^2/2+r/2+3/8}$. In contrast, Miller [M2] proposes what is called the Independent Model, with eigenvalue distribution decomposing as a sum of $(2\lfloor r/2 \rfloor + 1)$ point-masses and the eigenvalue distribution of the symmetry group $SO(\text{even})$ or $SO(\text{odd})$. In this case the rth derivative analogue of (1.1) and (1.2) is given by (1.1) for $SO(\text{even})$ symmetry and (1.2) for $SO(\text{odd})$ symmetry. There is both theoretical evidence [M1, Y] and numerical data [M2] that the 1- and 2-level densities of zeros follow Miller's Independent Model for L-functions associated with parameterised families of elliptic curves with r constructed points that generate (the infinite part of) the Mordell-Weil group. But there is no evidence to suggest that the Miller model should hold for quadratic twists, and in fact the exponent $3/2$ in (1.2) is supported by the shape of the value distribution of $L'(E_d, 1)$ (see Figure 4.1) as well as by the results in Section 4.1. This illustrates that for odd twists the zero of $L(E_d, s)$ at $s = 1$ is apparently not independent — in contrast to a case of Young's [Y] where the zero came from a constructed rational point on the elliptic curve.

Finally there is a model due to A. Granville. Let E be a fixed elliptic curve given by the model $y^2 = x^3 + Ax + B$. Here we make a heuristic for the number of integral points (d, u, v, w) with $dw^2 = v(u^3 + Auv^2 + Bv^3)$ and $D < |d| < 2D$ and $X < |u|, |v| < 2X$. There are about $\approx X^2$ such (u, v)-pairs, and each leads to a right-hand side which is of size X^4. The number of integers that are of size X^4 and are d times a square with $D < |d| < 2D$ is $\approx D\sqrt{X^4/D}$, and thus the probability that an integer of size X^4 is of this form is $\approx \sqrt{DX^4}/X^4$. Multiplying this by our $\approx X^2$ possibilities for (u, v), we get a total of $\approx \sqrt{D}$ integral solutions, independent of X. Summing this dyadically over X, we get $\approx \sqrt{D}\log Y$ total solutions up to Y, and switching to logarithmic heights, we get[9] that the number of points of height less than H on the first D twists of E is $\approx H\sqrt{D}$. We then note (under GRH) that E_d has regulator at most size $|d|^{1/2+\varepsilon}$; if E_d is of rank 3, since a random 3-dimensional lattice of this covolume should have a vector whose length is of size $(|d|^{1/2+\varepsilon})^{1/3}$, we then expect a point of height less than $|d|^{1/6+\varepsilon}$ on E_d. From the above with $H = |d|^{1/6+\varepsilon}$, we expect no more than about $D^{1/2+1/6+\varepsilon}$ such twists up to D. Note that this only counts generators on E_d, and not all rational points on the twisted curve; up to height h, a rank r curve should (asymptotically) have $h^{r/2}/\sqrt{R_d}$ rational points. With $h = R_d \approx |d|^{1/2}$ and $r = 3$, we get $|d|^{1/2}$ points up to height $|d|^{1/2}$ on each rank 3 twist, and taking $H = |d|^{1/2}$ in the above accounting, we get no more than $D^{1/2+\varepsilon}$ rank 3 twists up to D.

The prediction of $\approx H\sqrt{D}$ such (d, u, v, w)-tuples can be proved via a sieve argument for small H, but is more dubious for large H. Indeed, the above-noted count of $h^{r/2}/\sqrt{R}$ rational points up to height h on a rank r curve

---

[9]This is a simplification; Granville uses congruences rather than this crude probabilistic method, and gets $H^\eta\sqrt{D}$ where η is the number of rational factors of the cubic polynomial $x^3 + Ax + B$. So our case is when the cubic is irreducible.

outgrows the linear prediction (as $H \to \infty$) with $r = 3$. However, we only need H to be a small power of D, and it is unclear how far the heuristic can be pushed.

4   Data

We now give tables and graphs that concern the above heuristics and conjectures. In our first graph (Figure 4.1), we plot the L' values for odd twists of $X_0(11)$ with $|d| < 10^6$. We are most concerned with the behaviour as $L' \to 0$, so we zoom in on this point; there are about 300000 total curves, of which 760 have $L' = 0$.

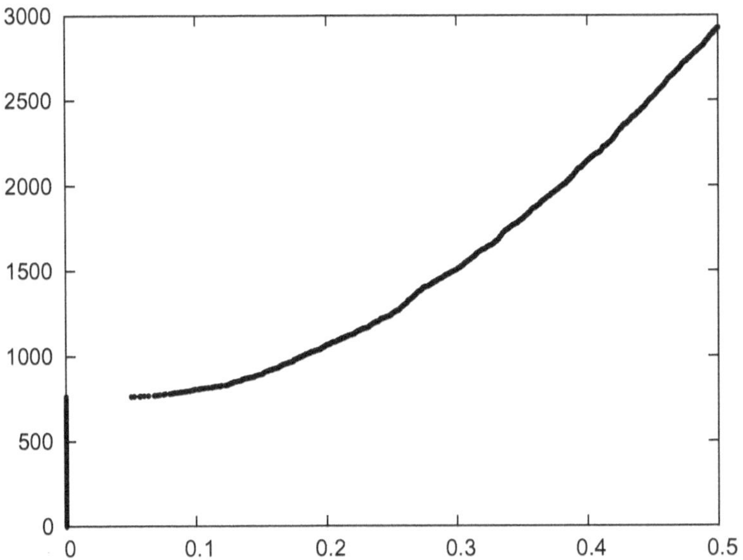

Figure 4.1: Cumulative L'-distribution for odd twists of $X_0(11)$ for $|d| < 10^6$.

From this graph, it looks as though there is an abrupt cutoff. We find that the smallest nonzero value of $L'(E_d, 1)$ is about 0.051 for $d = 477121$. However, it might be superior to look at the distribution of $L'(E_d, 1)/\big(\log|d|\big)$, as the average value of $L'(E_d, 1)$ is proportional to $\log|d|$ (see [BFH, I, MM]). This changes the picture quantitatively (see Figure 4.2), as the gap size becomes comparable to that of the L-distribution at the top of the graph.[10]

We compare the situation between even and odd twists. For $|d| < 10^6$ there are about 30 times more even twists with $L(E_d, 1) = 0$ than odd twists with $L'(E_d, 1) = 0$; however this factor of 30 is dependent on our cutoff of 10^6, and as we note below, it is not clear what happens asymptotically. If we

---

[10]It can be noted that $\log|d|$ is about size $|d|^{1/6}$ for our d, and thus it becomes difficult to distinguish in our data between a logarithm and a power of d.

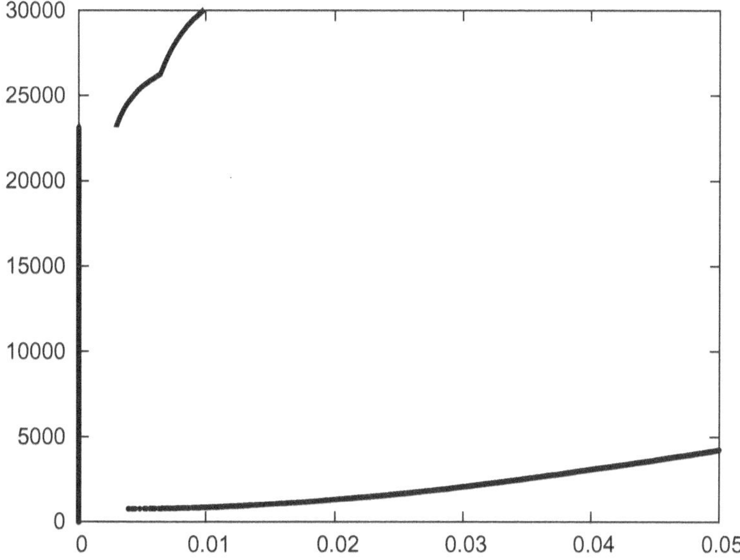

Figure 4.2: Cumulative distributions for L (top) andnormalised L' for $|d| < 10^6$.

restricted our range of d to a shorter interval, say $9 \cdot 10^5 < |d| < 10^6$, then the upper graph of L-values would be close to steplike, since the size of d is the only continuous variable in the BSD formula. However, the lower graph would still be rather smooth, as the regulator cannot be modelled as a discrete variable in the rank 1 case.

We let $S_0^-(X)$ be the subset of $S^-(X)$ with $L'(E_d, 1) = 0$, suppose that $\#S_0^-(X) \sim cX^A(\log X)^B$ and try to fit the data to get the exponent A. For $X_0(11)$ there are 760 odd twists with $L' = 0$ with $|d| < 10^6$. The best-fit exponent is $A = 0.86$, though if we just look at the last 380 curves, we get $A = 0.82$. The computations of Elkies[11] [E] for $X_0(32)$ go up to 10^7, and give $A = 0.84$ overall with $A = 0.80$ for the last half of the data; of course, we are ignoring log-factors. For $X_0(14)$ we get $A = 0.94$ and for $X_0(15)$ we get $A = 0.95$. These might seem large, but Elkies has $A = 0.93$ at 10^6 before it drops to $A = 0.86$ at 10^7. Also, since $X_0(14)$, $X_0(15)$, and $X_0(32)$ all have nontrivial 2-torsion while $X_0(11)$ does not, we might expect the exponent of the logarithm to be larger for them, which could lead to a larger observed value of A across the range of our dataset. For comparison with the even twist case, the dataset of Rubinstein [R] for the number of rank 2 imaginary quadratic twists of $X_0(11)$ has best-fit exponents of about $0.89, 0.86, 0.84$ up to $10^6, 10^7, 10^8$, while we expect the exponent to be 0.75.

To get a dataset of twists with points of small height, we looked at the dth twist of $y^2 = x^3 - 1$ for $d = t^3 - 1$; the curve $dy^2 = x^3 - 1$ will have the point $(t, 1)$

---

[11]He divides even fundamental discriminants by 4, so has different curve counts.

whose height is of size of $\log d$. As mentioned above, if (1.3) holds, we would expect such curves to have large values of $\#\text{III}_d$. Though we get some large examples like $t = 624$ and $d = 242970623$ for which $\#\text{III}_d = 47^2$, this idea does not always work so well. For instance, with $t = 810$ and $d = 531440999$ we have $\#\text{III}_d = 1$, where here we have $L'(E_d, 1) \approx 0.0315$; similarly $t = 902$ and $d = 733870807$ has $\#\text{III}_d = 1$, though in this case $L'(E_d, 1) \approx 0.0546$ is not quite so small. Relatedly, the results of Delaunay and Duquesne [DD] for curves connected to the simplest cubic fields show $\#\text{III} = 1$ to occur often.

More extensive experiments with techniques similar to those of Elkies are planned — indeed, it would be nice to have data for the odd twists comparable to that which [CKRS2] has for even twists. Our experiments for odd twists have simply computed the value of L' for every twist up to X and so takes X^2 total time, while the method of Elkies takes $X^{3/2}$ time, as does[12] the computation of [CKRS2].

4.1   Quadratic twists in arithmetic progressions

We note that the computations of Elkies [E] already give indirect evidence that (1.2) is probably correct. While Elkies notes a strange discrepancy in the counts E_d with rank 3 for d modulo 16, in fact, as explained in the last section of [CKRS], we expect such discrepancies for all (prime) moduli p whose Frobenius trace a_p is nonzero. In particular, of the d with $E_d \in S_0^-(X)$ we expect that the number of nonzero quadratic residues mod p is not the same as the number of quadratic nonresidues. The derivation in [CKRS] gives a ratio of $\left(\frac{p+1+a_p}{p+1-a_p}\right)^k$ where the exponent $k = -1/2$ is taken to be the rightmost pole of the distribution function; in the rank 1 case, the corresponding calculation of [Sn1] implies that we should take $k = -3/2$. This is a reasonably testable prediction, given that the dataset of Elkies has 8740 curves. In Table 1 we give the results for some primes that are 1 mod 4; since $a_p = 0$ for other odd primes the ratio should be 1, and indeed it is always quite close. Here the R and N columns count the d for which E_d has rank 3 and d is respectively a nonzero quadratic residue and a quadratic nonresidue mod p, while the E column calculates their experimentally-determined ratio, and C is the conjectured ratio from the above with $k = -3/2$.

Note that the fit is not as tight for small primes; indeed this also shows up in the even rank case, even when accounting for a secondary term as in [CPRW]. Given our dataset size, the confidence interval width for the experimental value is about 0.1 across most of our data range. If we take all the primes up to 1000 and do a fit for the best k, we get a result of -1.41, which is reasonably close to our expected value of $-3/2$. This gives us a modicum of confidence that (1.2) is correct; we hope a consideration of the secondary term will give an even better fit.

---

[12]With convolution techniques this reduces to essentially linear time, which is one reason why we seek to improve on [E] via p-adic computations and Θ-series.

Table 1: Residuosity effects in arithmetic progressionsfor rank 3 quadratic twists for the congruentnumber curve (data from Elkies)

p	R	N	E	C
5	4240	1951	2.17	2.83
13	1827	5580	0.33	0.25
17	3186	4197	0.76	0.72
29	5873	2249	2.61	2.83
37	4451	3820	1.17	1.17
41	2711	5411	0.50	0.48
53	2672	5723	0.47	0.45
61	5239	3245	1.61	1.63
73	4696	3688	1.27	1.28
89	3648	4828	0.76	0.72
97	2958	5526	0.54	0.57
929	4836	3876	1.25	1.16
937	4679	4035	1.16	1.13
941	4807	3922	1.23	1.20
953	4196	4524	0.93	0.92
977	4791	3929	1.22	1.21
997	4019	4712	0.85	0.83

4.2   Beyond twists

To go further, we can look at generic elliptic curves (rather than just twists); for this the database of Stein and Watkins [SW] is useful. We suspect a bound like $L'(E, 1) \gg 1/|\Delta|^{\theta/6}$ in analogy with the prediction (1.3) of $L'(E_d, 1) \gg 1/|d|^{\theta}$ for quadratic twists.[13]   However, as above, we really have no idea how to generate a good value of θ. The Stein-Watkins database (ECDB) has 11372286 curves of prime conductor less than 10^{10} (we make the choice of prime conductor so as to exclude twists from our data; looking at other curves does not change the result too much), of which 5253162 have analytic rank 1. The minimal L'-value for these curves is about 0.193 for the curve[14] $[0, 0, 1, -76931443, -259719125220]$ of conductor 8519438341. We get[15] 423944

---

[13]This analogy comes from the fact that the discriminant grows like d^6 in quadratic families, and our impression is that the discriminant is better than the conductor as a measure of the likelihood that the L-derivative vanishes. Actually we might suspect the real period to be the most significant datum in general, but it should be approximately $|\Delta|^{1/12}$ up to log-factors. In any case, considering the conductor is more difficult, even with the ABC conjecture.

[14]Here and below a curve $y^2 + a_1xy + a_3y = x^3 + a_2x^2 + a_4x + a_6$ is denoted by $[a_1, a_2, a_3, a_4, a_6]$.

[15]The usual caveats about not being able to prove that a curve actually has analytic rank r when $r \geq 4$ apply here.

curves of analytic rank 3, and 1296 of analytic rank 5. In Figure 4.3 we again see fewer curves with small normalised L'-value with the normalised gap for L' about as big as that for L.

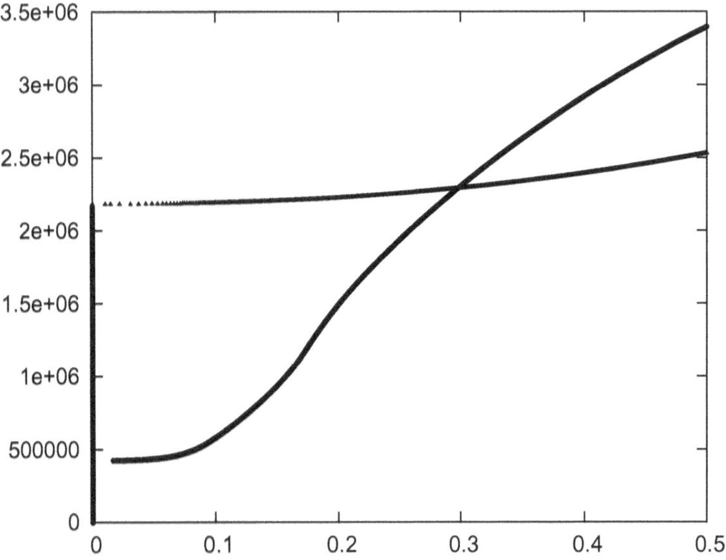

Figure 4.3: Cumulative L and normalised-L' distributionfor ECDB curves. The plot going from thelower-left to the upper-right is that for L'.

It was noted to us by N. D. Elkies that the small values of L' correspond to curves with large cancellation between c_4^3 and c_6^2. See Table 2 for the smallest values of L' in the database. For the even rank case, the smallest 85 L-values all come from Neumann-Setzer [N, Se] curves (with conductor of the form u^2+64), with the next smallest coming from $[1, 1, 1, -2413424773, -45636080008772]$ of conductor 6375846313; these thus similarly exhibit large cancellation between c_4^3 and c_6^2. Indeed, many of the curves come from families similar to those investigated by Delaunay and Duquesne [DD].

Following a suggestion of A. Venkatesh, we might speculate whether all the small L' values (possibly including $L' = 0$) essentially come from a small number of parametrised families. However, we can make a heuristical argument against the analogous claim that all rank 2 curves should come from parametrised families. A heuristic of Watkins [W] gives that there should be at least $X^{19/24-\varepsilon}$ curves of analytic rank 2 with conductor less than X, whereas we expect[16] there only to be about $X^{2/3+\varepsilon}$ curves with two small generators.

We can go to curves of larger rank and look at the distribution of $L''(E, 1)/2!$

---

[16]This type of heuristic appears (though not explicitly) in the work of Elkies and Watkins [EW]. They only consider small generators that are integral, but by passing to rationality we only lose logarithmic factors.

Table 2: Small L'-values for prime conductor curves in the Stein-Watkins ECDB

L'	conductor	equation
0.193	8519438341	$[0, 0, 1, -76931443, -259719125220]$
0.217	8072290789	$[0, -1, 1, -168735150, 843694875000]$
0.218	7807742161	$[1, 0, 0, -162115427, 794469530026]$
0.219	7598316169	$[1, -1, 1, -157763487, 762746660718]$
0.219	972431659	$[1, -1, 0, -42359524, -106103907983]$
0.220	7344220789	$[1, -1, 1, -153528564, 732242039802]$
0.225	6436262197	$[1, -1, 1, -133616676, 594515948970]$
0.226	6347138731	$[0, 1, 1, -131764782, 582122479302]$
0.226	2829273949	$[1, -1, 1, -119862711, -505066414494]$
0.229	5907969559	$[1, -1, 1, -122639979, 522783273972]$

and $L'''(E, 1)/3!$ for curves of (analytic) rank 2 and 3 in the database. If we ignore various examples of small conductor, the smallest value of $L''(E, 1)/2!$ for a curve of larger conductor is about 1.554 for the curve $[0, 0, 1, -2664919573, -529951013063110]$ of conductor 6264757621, where again we see the large cancellation between c_4^3 and c_6^2. For rank 3 the smallest value of $L'''(E, 1)/3!$ for curves of larger conductor is about 8.089 for the curve $[0, 0, 1, -7990342, 8693530176]$ whose conductor is 1531408357. Though there is large cancellation between c_4^3 and c_6^2 here, it is not as noticeable as in the cases above; however, the large cancellation appears again for the next-notable curve $[0, 0, 1, -217363231, 1233466148550]$ of conductor 6352778197 for which $L'''(E, 1)/3! \approx 8.24$. As noted above, it is better to divide the $L^{(r)}$-values through by the expected average value, which is proportional to $(\log N)^r$, before making these comparisons; upon doing this, the listed curves of conductor 6264757621 and conductor 6352778197 have the smallest respective values.

5 Conclusion

Via analogies with random matrix theory, we have given a link (as in the case of rank 2 quadratic twists) between the distribution of L'-values and the number of rank 3 quadratic twists, but are unable to gain much insight into solving the discretisation problem. Although we might expect a smooth distribution function for $L'(E_d, 1)$ (especially as it is an analytic and not an arithmetic object), there is some evidence of a rather abrupt cutoff in the distribution. This has led some of the authors of this paper to conjecture (1.3) in a universal form, while others remain more skeptical and propose something along the lines (with a smooth cutoff function) of (1.4) instead.[17] We have also

---

[17] It may be noted that (1.3) has been referred to as the "Saturday Night Conjecture" due its formulation on a Saturday night at the Isaac Newton Institute.

discussed various methods for modelling the number of rank 3 quadratic twists of a given elliptic curve. However, currently we do not have enough data to feel confident in eliminating any of the suggestions.

6   Acknowledgments

This work was largely done during the "Random Matrix Approaches in Number Theory" programme at the Isaac Newton Institute (INI), whose hospitality we enjoyed. The first author was partially supported by Focused Research Group grant DMS-0244660 from the National Science Foundation (NSF/USA), the second author by NSF grant DMS-0138597 and a Natural Sciences and Engineering Research Council grant (Canada), the third author by a Dorothy Hodgkin Fellowship from the Royal Society and Engineering and Physical Sciences Research Council (EPSRC/UK) grants GR/T00825/01 and GR/T00832/01, and the fourth author by an INI Fellowship (EPSRC grant GR/N09176/01) and EPSRC grant GR/T00658/01.

References

[BSD]   B. J. Birch, H. P. F. Swinnerton-Dyer, *Notes on elliptic curves. I. II.* J. reine angew. Math. **212** (1963), 7–25, **218** (1965), 79–108.

[BFH]   D. Bump, S. Friedberg, J. Hoffstein, *Nonvanishing theorems for L-functions of modular forms and their derivatives.* Invent. Math. **102** (1990), no. 3, 543–618.

[CKRS]   J. B. Conrey, J. P. Keating, M. O. Rubinstein, N. C. Snaith, *On the frequency of vanishing of quadratic twists of modular L-functions.* In *Number theory for the millennium, I* (Urbana, IL, 2000), edited by M. A. Bennett, B. C. Berndt, N. Boston, H. G. Diamond, A. J. Hildebrand and W. Philipp, A K Peters, Natick, MA (2002), 301–315. Available online at `arxiv.org/math.NT/0012043`

[CKRS2]   J. B. Conrey, J. P. Keating, M. O. Rubinstein, N. C. Snaith, *Random Matrix Theory and the Fourier Coefficients of Half-Integral Weight Forms.* Available online at `arxiv.org/math.NT/0412083`

[CPRW]   J. B. Conrey, A. Pokharel, M. O. Rubinstein, M. Watkins, *Secondary terms in the number of vanishings of quadratic twists of elliptic curve L-functions.* In this volume. Also available online at `arxiv.org/math.NT/0509059`

[D]   C. Delaunay, *Heuristics on Tate-Shafarevitch Groups of Elliptic Curves Defined over* **Q**. Experiment. Math. **10** (2001), no. 2, 191–196.

[DD]   C. Delaunay, S. Duquesne, *Numerical investigations related to the derivatives of the L-series of certain elliptic curves.* Experiment. Math. **12** (2003), no. 3, 311–317.

[E]   N. D. Elkies, *Heegner point computations.* In *Algorithmic Number Theory*, Proceedings of the First International Symposium (ANTS-I) held at Cornell University, Ithaca, New York, May 6–9, 1994. Edited by L. M. Adleman and M.-D. Huang. Lecture Notes in Computer Science, **877**. Springer-Verlag, Berlin (1994), 122–133.
N. D. Elkies, *Curves $Dy^2 = x^3 - x$ of odd analytic rank.* In *Algorithmic number theory*, Proceedings of the 5th International Symposium (ANTS-V) held at the University of Sydney, Sydney, July 7–12, 2002. Edited by C. Fieker and D. R. Kohel. Lecture Notes in Computer Science, **2369**. Springer-Verlag, Berlin (2002), 244–251. Available online at `arxiv.org/math.NT/0208056`

[EW]   N. D. Elkies, M. Watkins, *Elliptic curves of large rank and small conductor.* In *Algorithmic number theory,* Proceedings of the 6th International Symposium (ANTS-VI) held at the University of Vermont, Burlington, VT, June 13–18, 2004. Edited by D. Buell. Lecture Notes in Computer Science, **3076**. Springer-Verlag, Berlin (2004), 42–56. Available online at `arxiv.org/math.NT/0403374`

[GZ]   B. H. Gross, D. B. Zagier, *Heegner points and derivatives of L-series.* Invent. Math. **84** (1986), no. 2, 225–320.

[I]   H. Iwaniec, *On the order of vanishing of modular L-functions at the critical point.* Sém. Théor. Nombres Bordeaux (2) **2** (1990), no. 2, 365–376.

[L]   S. Lang, *Number theory. III. Diophantine geometry.* Encyclopaedia of Mathematical Sciences, **60**. Springer-Verlag, Berlin, 1991. xiv+296 pp.

[M1]   S. J. Miller, *One- and two-level densities for rational families of elliptic curves: evidence for the underlying group symmetries.* Compos. Math. **140** (2004), no. 4, 952–992. Preprint version: `arxiv.org/math.NT/0310159`

[M2]   S. J. Miller, *Investigations of Zeros Near the Central Point of Elliptic Curve L-functions.* Preprint, `arxiv.org/math.NT/0508150`

[MM]   M. R. Murty, V. K. Murty, *Mean values of derivatives of modular L-series.* Ann. of Math. (2) **133** (1991), no. 3, 447–475.

[N]   O. Neumann, *Elliptische Kurven mit vorgeschriebenem Reduktionsverhalten. I, II.* (German) [Elliptic curves with prescribed reduction behaviour]. Math. Nachr. **49** (1971), 107–123, **56** (1973), 269–280.

[RS]   K. Rubin, A. Silverberg, *Rank frequencies for quadratic twists of elliptic curves*, Exper. Math. **10** (2001), 559–569.

[RS2]   K. Rubin, A. Silverberg, *Ranks of elliptic curves.* Bull. Amer. Math. Soc. (N.S.) **39** (2002), no. 4, 455–474 (electronic), online at `www.ams.org/bull/2002-39-04/S0273-0979-02-00952-7/home.html`

[R]   M. O. Rubinstein, *Online data.* Currently available from `www.math.uwaterloo.ca/~mrubinst`

[Se]   B. Setzer, *Elliptic Curves of prime conductor.* J. London Math. Soc. (2), **10** (1975), 367–378.

[Si]   J. H. Silverman, *Lower bounds for height functions*. Duke Math. J. **51** (1984), no. 2, 395–403.

[Sn1]   N. C. Snaith, *Derivatives of random matrix characteristic polynomials with applications to elliptic curves*. J. Phys. A: Math. Gen. **38** (2005) 10345-10360, `arxiv.org/math.NT/0508256`

[Sn2]   N. C. Snaith, *The derivative of $SO(2N+1)$ characteristic polynomials and rank 3 elliptic curves*. In this volume.

[SW]   W. A. Stein, M. Watkins, *A Database of Elliptic Curves—First Report*. In *Algorithmic number theory*, Proceedings of the 5th International Symposium (ANTS-V) held at the University of Sydney, Sydney, July 7–12, 2002. Edited by C. Fieker and D. R. Kohel. Lecture Notes in Computer Science, **2369**. Springer-Verlag, Berlin (2002), 267–275.

[W]   M. Watkins, *Some heuristics about elliptic curves*. Draft, available from `www.maths.bris.ac.uk/~mamjw/heur.ps`

[Y]   M. P. Young, *Low-lying zeros of families of elliptic curves*. Preprint. `arxiv.org/math.NT/0406330`

J.B. Conrey
American Institute of Mathematics
360 Portage Avenue
Palo Alto, CA 94306
USA

J.B. Conrey, N.C. Snaith, M. Watkins School of Mathematics
University of Bristol
Bristol BS8 1TW
UK

M.O. Rubinstein
Pure Mathematics
University of Waterloo
200 University Ave W
Waterloo, ON, Canada
N2L 3G1

Secondary terms in the number of vanishings of quadratic twists of elliptic curve L-functions

J. Brian Conrey, Atul Pokharel, Michael O. Rubinstein and Mark Watkins

Abstract

We examine the number of vanishings of quadratic twists of the L-function associated to an elliptic curve. Applying a conjecture for the full asymptotics of the moments of critical L-values we obtain a conjecture for the first two terms in the ratio of the number of vanishings of twists sorted according to arithmetic progressions.

1 Introduction

Let E be an elliptic curve over $\mathbb{Q}$ with associated L-function given by

$$L_E(s) = \sum_{n=1}^{\infty} \frac{a_n}{n^s} = \prod_{p|\Delta} \left(1 - a_p p^{-s}\right)^{-1} \prod_{p\nmid\Delta} \left(1 - a_p p^{-s} + p^{1-2s}\right)^{-1} \qquad (1.1)$$

$$= \prod_p \mathcal{L}_p(1/p^s), \qquad \Re(s) > 3/2. \qquad (1.2)$$

Here, Δ is the discriminant of E, and $a_p = p + 1 - \#E(\mathbb{F}_p)$, with $\#E(\mathbb{F}_p)$ the number of points, including the point at infinity, of E over $\mathbb{F}_p$. $L_E(s)$ has analytic continuation to $\mathbb{C}$ and satisfies a functional equation [12] [11] [1] of the form

$$\left(\frac{2\pi}{\sqrt{Q}}\right)^{-s} \Gamma(s) L_E(s) = w_E \left(\frac{2\pi}{\sqrt{Q}}\right)^{s-2} \Gamma(2-s) L_E(2-s), \qquad (1.3)$$

where Q is the conductor of the elliptic curve E and $w_E = \pm 1$.

Let

$$L_E(s, \chi_d) = \sum_{n=1}^{\infty} \frac{a_n \chi_d(n)}{n^s} \qquad (1.4)$$

be the L-function of the elliptic curve E_d, the quadratic twist of E by the fundamental discriminant d. If $(d, Q) = 1$, then $L_E(s, \chi_d)$ satisfies the functional equation

$$\left(\frac{2\pi}{\sqrt{Q}|d|}\right)^{-s} \Gamma(s) L_E(s, \chi_d) = \chi_d(-Q) w_E \left(\frac{2\pi}{\sqrt{Q}|d|}\right)^{s-2} \Gamma(2-s) L_E(2-s, \chi_d). \qquad (1.5)$$

In [5] and [6] conjectures, modeled after corresponding theorems in random matrix theory, are stated concerning the distribution of values of $L_E(1, \chi_d)$ with an application made to counting the number of vanishings of $L_E(1, \chi_d)$. We focus on the case $w_E \chi_d(-Q) = 1$, since otherwise $L_E(1, \chi_d)$ is trivially equal to zero. One quantity studied concerns the ratio of the number of vanishings sorted according to residue classes mod q for a fixed prime $q \nmid Q$. Let

$$R_q(X) = \frac{\displaystyle\sum_{\substack{|d| < X, w_E \chi_d(-Q)=1 \\ L_E(1,\chi_d)=0 \\ \chi_d(q)=1}} 1}{\displaystyle\sum_{\substack{|d| < X, w_E \chi_d(-Q)=1 \\ L_E(1,\chi_d)=0 \\ \chi_d(q)=-1}} 1} \tag{1.6}$$

be the ratio of the number of vanishings of $L_E(1, \chi_d)$ sorted according to whether $\chi_d(q) = 1$ or -1.

By looking at this ratio, certain elusive and mysterious quantities that appear in the asymptotics for both the numerator and denominator cancel each other out and one is left with a precise prediction for its limit. Let

$$R_q = \left(\frac{q + 1 - a_q}{q + 1 + a_q} \right)^{1/2}. \tag{1.7}$$

A conjecture from [5] asserts that, for $q \nmid Q$,

$$\lim_{X \to \infty} R_q(X) = R_q. \tag{1.8}$$

It is believed that this continues to hold if the set of quadratic twists is restricted to subsets such as $d < 0$ or $d > 0$, or to $|d|$ prime, though in the latter case we must be sure to rule out there being no vanishings at all due to arithmetic reasons [7].

Numerical evidence for three elliptic curves is presented in [5] and confirms this prediction. However, even taking X of size roughly 10^9 (and, in that paper, $d < 0$ and $|d|$ prime), the numeric value of the ratio was found in that paper to agree with the predicted value to about two decimal places. In other cases, when a_q of $L_E(S)$ in (1.1) equals 0, the numeric value of $R_q(X)$ compared to the predicted limit R_q to three or more decimal places.

In this paper we examine secondary terms in the above conjecture applying new conjectures [4] for the full asymptotics of the moments of $L_E(1, \chi_d)$. We obtain a conjectural formula for the next to leading term in the asymptotics for $R_q(X)$. It is of size $O(1/\log(X))$ and explains the slow convergence to the limit R_q. We also explain in Section 3 the tighter fit when $a_q = 0$.

While the main term, R_q, in the above conjecture is robust and does not depend heavily on the set of d's considered, the secondary terms are more sensitive, for example, to the residue classes of d modulo the primes that divide Q. Therefore, for simplicity we focus on the following dense collection of fundamental discriminants d. Assume that Q is squarefree and let

$$S^-(X) = S_E^-(X) = \{-X \leqslant d < 0; \chi_d(p) = -a_p \text{ for all } p \mid Q\} \tag{1.9}$$

For curves of prime conductor Q we also consider the set of fundamental discriminants

$$S^+(X) = S_E^+(X) = \{0 < d \leqslant X; \chi_d(Q) = a_Q\}. \tag{1.10}$$

These sets of discriminants are also chosen because they allow us to efficiently compute $L_E(1, \chi_d)$ using a relationship to the coefficients of certain modular forms of weight $3/2$ that has been worked out explicitly for many examples by Tornaria and Rodiguez-Villegas [9] (see [6] for more details). The sets $S^\pm(X)$ restrict d according to certain residue classes mod Q in the case that Q is odd and squarefree, and $4Q$ in the case that Q is even and squarefree.

2 Moments of $L_E(1, \chi_d)$

Let

$$M_E^\pm(X, k) = \frac{1}{|S^\pm(X)|} \sum_{d \in S^\pm(X)} L_E(1, \chi_d)^k. \tag{2.1}$$

be the kth moment of $L_E(1, \chi_d)$.

The conjecture of Conrey-Farmer-Keating-Rubinstein-Snaith [4, 4.4] says here that, for $k \geqslant 1$, $k \in \mathbb{Z}$,

$$M_E^\pm(X, k) = \frac{1}{X} \int_0^X \Upsilon_k^\pm(\log(t)) \, dt + O(X^{-\frac{1}{2}+\varepsilon}) \tag{2.2}$$

as $X \to \infty$, where Υ_k is the polynomial of degree $k(k-1)/2$ given by the k-fold residue

$$\Upsilon_k^\pm(x) = \frac{(-1)^{k(k-1)/2} 2^k}{k!} \frac{1}{(2\pi i)^k}$$
$$\times \oint \cdots \oint \frac{F_k^\pm(z_1, \ldots, z_k) \Delta(z_1^2, \ldots, z_k^2)^2}{\prod_{j=1}^k z_j^{2k-1}} e^{x \sum_{j=1}^k z_j} \, dz_1 \ldots dz_k, \tag{2.3}$$

where the contours above enclose the poles at $z_j = 0$,

$$F_k^\pm(z_1, \ldots, z_k) = A_k^\pm(z_1, \ldots, z_k) \prod_{j=1}^k \left(\frac{\Gamma(1+z_j)}{\Gamma(1-z_j)} \left(\frac{Q}{4\pi^2} \right)^{z_j} \right)^{\frac{1}{2}} \prod_{1 \leqslant i < j \leqslant k} \zeta(1+z_i+z_j), \tag{2.4}$$

and Δ is the Vandermonde. The factor $A_k^\pm$, which depends on E, is the Euler product which is absolutely convergent for $\sum_{j=1}^k |z_j| < 1/2$,

$$A_k^\pm(z_1, \ldots, z_k) = \prod_p F_{k,p}^\pm(z_1, \ldots, z_k) \prod_{1 \leqslant i < j \leqslant k} \left(1 - \frac{1}{p^{1+z_i+z_j}} \right) \tag{2.5}$$

with, for $p \nmid Q$,

$$F_{k,p}^\pm = \left(1 + \frac{1}{p} \right)^{-1} \left(\frac{1}{p} + \frac{1}{2} \left(\prod_{j=1}^k \mathcal{L}_p \left(\frac{1}{p^{1+z_j}} \right) + \prod_{j=1}^k \mathcal{L}_p \left(\frac{-1}{p^{1+z_j}} \right) \right) \right). \tag{2.6}$$

and, for $p \mid Q$,

$$F_{k,p}^{\pm} = \prod_{j=1}^{k} \mathcal{L}_p\left(\frac{\pm a_p}{p^{1+z_j}}\right). \tag{2.7}$$

Because we are limiting ourselves to Q squarefree (Q prime in the S^+ case), we have $a_p = \pm 1$ when $p \mid Q$ and so

$$F_{k,p}^{\pm} = \begin{cases} \prod_{j=1}^{k}(1 + p^{-1-z_j})^{-1} & \text{in the } S^- \text{ case, for } p \mid Q \\ \prod_{j=1}^{k}(1 - p^{-1-z_j})^{-1} & \text{in the } S^+ \text{ case, for } p = Q. \end{cases} \tag{2.8}$$

The r.h.s. of (2.2) is [4] asymptotically, as $X \to \infty$,

$$M_E^{\pm}(X, k) \sim A^{\pm}(k) M_O(\lfloor \log X \rfloor, k) \tag{2.9}$$

where

$$A^{\pm}(k) = \tag{2.10}$$

$$\prod_{p \nmid Q} (1 - p^{-1})^{k(k-1)/2} \left(\frac{p}{p+1}\right) \left(\frac{1}{p} + \frac{1}{2}\left(\mathcal{L}_p(1/p)^k + \mathcal{L}_p(-1/p)^k\right)\right)$$

$$\times \prod_{p \mid Q} \left(1 - p^{-1}\right)^{k(k-1)/2} \mathcal{L}_p(\pm a_p/p)^k$$

with

$$M_O(N, k) = 2^{2Nk} \prod_{j=1}^{N} \frac{\Gamma(N + j - 1)\Gamma(k + j - 1/2)}{\Gamma(j - 1/2)\Gamma(k + j + N - 1)}. \tag{2.11}$$

The leading asymptotics given above for the moments of $L_E(1, \chi_d)$ was first made in [8] and [2], though the arithmetic factor was off for primes dividing Q. One nice thing about (2.9) is that it makes sense for complex values of k and in [8] was conjectured to hold for $\Re k > -1/2$.

In [5] it is shown how the conjectured asymptotics for moments can be used to obtain information concerning the distribution of values of $L_E(1, \chi_d)$. That paper discusses the importance of the first pole of the r.h.s. of (2.11) at $k = -1/2$ in analyzing the number of vanishings of $L_E(1, \chi_d)$.

3   Vanishings of $L_E(1, \chi_d)$ in progressions

We fix a prime $q \nmid Q$ and restrict d further according to residue classes mod q as follows. For $\lambda = \pm 1$ we set

$$S^{\pm}(X; q, \lambda) = \{d \in S^{\pm}(X); \chi_d(q) = \lambda\} \tag{3.1}$$

Let

$$R_q^{\pm}(X) = \frac{\sum_{\substack{d \in S^{\pm}(X;q,1) \\ L_E(1,\chi_d)=0}} 1}{\sum_{\substack{d \in S^{\pm}(X;q,-1) \\ L_E(1,\chi_d)=0}} 1} \tag{3.2}$$

denote the number of ratio of the number of vanishings of $L_E(1, \chi_d)$, with $d \in S^\pm$, sorted according to residue classes mod q.

To study this ratio we need to look at the moments:

$$M_E^\pm(X, k; q, \lambda) = \frac{1}{|S^\pm(X; q, \lambda)|} \sum_{d \in S^\pm(X; q, \lambda)} L_E(1, \chi_d)^k. \tag{3.3}$$

The conjecture in [4] then gives

$$M_E^\pm(X, k; q, \lambda) = \frac{1}{X} \int_0^X \Upsilon_{k,q,\lambda}^\pm \left(\log(t)\right) dt + O(X^{-\frac{1}{2}+\varepsilon}) \tag{3.4}$$

where $\Upsilon_{k,q,\lambda}^\pm(x)$ is given by the same formula as in (2.3) but with a slight but important modification: the local factor corresponding to the prime q, $F_{k,q}^\pm$, gets replaced by

$$F_{k,q,\lambda}^\pm = \prod_{j=1}^k (1 - \lambda a_q q^{-1-z_j} + q^{-1-2z_j})^{-1}. \tag{3.5}$$

Similarly, in (2.10), the local factor

$$\left(\frac{q}{q+1}\right)\left(\frac{1}{q} + \frac{1}{2}\left(\mathcal{L}_q(1/q)^k + \mathcal{L}_q(-1/q)^k\right)\right) \tag{3.6}$$

at the prime q gets replaced by

$$L_q(\lambda/q)^k = (1 - \lambda a_q q^{-1} + q^{-1})^{-k}. \tag{3.7}$$

From this we immediately surmise several things. First, $R_q^\pm(X)$ which is conjectured to be, asymptotically, equal to the ratio of the residues of the two moments (3.4), corresponding to $\lambda = 1$ and -1, at the pole $k = -1/2$ should thus equal, up to leading order,

$$\left(\frac{q+1-a_q}{q+1+a_q}\right)^{1/2}. \tag{3.8}$$

Second, when $a_q = 0$, the complete asymptotic expansion for both moments are identical up to the conjectured error of size $O(X^{-1/2+\varepsilon})$. The reason for this is that, in (3.5), if $a_q = 0$, there is no dependence on λ. Indulging in conjectural bravado, we predict that when $a_q = 0$

$$R_q^\pm(X) = 1 + O(X^{-1/2+\varepsilon}) \tag{3.9}$$

and similarly for $R_q(X)$ in (1.6). This fits well with our numeric data. See section 6 and also Table 1 in [5] .

Third, from this formula for the moments we are able to work out, in principle, arbitrarily many terms in the asymptotic expansion of $R_q^\pm(X)$. Below, we describe the next to leading term in detail. It is of size $O(1/\log(X))$. The lower terms in the asymptotics of $R_q^\pm(X)$ do depend on whether we are looking at $S^+(X)$ as opposed to $S^-(X)$. This arises from the fact that the local factors $F_{k,p}^\pm$ for $p \mid Q$ in equation (2.8) depend on whether we are looking at S^+ or S^-. While this does not affect the main term R_q, it does show up in the secondary terms.

4   Evaluating the first two terms of $M_E^\pm(X, k; q, \lambda)$

To evaluate the residue that defines $\Upsilon_{k,q,\lambda}^\pm$ we need to examine the multiple Laurent series about $z_j = 0$ of the corresponding integrand. In the numerator, we must evaluate the coefficient of $\prod_{j=1}^k z_j^{2k-2}$ of degree $2k(k-1)$. Now $\Delta(z_1^2, \ldots, z_k^2)^2$ is a homogeneous polynomial consisting of terms of degree $4\binom{k}{2} = 2k(k-1)$. However, the poles of $\prod_{1 \leqslant i < j \leqslant k} \zeta(1 + z_i + z_j)$ cancel $\binom{k}{2}$ factors of the Vandermonde. Therefore, in computing the residue, we only need to take terms from the series for $e^{x \sum_{j=1}^k z_j}$ up to degree $\binom{k}{2}$. From this we see that $\Upsilon_{k,q,\lambda}^\pm(x)$ is a polynomial in x of degree $\binom{k}{2}$.

To obtain the leading two terms of $\Upsilon_{k,q,\lambda}^\pm(x)$, i.e. those of degree $\binom{k}{2}$ and $\binom{k}{2} - 1$ in x, we need to evaluate the constant and linear terms in the multiple Maclaurin series of the function

$$
h_k^\pm(z; q, \lambda) = A_k^\pm(z_1, \ldots, z_k; q, \lambda) \prod_{j=1}^k \left(\frac{\Gamma(1+z_j)}{\Gamma(1-z_j)} \left(\frac{Q}{4\pi^2} \right)^{z_j} \right)^{\frac{1}{2}} \tag{4.1}
$$
$$
\times \prod_{1 \leqslant i < j \leqslant k} \zeta(1 + z_i + z_j)(z_i + z_j).
$$

Here $A_k^\pm(z_1, \ldots, z_k; q, \lambda)$ is the same as the function $A_k^\pm(z_1, \ldots, z_k)$ but with the local factor $F_{k,q}^\pm$ replaced by $F_{k,q,\lambda}^\pm$.

For example, the term involving $x^{k(k-1)/2}$ of $\Upsilon_{k,q,\lambda}^\pm(x)$ is equal to

$$
h_k^\pm(0; q, \lambda) \frac{(-1)^{k(k-1)/2} 2^k}{k!} \frac{1}{(2\pi i)^k} \tag{4.2}
$$
$$
\times \oint \cdots \oint \frac{\Delta(z_1^2, \ldots, z_k^2)^2}{\prod_{j=1}^k z_j^{2k-1}} \frac{e^{x \sum_{j=1}^k z_j}}{\prod_{1 \leqslant i < j \leqslant k}(z_i + z_j)} dz_1 \ldots dz_k.
$$

It is shown in [3] that the above equals

$$
h_k^\pm(0; q, \lambda) g_k(O^+) x^{k(k-1)/2} \tag{4.3}
$$

where

$$
g_k(O^+) = 2^{k(k+1)/2} \prod_{j=1}^{k-1} \frac{j!}{2j!}. \tag{4.4}
$$

We also have

$$
h_k^\pm(0; q, \lambda) = A_k^\pm(0, \ldots, 0; q, \lambda). \tag{4.5}
$$

To compute the leading two terms of the moments we prefer to write

$$
h_k^\pm(z; q, \lambda) = \exp(\log h_k^\pm(z; q, \lambda)) \tag{4.6}
$$

and evaluate the constant and linear terms of

$$
\log h_k^\pm(z; q, \lambda) = \alpha_k^\pm(q, \lambda) + \beta_k^\pm(q, \lambda) \sum z_j + \ldots. \tag{4.7}
$$

Notice that the linear terms all share the same coefficient because $h_k^{\pm}(z; q, \lambda)$ is symmetric in the z_j's.

The constant term can be pulled out of the integral as $e^{\alpha_k^{\pm}(q,\lambda)} = h_k^{\pm}(0; q, \lambda)$. The linear terms can be absorbed into the $\exp(x \sum_{j=1}^{k} z_j)$. Dropping the terms of degree two or higher in $\log h_k^{\pm}(z; q, \lambda)$ we can evaluate the residue using (4.3):

$$h_k^{\pm}(0; q, \lambda) g_k(O^+)(x + \beta_k^{\pm}(q, \lambda))^{k(k-1)/2} \tag{4.8}$$

and thus find that

$$\Upsilon_{k,q,\lambda}^{\pm}(x) = h_k^{\pm}(0; q, \lambda) g_k(O^+)(x^{\frac{k(k-1)}{2}} + \frac{k(k-1)}{2}\beta_k^{\pm}(q, \lambda) x^{\frac{k(k-1)}{2}-1} + \ldots). \tag{4.9}$$

Inserting (4.9) into (3.4) and integrating, we obtain

$$\begin{aligned}
M_E^{\pm}(X, k; q, \lambda) &= \frac{h_k^{\pm}(0; q, \lambda) g_k(O^+)}{X} \\
&\times \int_0^X \left(\log(t)^{\frac{k(k-1)}{2}} + \frac{k(k-1)\beta_k^{\pm}(q, \lambda)}{2} \log(t)^{\frac{k(k-1)}{2}-1} \right) dt \\
&\quad + O(\log(X)^{\frac{k(k-1)}{2}-2})
\end{aligned} \tag{4.10}$$

and hence

$$\begin{aligned}
M_E^{\pm}(X, k; q, \lambda) &= h_k^{\pm}(0; q, \lambda) g_k(O^+) \log(X)^{\frac{k(k-1)}{2}} \\
&\times \left(1 + \frac{k(k-1)}{2\log(X)}(\beta_k^{\pm}(q, \lambda) - 1) \right) + O(\log(X)^{\frac{k(k-1)}{2}-2}).
\end{aligned} \tag{4.11}$$

Therefore, the remaining work is to compute above the coefficient $\beta_k^{\pm}(q, \lambda)$. To do so we evaluate individually the linear terms in the Maclaurin expansions of:

$$\frac{1}{2} \log \prod_{j=1}^{k} \left(\frac{\Gamma(1 + z_j)}{\Gamma(1 - z_j)} \left(\frac{Q}{4\pi^2} \right)^{z_j} \right), \tag{4.12}$$

$$\log \prod_{1 \leqslant i < j \leqslant k} \zeta(1 + z_i + z_j)(z_i + z_j), \tag{4.13}$$

and

$$\log A_k^{\pm}(z_1, \ldots, z_k; q, \lambda). \tag{4.14}$$

First, $\log \Gamma(1 + z) = -\gamma z + \frac{\pi^2}{12} z^2 + \ldots$ hence

$$\frac{1}{2} \log \left(\frac{\Gamma(1 + z)}{\Gamma(1 - z)} \left(\frac{Q}{4\pi^2} \right)^z \right) = (-\gamma + \log(Q^{1/2}/(2\pi)))z + \ldots \tag{4.15}$$

and so (4.12) equals

$$(-\gamma + \log(Q^{1/2}/(2\pi))) \sum z_j + \ldots. \tag{4.16}$$

Next,
$$\zeta(1 + z_i + z_j)(z_i + z_j) = 1 + \gamma(z_i + z_j) + \dots \qquad (4.17)$$

so
$$\prod_{1 \leqslant i < j \leqslant k} \zeta(1 + z_i + z_j)(z_i + z_j) \;=\; 1 + \gamma \sum_{1 \leqslant i < j \leqslant k} (z_i + z_j) + \dots$$
$$=\; 1 + (k-1)\gamma \sum z_j + \dots \qquad (4.18)$$

Therefore, (4.13) equals
$$(k-1)\gamma \sum z_j + \dots . \qquad (4.19)$$

We now turn to (4.14). The function $A_k^{\pm}(z_1, \dots, z_k; q, \lambda)$ is given by (2.5) except that the local factor at $p = q$, namely $F_{k,q}^{\pm}$, gets replaced by (3.5). To find the coefficient of $\sum z_j$ in the Maclaurin series for

$$\prod_{1 \leq i < j \leq k} \left(1 - \frac{1}{p^{1+z_i+z_j}}\right) \qquad (4.20)$$

we can, because the above is symmetric in the z_j's, differentiate with respect to z_1 and set all z_j equal to 0. We thus find that the coefficient of $\sum z_j$ equals

$$\frac{(k-1)\log p}{p-1}. \qquad (4.21)$$

Next we consider the contribution from the local factor when $p = q$:

$$\log F_{k,q,\lambda}^{\pm} = -\sum_{j=1}^{k} \log(1 - \lambda a_q q^{-1-z_j} + q^{-1-2z_j}). \qquad (4.22)$$

Differentiating w.r.t. z_1 and setting all $z_j = 0$ we find that the coefficient of $\sum z_j$ in the Maclaurin series for $\log F_{k,q,\lambda}^{\pm}$ equals

$$\frac{\log q(\lambda a_q - 2)}{\lambda a_q - q - 1}. \qquad (4.23)$$

Finally, we consider the local factor when $p \neq q$. If $p \mid Q$, we have, on taking the logarithm of (2.8), differentiating w.r.t. z_1, setting all $z_j = 0$, that the coefficient of $\sum z_j$ in the series for $\log F_{k,p}^{\pm}$ equals

$$\begin{cases} \log(p)/(1+p) & \text{in the } S^- \text{ case} \\ \log(p)/(1-p) & \text{in the } S^+ \text{ case.} \end{cases} \qquad (4.24)$$

If $p \nmid Q$, taking the logarithm of (2.6), differentiating w.r.t. z_1, and letting $z_j = 0$, we get the coefficient of $\sum z_j$ equal to

$$\log(p) \left(\frac{(2 - a_p)f_1(p)^{-k-1} + (2 + a_p)f_2(p)^{-k-1}}{2 + p\left(f_1(p)^{-k} + f_2(p)^{-k}\right)} \right) \qquad (4.25)$$

where

$$f_1(p) = 1 - a_p/p + 1/p$$
$$f_2(p) = 1 + a_p/p + 1/p. \tag{4.26}$$

Hence, adding all the coefficients of $\sum z_j$ we find that $\beta_k^\pm(q, \lambda)$ in (4.7), and hence in (4.9), equals

$$(k - 2)\gamma + \log(Q^{1/2}/(2\pi)) + \sum_p \beta_k(p) \tag{4.27}$$

where

$$\beta_k(p) = \frac{(k - 1)\log p}{p - 1} + \begin{cases} \frac{\log(q)(\lambda a_q - 2)}{\lambda a_q - q - 1} & \text{if } p = q \\ \log(p)\left(\frac{(2 - a_p)f_1(p)^{-k-1} + (2 + a_p)f_2(p)^{-k-1}}{2 + p\left(f_1(p)^{-k} + f_2(p)^{-k}\right)} \right) & \text{if } p \neq q,\ p \nmid Q \\ \log(p)/(1 + p) & \text{if } p \mid Q, \text{ in the } S^- \text{ case} \\ \log(p)/(1 - p) & \text{if } p \mid Q, \text{ in the } S^+ \text{ case.} \end{cases} \tag{4.28}$$

Notice that the only dependence in $\beta_k^\pm(q, \lambda)$ on q is in the term

$$\beta_k(q) = \frac{(k - 1)\log q}{q - 1} + \frac{\log(q)(\lambda a_q - 2)}{\lambda a_q - q - 1}. \tag{4.29}$$

5 Conjecture for the first two terms in $R_q^\pm(X)$

Dividing $M_E^\pm(X, k; q, 1)$ by $M_E^\pm(X, k; q, -1)$, using equation (4.11)

$$\frac{M_E^\pm(X, k; q, 1)}{M_E^\pm(X, k; q, -1)} = \frac{h_k^\pm(0; q, 1)}{h_k^\pm(0; q, -1)} \frac{\left(1 + \frac{k(k-1)}{2\log(X)}(\beta_k^\pm(q, 1) - 1)\right)}{\left(1 + \frac{k(k-1)}{2\log(X)}(\beta_k^\pm(q, -1) - 1)\right)} + O(\log(X)^{-2}). \tag{5.1}$$

The first factor $\frac{h_k^\pm(0;q,1)}{h_k^\pm(0;q,-1)}$ equals

$$\left(\frac{q + 1 - a_q}{q + 1 + a_q}\right)^{-k}. \tag{5.2}$$

Interpolating to $k = -1/2$ gives our conjecture:

Conjecture 1. *For $q \nmid Q$*

$$R_q^\pm(X) = R_q \frac{1 + \frac{3}{8\log(X)}(\beta_{-\frac{1}{2}}^\pm(q, 1) - 1)}{1 + \frac{3}{8\log(X)}(\beta_{-\frac{1}{2}}^\pm(q, -1) - 1)} + O(\log(X)^{-2}) \tag{5.3}$$

where $\beta_{-\frac{1}{2}}^\pm(q, \lambda)$ is given explicitly by equation (4.27). The implied constant in the remainder term depends on E and q, and thus also on a_q.

6   Numerical Data

We verify the conjecture described above for over two thousand elliptic curves and the sets $S_E^\pm(X)$, with $X = 10^8$. Altogether we have 2398 datasets. The curves in question and the method for computing $L_E(1, \chi_d)$ are detailed in [6]. Tables of L-values can be obtained from [10].

We first depict in Figure 6.1 the distribution of the remainder in comparing $R_q^\pm(X)$ to the conjectured first and second order approximations. More precisely, for our 2398 datasets, we examine the distribution of values of

$$R_q^\pm(X) - R_q \tag{6.1}$$

and of

$$R_q^\pm(X) - R_q \frac{1 + \frac{3}{8\log(X)}(\beta_{-\frac{1}{2}}^\pm(q, 1) - 1)}{1 + \frac{3}{8\log(X)}(\beta_{-\frac{1}{2}}^\pm(q, -1) - 1)} \tag{6.2}$$

with $X = 10^8$, $q \leqslant 3571$. We break up the horizontal axis into small bins of size .0002 and count how often the values fall within a given bin. The difference in (6.2) has smaller variance reflecting an overall better fit of the second order approximation compared with the first. These distributions are not Gaussian. There are yet further lower terms and these are given by complicated sums involving the Dirichlet coefficients of $L_E(s)$, and q.

In the first plot of Figure 6.2 we depict, for one hundred of our datasets, the raw data for the values given by equation (6.1). The horizontal axis is q. For each q on the horizontal axis there are 100 points corresponding to the 100 values, one for each dataset, of $R_q^\pm(X) - R_q$, with $X = 10^8$. We see the values fluctuating about zero, most of the time agreeing to within about .02. The convergence in X is predicted from the secondary term to be logarithmically slow and one gets a better fit by including the second order term.

This is depicted in the second plot of Figure 6.2 which shows the difference given in (6.2). again with $X = 10^8$, and the same one hundred elliptic curves E. We see an improvement to the first plot which uses just the main term. We only depict data for 100 datasets in these plots since otherwise there would be too many data points leading to a thick black mess.

Finally, a sequence of plots shows the dependence of the remainder term in the first and second order approximations on q and a_q. Given an integer n, we display, in Figure 6.3 q v.s. $R_q^\pm(10^8) - R_q$ for the subset of our elliptic curves satisfying $a_q = n$. For each of $n = -20, -9, -6, -4, -3, -2, -1, 0, 1, 2, 3, 4, 6, 9, 20$ there is one plot. Figure 6.4 does the same but for the values given by equation (6.2).

We notice several things. Overall, the plots in the Figure 6.4 are more symmetric about the horizontal axis reflecting a tighter fit by including the second order term. For smaller q however, incorporating the second order term leads to a correction that tends to overshoot. Compare for example the fourth plot in Figures 6.3 and 6.4. Presumably, the third and further

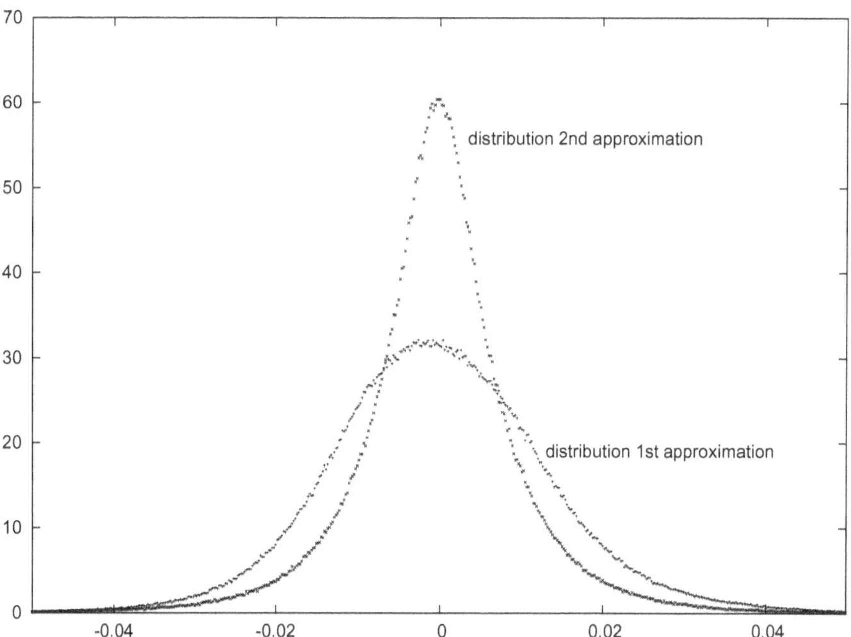

Figure 6.1: Distribution first approximation v.s. second approximation for ratio of vanishings

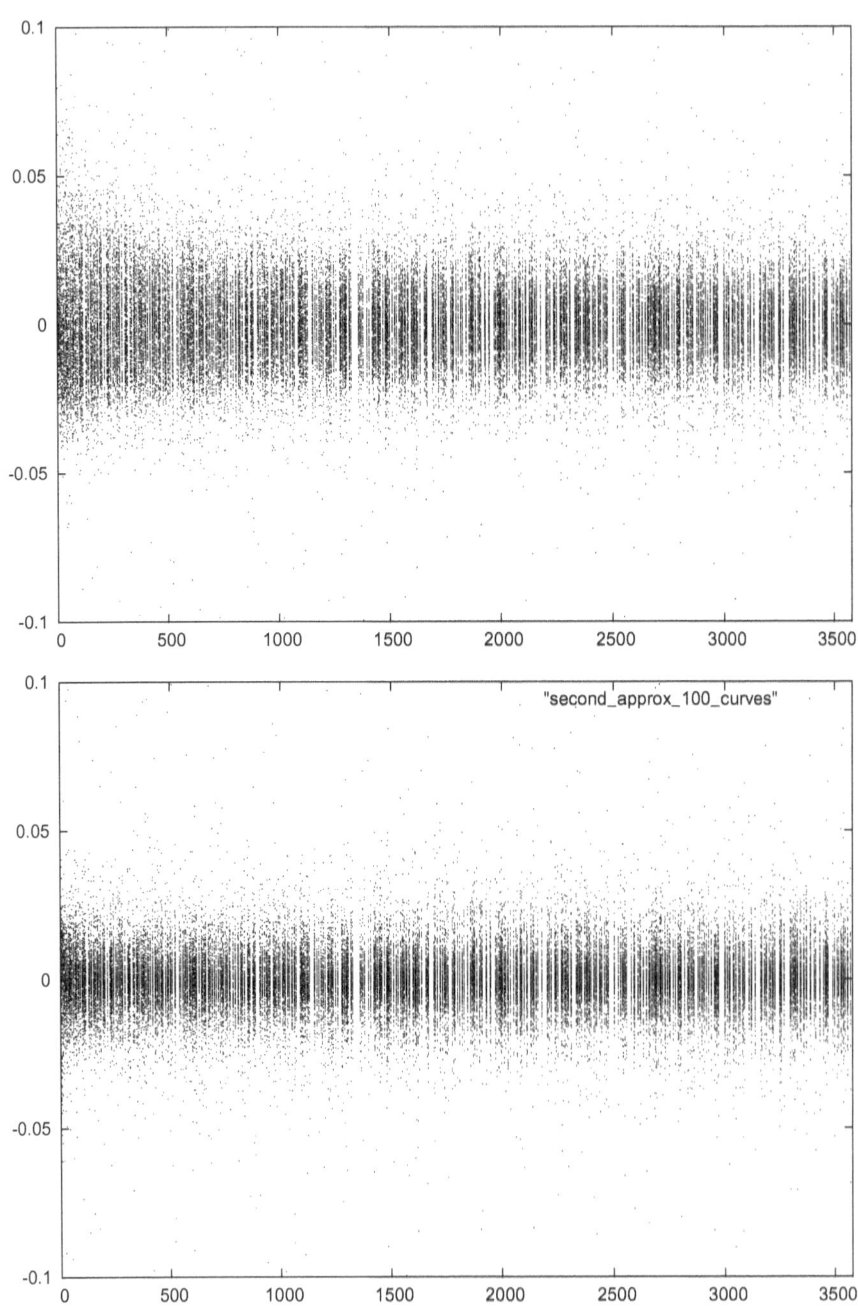

Figure 6.2: A plot for one hundred datasets of $R_q^{\pm}(10^8) - R_q$, top plot, and of (6.2), bottom plot, for $2 \leqslant q \leqslant 3571$.

order terms, while of size $O(\log(X)^{-2})$ can have relatively large constants for smaller q requiring one to take X larger than 10^8 to see an improvement from the second order term.

This is also reflected in Tables 1– 2 which lists for two elliptic curves and the sets $S^+(10^8)$ and $S^-(10^8)$ the numeric values of (6.1) and (6.2) for $q \leqslant 179$.

Acknowledgements

We wish to thank the Newton Institute in Cambridge where some of this research was carried out. We also thank Gonzalo Tornaria and Fernando Rodriguez-Villegas who supplied us with a table of weight three halves forms that were used to compute the L-values studied in this paper.

References

[1] C. Breuil, B. Conrad, F. Diamond, and R. Taylor, *J. Amer. Math. Soc.* **14**:843–939, 2001, no. 4.

[2] J.B. Conrey and D.W. Farmer, Mean values of L-functions and symmetry, *Int. Math. Res. Notices*, **17**:883–908, 2000. arXiv:math.nt/9912107.

[3] J.B. Conrey, D.W. Farmer, J.P. Keating, M.O. Rubinstein, and N.C. Snaith, Autocorrelation of random matrix polynomials, *Comm. Math. Phys.*:365–395, 2003, no. 3. arXiv:math-ph/02080077.

[4] J.B. Conrey, D.W. Farmer, J.P. Keating, M.O. Rubinstein, and N.C. Snaith, Integral moments of L-functions, *Proceedings of the London Mathematical Society*, **91**, 33–104. arXiv:math.nt/0206018.

[5] J.B. Conrey, J.P. Keating, M.O. Rubinstein, and N.C. Snaith, On the frequency of vanishing of quadratic twists of modular L-functions, In *Number Theory for the Millennium I: Proceedings of the Millennial Conference on Number Theory*, editor, M.A. Bennett et al., pages 301–315. A K Peters, Ltd, Natick, 2002. arXiv:math.nt/0012043.

[6] J.B. Conrey, J.P. Keating, M.O. Rubinstein, and N.C. Snaith, Random Matrix Theory and the Fourier Coefficients of Half-Integral Weight Forms, *Experimental Mathematics*, **15** 2006, no. 1. arXiv:math.nt/0412083

[7] C. Delaunay, Note on the frequency of vanishing of L-functions of elliptic curves in a family of quadratic twists, in this volume.

[8] J.P. Keating and N.C. Snaith, Random matrix theory and L-functions at $s = 1/2$, *Commun. Math. Phys*, **214**:91–110, 2000.

[9] F. Rodriguez-Villegas and G. Tornaria, private communication.

[10] M. Rubinstein, *The L-function database*, Available at `www.math.uwaterloo.ca/~mrubinst`.

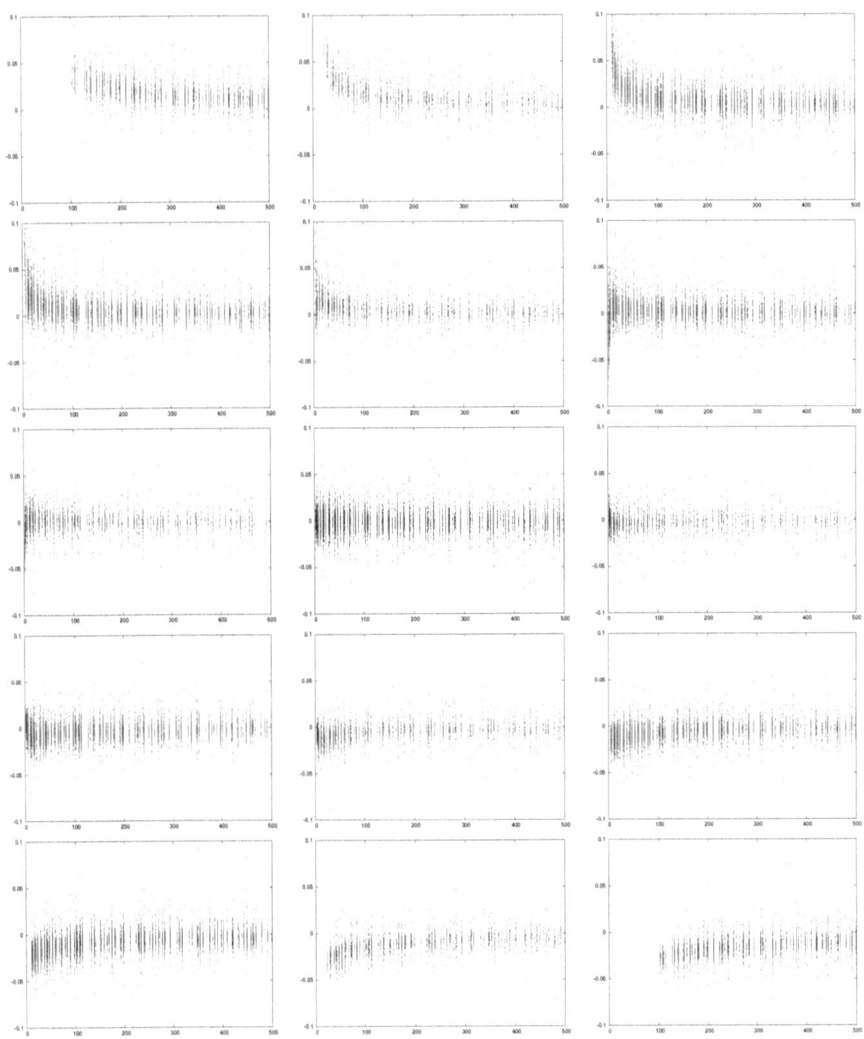

Figure 6.3:   Left to right, top to bottom: $n = -20, -9, -6, -4, -3, -2, -1, 0, 1, 2, 3, 4, 6, 9, 20$.   Values of $R_q^{\pm}(X) - R_q$, with $X = 10^8$, $2 \leqslant q < 500$, for the subset of our elliptic curves satisfying $a_q = n$. The blank white area on the left of the plots for larger n reflects Hasse's theorem that $|a_q| < 2q^{1/2}$ which restricts how small q can be given $a_q = n$.

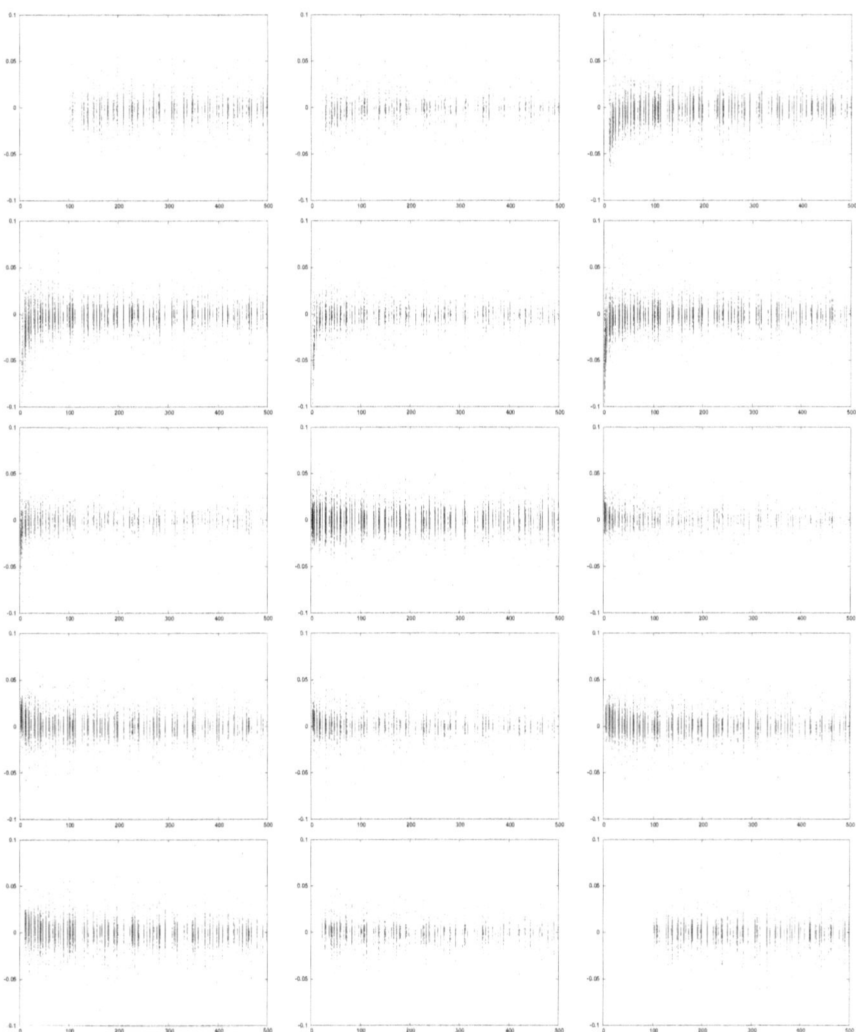

Figure 6.4: Left to right, top to bottom: $n = -20, -9, -6, -4, -3, -2, -1, 0, 1, 2, 3, 4, 6, 9, 20$ Values of (6.2), with $X = 10^8$, $2 \leqslant q < 500$, for the subset of our elliptic curves satisfying $a_q = n$.

q	a_q	(6.1), R^- case	(6.2), R^- case	(6.1), R^+ case	(6.2), R^+ case
2	-2	-0.0770803072	-0.1058493733	-0.0586746787	-0.0877402111
3	-1	-0.0226715635	-0.0314020531	-0.0112745015	-0.0200944948
5	1	0.0039386614	0.0110670332	0.0036670414	0.0108679937
7	-2	-0.0086677613	-0.0320476479	0.0122162834	-0.0114052128
13	4	-0.0117312471	0.0114581936	-0.0109800729	0.0124435613
17	-2	0.0068671146	-0.0078374991	0.0156420190	0.0007858160
19	0	0.0018786796	0.0018786796	0.0017548761	0.0017548761
23	-1	0.0065085545	0.0007253864	0.0087254527	0.0028829043
29	0	0.0015867409	0.0015867409	0.0024574134	0.0024574134
31	7	-0.0203976628	0.0065021478	-0.0212844047	0.0058867043
37	3	-0.0076213530	0.0038881303	-0.0081586993	0.0034679279
41	-8	0.0293718254	-0.0104233512	0.0370003139	-0.0032097869
43	-6	0.0200767559	-0.0066399665	0.0230632720	-0.0039304770
47	8	-0.0166158276	0.0077120067	-0.0181946828	0.0063789626
53	-6	0.0175200151	-0.0048911726	0.0194053316	-0.0032378110
59	5	-0.0095451504	0.0043844494	-0.0127090647	0.0013621363
61	12	-0.0229944549	0.0068341556	-0.0279181705	0.0022108579
67	-7	0.0114509369	-0.0104875891	0.0227073168	0.0005417642
71	-3	0.0078736247	-0.0004772247	0.0051206275	-0.0033160932
73	4	-0.0037492048	0.0060879152	-0.0119406010	-0.0020032563
79	-10	0.0300180540	0.0013488112	0.0296738495	0.0007070253
83	-6	0.0142507227	-0.0012053860	0.0124985709	-0.0031170117
89	15	-0.0230738419	0.0057929377	-0.0246777538	0.0044799769
97	-7	0.0105905604	-0.0054712607	0.0154867447	-0.0007408496
101	2	-0.0037100582	0.0002953972	-0.0044847165	-0.0004383257
103	-16	0.0324024693	-0.0068711726	0.0357260869	-0.0039571170
107	18	-0.0228240764	0.0073200274	-0.0245602341	0.0058874808
109	10	-0.0097574184	0.0078543625	-0.0133419792	0.0044484844
113	9	-0.0120886539	0.0035056429	-0.0113667336	0.0043859550
127	8	-0.0093873089	0.0034881040	-0.0081483592	0.0048580252
131	-18	0.0320681832	-0.0038139100	0.0371594888	0.0009037228
137	-7	0.0117897817	-0.0002445226	0.0086451554	-0.0035131214
139	10	-0.0148514126	0.0000259176	-0.0112784046	0.0037500975
149	-10	0.0140952751	-0.0023344544	0.0172405748	0.0006412396
151	2	-0.0041170706	-0.0011557351	-0.0070016068	-0.0040099902
157	-7	0.0108322334	0.0000925632	0.0097641977	-0.0010860401
163	4	-0.0014750980	0.0040361356	-0.0066512858	-0.0010837710
167	-12	0.0171132732	-0.0010302790	0.0222297420	0.0038987403
173	-6	0.0054181738	-0.0030119338	0.0036566390	-0.0048601622
179	-15	0.0177416502	-0.0040658274	0.0261766468	0.0041434818

Table 1: The values of $R_q^{\pm}(10^8)$, for the elliptic curve 11_A of conductor 11 given by $y^2 + y = x^3 - x^2 - 10x - 20$, compared to the conjectured first order approximation (6.1) and second order approximation (6.2).

q	a_q	(6.1), R^- case	(6.2), R^- case	(6.1), R^+ case	(6.2), R^+ case
2	0	0.0001964177	0.0001964177	0.0025336244	0.0025336244
3	0	-0.0007380207	-0.0007380207	-0.0025236647	-0.0025236647
5	4	-0.0128879806	0.0109510354	-0.0166316058	0.0072258354
7	0	-0.0048614428	-0.0048614428	-0.0014203548	-0.0014203548
11	3	-0.0076239866	0.0095824910	-0.0101221542	0.0070977143
13	6	-0.0212338218	0.0089386380	-0.0276990384	0.0024967032
17	-1	0.0033655021	-0.0029005302	0.0086797465	0.0024087738
19	-1	0.0055745934	-0.0003223680	0.0020465484	-0.0038550619
23	-2	0.0074744917	-0.0036255406	0.0079256468	-0.0031831583
29	0	0.0004190042	0.0004190042	-0.0010879108	-0.0010879108
31	4	-0.0108662407	0.0041956843	-0.0096223973	0.0054512748
37	3	-0.0067227670	0.0037940655	-0.0162107316	-0.0056856756
41	5	-0.0109118090	0.0049138186	-0.0164777387	-0.0006397725
43	-10	0.0406071465	-0.0060036473	0.0409949651	-0.0056532348
47	-6	0.0284021024	0.0057897746	0.0209827487	-0.0016475471
53	-10	0.0361234610	-0.0017568821	0.0423409405	0.0044303004
59	4	-0.0054935724	0.0048495607	-0.0148985734	-0.0045473511
61	-8	0.0227634479	-0.0025651538	0.0253866588	0.0000379053
67	-8	0.0217284008	-0.0016029354	0.0249365465	0.0015866634
71	-15	0.0398795640	-0.0080932079	0.0531538377	0.0051425339
73	2	-0.0003657281	0.0042519609	-0.0019954011	0.0026259102
79	-13	0.0270702549	-0.0087950276	0.0328555729	-0.0030383756
83	5	-0.0120289758	-0.0018576129	-0.0140337206	-0.0038544019
89	9	-0.0117002661	0.0050406278	-0.0159501141	0.0008038275
97	7	-0.0121449601	0.0003884458	-0.0126491435	-0.0001059465
101	10	-0.0162655200	0.0006799944	-0.0166873803	0.0002713400
103	11	-0.0154514081	0.0027879315	-0.0155096044	0.0027439391
107	-15	0.0298791131	-0.0020491232	0.0346054275	0.0026517125
109	-7	0.0131301691	-0.0001660211	0.0138662913	0.0005595788
113	14	-0.0219346950	-0.0006197951	-0.0199798581	0.0013516122
127	17	-0.0231978866	0.0002623636	-0.0235951007	-0.0001166344
131	-6	0.0075864820	-0.0020988314	0.0132703492	0.0035773840
137	-6	0.0049307893	-0.0044030816	0.0085067787	-0.0008344650
139	14	-0.0179638452	0.0005718014	-0.0220919830	-0.0035419036
149	19	-0.0184587534	0.0048182811	-0.0206858659	0.0026092450
151	-14	0.0157624561	-0.0057016072	0.0217272592	0.0002461455
157	-14	0.0258394912	0.0051129620	0.0236949594	0.0029519710
163	-8	0.0115026664	0.0005637031	0.0044174198	-0.0065301910
167	21	-0.0224356707	0.0011192167	-0.0284090909	-0.0048359139
173	-6	0.0056158047	-0.0020804090	0.0044893748	-0.0032129134
179	0	0.0018844544	0.0018844544	-0.0007004350	-0.0007004350

Table 2: The values of $R_q^\pm(10^8)$, for the elliptic curve 307_A of conductor 307 given by $y^2 + y = x^3 - x - 9$, compared to the conjectured first order approximation (6.1) and second order approximation (6.2).

[11] R. Taylor and A. Wiles Ring-theoretic properties of certain Hecke algebras *Ann. of Math.*, (2) **141**:553–572 (1995), no. 3.

[12] A. Wiles, Modular elliptic curves and Fermat's last theorem, *Ann. of Math.*, (2) **141**:443–551 (1995), no. 3.

J.B. Conrey
American Institute of Mathematics
360 Portage Avenue
Palo Alto, CA 94306
USA

School of Mathematics
University of Bristol
Bristol BS8 1TW
UK

A. Pocharel
Department of Mathematics
Princeton University
Princeton, NJ 08544
USA

M.O. Rubinstein
Pure Mathematics
University of Waterloo
200 University Ave W
Waterloo, ON, Canada
N2L 3G1

M. Watkins
School of Mathematics
University of Bristol
Bristol BS8 1TW
UK

Fudge Factors in the Birch and Swinnerton-Dyer Conjecture

Karl Rubin

The aim of this note is to describe how the "fudge factors" in the Birch and Swinnerton-Dyer conjecture vary in a family of quadratic twists (see Proposition 5, which follows directly from Tate's algorithm [T]). We illustrate with two examples.

Definition 1. *If E is an elliptic curve over $\mathbf{Q}$ and p is a prime, the* fudge factor *(or* Tamagawa factor*) $c_p(E)$ is defined by*

$$c_p(E) = [E(\mathbf{Q}_p) : E_0(\mathbf{Q}_p)]$$

where $E_0(\mathbf{Q}_p)$ is the subgroup of $E(\mathbf{Q}_p)$ consisting of those points whose reduction modulo p (on a minimal model of E) is nonsingular.

The fundamental method for computing the fudge factors is Tate's algorithm. This algorithm, originally described in a 1965 letter to Cassels, was published in [T] and essentially reproduced in §IV.9 of [S]. Standard number theoretic computer packages, such as PARI/GP (available at `http://pari.math.u-bordeaux.fr`), will compute these factors very efficiently.

Let $\Delta(E)$ denote the discriminant of a minimal model of E.

Proposition 2. *Suppose E is an elliptic curve over $\mathbf{Q}$.*

(i) *If E has good reduction at p, then $c_p(E) = 1$.*

(ii) *If E has split multiplicative reduction at p, then $c_p(E) = \mathrm{ord}_p(\Delta(E))$, i.e., $p^{c_p(E)}$ is the highest power of p dividing $\Delta(E)$.*

(iii) *If E has nonsplit multiplicative reduction at p, then $c_p(E) \leq 2$ and $c_p(E) \equiv \mathrm{ord}_p(\Delta(E)) \pmod{2}$.*

(iv) *If E has additive reduction at p, then $c_p(E) \leq 4$.*

Proof These are cases 1, 2a, 2b, and 3 through 10, respectively, of Tate's algorithm [T]. $\qquad\square$

Fix an elliptic curve E and a model of E of the form

$$y^2 = f(x)$$

---

[1] Supported by NSF grant DMS-0140378.

with a monic cubic polynomial $f(x) \in \mathbf{Z}[x]$, and let Δ denote the discriminant of this model. We may assume that the model is minimal at all primes $p > 2$, but this is not necessary for what follows.

Definition 3. *The quadratic twist of E by a nonzero rational number d is*

$$E_d : y^2 = d^3 f(x/d).$$

We will write simply $c_p(d)$ for $c_p(E_d)$. The purpose of this note is to describe how $c_p(d)$, and $\prod_p c_p(d)$, vary with d.

Lemma 4. *Suppose $d, d' \in \mathbf{Q}^\times$.*

  (i) *If d/d' is a square in $\mathbf{Q}$, then E_d is isomorphic to $E_{d'}$.*

  (ii) *If p is a prime and d/d' is a square in $\mathbf{Q}_p$, then $c_p(d) = c_p(d')$.*

Proof If $d' = dr^2$, then the map $(x, y) \mapsto (r^2 x, r^3 y)$ is an isomorphism from E_d to $E_{d'}$. If $r \in \mathbf{Q}^\times$, this proves (i). If $r \in \mathbf{Q}_p^\times$, this isomorphism identifies $E_d(\mathbf{Q}_p)$ with $E_{d'}(\mathbf{Q}_p)$ and by the definition of $c_p(d)$ we get $c_p(d) = c_p(d')$. $\quad\square$

By Lemma 4(i), every quadratic twist E_d of E is a twist by some (unique) squarefree integer. From now on we will assume that d is a squarefree integer.

Proposition 5. *Suppose p is a prime not dividing 2Δ. If $p \nmid d$ then $c_p(d) = 1$. If $p \mid d$, then*

$$c_p(d) = 1 + \#\{\text{roots of } f(x) \equiv 0 \ (\mathrm{mod} \ p) \text{ in } \mathbf{Z}/p\mathbf{Z}\} = 1, \ 2, \ \text{or } 4.$$

Proof If $p \nmid 2\Delta d$ then E_d has good reduction at p, so $c_p(d) = 1$. If $p \mid d$ but $p \nmid 2\Delta$ then we are in case 6 of Tate's algorithm [T]. $\quad\square$

Note that for every p not dividing 2Δ, the number of roots of $f(x)$ modulo p is at least as large as the number of roots of $f(x)$ in $\mathbf{Q}$. Thus if $p \mid d$ and $p \nmid 2\Delta$, then $c_p(d) \geq \#E(\mathbf{Q})[2]$.

If $p \mid 2\Delta$ the situation is more complicated. However, for those primes, to determine $c_p(d)$ for every d, Lemma 4(ii) shows that it is enough to compute $c_p(d)$ (using Tate's algorithm) for d in a set of representatives of $\mathbf{Q}_p^\times/(\mathbf{Q}_p^\times)^2$. Note that $\mathbf{Q}_p^\times/(\mathbf{Q}_p^\times)^2$ has order 4 if $p > 2$, and order 8 if $p = 2$.

Example 6. $E : y^2 = x^3 - x$

We have $\Delta = 64$, and $x^3 - x$ factors into linear factors over $\mathbf{Q}$, so Proposition 5 shows that for $p > 2$ we have

$$c_p(d) = \begin{cases} 1 & \text{if } p \nmid d, \\ 4 & \text{if } p \mid d. \end{cases} \tag{1.1}$$

Tate's algorithm (cases 4 and 7.2, respectively) gives

$$c_2(d) = \begin{cases} 2 & \text{if } 2 \nmid d, \\ 4 & \text{if } 2 \mid d. \end{cases} \tag{1.2}$$

(Alternatively, we can use PARI/GP to compute that

$$c_2(1) = c_2(3) = c_2(-1) = c_2(-3) = 2,$$
$$c_2(2) = c_2(6) = c_2(-2) = c_2(-6) = 4,$$

and then use Lemma 4(ii) to deduce (1.2).)

Combining (1.1) and (1.2) we conclude that

$$\prod_p c_p(d) = \begin{cases} 2^{2\omega(d)+1} & \text{if } d \text{ is odd,} \\ 2^{2\omega(d)} & \text{if } d \text{ is even,} \end{cases}$$

where $\omega(d)$ is the number of prime divisors of d.

Example 7. $E : y^2 + y = x^3 - x^2 - 10x - 20$

This is the modular curve $X_0(11)$, with discriminant -11^5. We will use the model (not minimal at 2)

$$y^2 = x^3 - 4x^2 - 160x - 1264$$

with discriminant $\Delta = -2^{12}11^5$. For $p \neq 2, 11$, Proposition 5 shows that

$$c_p(d) = \begin{cases} 1 & \text{if } p \nmid d, \\ 1 + \#\{\text{roots of } x^3 - 4x^2 - 160x - 1264 \bmod p\} & \text{if } p \mid d. \end{cases}$$

Since $x^3 - 4x^2 - 160x - 1264$ is irreducible over $\mathbf{Q}$, $c_p(d)$ can be 1, 2, or 4. More precisely, the Galois group of $x^3 - 4x^2 - 160x - 1264$ over $\mathbf{Q}$ is S_3, so the Cebotarev theorem shows that if D_k is the density of the set of primes p such that $x^3 - 4x^2 - 160x - 1264$ has k roots modulo p, then $D_0 = 1/3$, $D_1 = 1/2$, and $D_3 = 1/6$.

We also compute

d	1	3	-1	-3	2	6	-2	-6
$c_2(d)$	1	1	1	1	1	1	1	1

d	1	-1	11	-11
$c_{11}(d)$	5	1	4	2

Therefore by Lemma 4(ii), $c_2(d) = 1$ for every d, and

$$c_{11}(d) = \begin{cases} 5 & \text{if } d \text{ is a nonzero square modulo 11,} \\ 1 & \text{if } d \text{ is not a square modulo 11,} \\ 4 & \text{if } 11 \mid d \text{ and } \frac{d}{11} \text{ is a square modulo 11,} \\ 2 & \text{if } 11 \mid d \text{ and } \frac{d}{11} \text{ is not a square modulo 11.} \end{cases}$$

References

[S]   J. H. Silverman, Advanced Topics in the Arithmetic of Elliptic Curves, Graduate Texts in Mathematics **151**, New York: Springer-Verlag (1994).

[T]   J. Tate, Algorithm for determining the type of a singular fiber in an elliptic pencil. In: Modular functions of one variable (IV), *Lecture Notes in Math.* **476**, New York: Springer-Verlag (1975) 33–52.

Karl Rubin
Department of Mathematics,
Stanford University,
Stanford, CA
94305 USA

Department of Mathematics
UC Irvine
Irvine, CA
92697 USA

Rank distribution in a family of cubic twists

Mark Watkins

Abstract

In 1987, Zagier and Kramarz published a paper in which they presented evidence that a positive proportion of the even-signed cubic twists of the elliptic curve $X_0(27)$ should have positive rank. We extend their data, showing that it is more likely that the proportion goes to zero.

1   Introduction

Let E_m be the elliptic curve defined by the equation $x^3 + y^3 = m$, which is isomorphic to $y^2 = x^3 - 432m^2$. The case of $m = 1$ is the modular curve $X_0(27)$, and the cubefree positive m-values give the cubic twists.

These equations have a long history, dating back to Fermat. An early study was done by Sylvester [Syl] in 1879-80, and another voluminous study in 1951 by Selmer [Sel]. In between these two, Nagell [N, p.14] proved sundry results concerning non-solvability in many cases. In the late 1960s, Stephens [Ste1, Ste2] did numerical experiments with these curves with respect to the then-new Birch–Swinnerton-Dyer conjecture. Zagier and Kramarz [ZK] did a large numerical experiment in the 1980s, which led them to suggest that a positive proportion of the curves have rank 2 or greater. The best results in this regard appear to be due to Mai [M], who showed that, assuming the Parity Conjecture, for every $\varepsilon > 0$ at least $c_\varepsilon T^{2/3-\varepsilon}$ of the cubefree even twists up to T have rank 2. Elkies and Rogers [ER] have recently found that the curve

$$x^3 + y^3 = 13293998056584952174157235$$

has rank at least 11. We shall mainly be concerned with rank 2 cubic twists and in extending the numerical data of [ZK], showing that the purported positive proportion does not seem to persist. We also consider the distribution of the size of the Tate–Shafarevitch groups attached to these curves, comment on effects stemming from the arithmetic of m, consider similar questions for quartic twists of $X_0(32)$, and discuss random matrix models for these.

We briefly review how to compute the central L-value of E_m. The first consideration is the sign of the functional equation, which was computed by Birch and Stephens [BS]. This is defined by $\varepsilon_m = \prod_p \varepsilon_m(p)$ where for $p \neq 3$ we have that $\varepsilon_m(p) = \left(\frac{p}{3}\right)$ if $p|m$ and $\varepsilon_m(p) = +1$ if p does not divide m. For

$p = 3$, we have that $\varepsilon_m(3) = +1$ if $m \equiv 1, 3, 6, 8 \pmod 9$, and $\varepsilon_m(3) = -1$ otherwise. Next, there is the conductor $N_m = \prod_p N_m(p)$ where for $p \neq 3$ we have that $N_m(p) = p^2$ if $p|m$ and $N_m(p) = 1$ otherwise, while for $p = 3$ we have that $N_m(3) = 3^5$ if $3|m$, that $N_m(3) = 3^2$ if $m \equiv \pm 2 \pmod 9$, and $N_m(3) = 3^3$ otherwise. There are also Tamagawa numbers and considerations for the real period Ω; the effects of these are given in the Section 4 (see also Table 1 of [ZK]).

When $\varepsilon_m = +1$, the central L-value is given by

$$L(E_m, 1) = 2 \sum_n \frac{a_m(n)}{n} e^{-2\pi n/\sqrt{N_m}},$$

where the conductor N_m is defined as above, and the $a_m(n)$ can be computed as follows. For primes $p \not\equiv 1 \pmod 3$ and primes $p|3m$, we define $a_m(p) = 0$. Given a prime $p \equiv 1 \pmod 3$, the set

$$A_p = \{a|\ a \equiv 2 \pmod 3,\ a^2 + 3b^2 = 4p \text{ for some } b \in \mathbf{Z}\}$$

has 3 elements. For such a prime we define $a_1(p)$ to be the unique element in A_p for which $3|b$. We then define $a_m(p)$ uniquely by the conditions $a_m(p) \equiv m^{(p-1)/3} a_1(p) \pmod p$ and $a_m(p) \in A_p$ (this second condition is equivalent to $|a_m(p)| < 2\sqrt{p}$ for $p > 13$ and not $p \geq 13$ as [ZK] claims). Having defined $a_m(p)$ for all primes p, we extend it to prime powers via the Hecke relations, and then to all positive integers via multiplicativity. In order to approximate $L(E_m, 1)$ well, we need to use about $C\sqrt{N_m}$ coefficients for some constant C. When $\varepsilon_m = -1$, the series for $L'(E_m, 1)$ has the exponential function replaced by an exponential integral — we did not deal with this case ([ZK] considered it for $m \leq 20000$) since the exponential homomorphism can be computed rapidly more readily than the exponential integral — for the latter, local power series would likely be useful. Lieman [L] has shown that the values of $L(E_m, 1)$ are the coefficients of a metaplectic form as was suggested in [ZK, §3.1], but this does not seem useful for computational purposes. We did not try to use the conditions given by Rodriguez-Villegas and Zagier [RVZ], and cannot comment on their computational efficacy.

2   Numerical data

Applying the above method for the cubefree $m \leq 10^7$ with $\varepsilon_m = +1$, we find that about 17.7% of the twists have vanishing central L-value. This is to be compared to 23.3% for the $m \leq 70000$, and 20.5% for $m \leq 10^6$. If we take the best linear fit to a log-log regression, we find that the number of twists up to x with vanishing central L-value appears to grow like $x^{0.935}$. Heuristic models involving the expected size of III as mentioned in [ZK, §3.2] imply that the growth should be more like $x^{5/6}$. Stronger models such as those in [CKRS] imply this should be more like $Bx^{5/6}(\log x)^C$ for some constants B and C; in the last section we make remarks about what random matrix theory predicts for C.

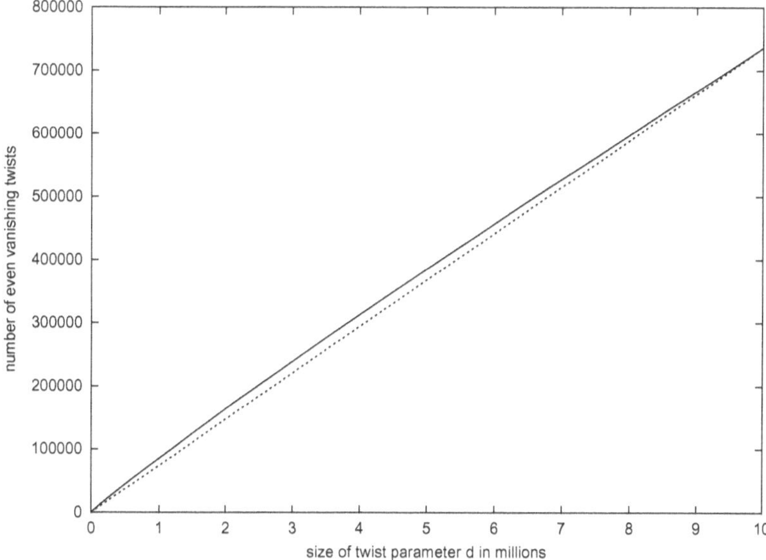

Figure 2.1: Number of even vanishing cubic twists of $X_0(27)$ compared to a (dotted) straight line.

There is also the question of arithmetic effects of m. Only 6.1% of the prime m in the above range $m \le 10^7$ have vanishing central L-value, while 11.3% of the m with two prime factors do, and 17.1% of the m with three prime factors. The number grows to 24.5% for four prime factors, and 35.3% for five prime factors, and is 51.4% for six or more prime factors. However, each of these percentages is about 20% lower than the comparative value when considering only the $m \le 10^6$. So even if we restrict to prime m we expect that the proportion of twists with vanishing central L-value tends to zero. Note in this context that 3-descent can tell us much about the rank when we limit the number of prime factors of m (see [C]). For instance, when m is prime and E_m has even functional equation, we know that $m \equiv 1, 2, 5 \pmod 9$, and the rank is zero in the latter two cases. Thus the 6.1% of above might be re-interpreted as 18.3% of the cases where descent considerations do not force the rank to be zero. Using the results of [N], we could similarly derive such results when m has two prime factors. Also, one can recall that Elkies (see [E1]) has proven that the rank is exactly 1 for primes $m \equiv 4, 7 \pmod 9$; here in fact the conjecture is that the same is true for $m \equiv 8 \pmod 9$. We return to such considerations below when we discuss random matrix models.

We next make some comments about how often various $|\text{Ш}|$-values occur. Zagier and Kramarz found that 26.3% of the even twists for $m \le 70000$ have rank 0 and trivial Ш, while we find the percentage to be 18.8% for $m \le 10^6$ and 14.1% for $m \le 10^7$. Indeed, already in [ZK] this percentage was noted to be diminishing. More interesting might be how often a given prime divides |Ш|, under the restriction to rank 0 twists. For instance, 32.4% of the even rank 0 twists with $m \le 70000$ have 3 dividing |Ш|. This number increases to 40.1% for $m \le 10^6$, and is 45.3% for $m \le 10^7$. The heuristics of Delaunay [De] imply a number more like 36.1%. There is a strong arithmetic impact from m, as for prime m the percentage for $m \le 10^7$ is only 5.8%. However, this last datum should probably be considered anomalous because of the special rôle that 3 plays for cubic twists.

Similarly, 2 divides |Ш| about 45.7% of the time for even rank 0 twists with $m \le 10^7$, while only 42.1% of the time for $m \le 10^6$ and 35.5% of the time for $m \le 70000$. Here Delaunay predicts 58.1%. Here prime m are *more* likely to cause 2-divisibility of |Ш|, with the percentage here for $m \le 10^7$ being 53.5%. As [ZK] notes, the expectation is that |Ш| should be of size $m^{1/3} \approx N_m^{1/6}$ for these cubic twists, larger than the expected $N_m^{1/12}$ in the general case. For 5-divisibility of |Ш|, the percentage increases from 3.6% to 5.9% to 8.0% as the m-range increases. It seems unlikely that these percentages (for $p \ne 3$) will climb all the way to 100%, and without a better guess, one could posit that they are tending toward the number suggested by the Delaunay heuristic. In Table 1, the "$r > 0$" column counts percentages of curves for which the central L-value vanishes, while the other four columns denote how often a given prime divides the |Ш|-value of a nonvanishing twist.

Table 1: Data for cubic twists

	$r > 0$	$p = 2$	$p = 3$	$p = 5$	$p = 7$
$m \le 10^5$	22.9	37.3	33.7	3.9	1.2
$m \le 10^6$	20.5	42.1	40.1	5.9	2.4
$m \le 10^7$	17.7	45.7	45.3	8.0	3.7
prime $m \le 10^7$	6.1	53.5	5.8	14.5	8.2
Delaunay		58.3	36.1	20.7	14.5

M. O. Rubinstein pointed out to us that in his data for quadratic twists, the |Ш| tend to the Delaunay number more readily upon including *all* even rank twists, instead of just the ones of rank 0. Indeed, as we expect that the high rank twists should form an asymptotically negligible set, there is perhaps no reason not to include them in our data. Furthermore, additionally restricting to prime twists also tends to speed convergence toward the number given by Delaunay. Upon implementing these two ideas, we get numbers of 56.3% for 2-divisibility, 19.7% for 5-divisibility, and 13.8% for 7-divisibility, which are fairly close to the percentages predicted by Delaunay. For 3-divisibility we have only 11.6%, as the existence of 3-isogenies for our curves appears to have a definite impact (Rubinstein reports similar phenomena for quadratic twists).

One can do a similar experiment with quartic twists of $X_0^\iota(32) : y^2 = x^3 - x$ or sextic twists of $X_0(27)$. We only looked at the former (where the ι indicates isogenous). For the computation of the sign of the functional equation in these cases, see [ST]. Note that [ZK] looked at the quadratic twists of $X_0^\iota(32)$ given by $y^2 = x^3 - m^2 x$ with $m \equiv 1 \pmod{16}$ for $m \le 500000$, and they found that the percentage of vanishing twists is dropping fairly rapidly, it being 15.2% for $m \le 50000$ and 10.6% for $m \le 500000$. For the quartic twists of $X_0^\iota(32)$ we are looking at $y^2 = x^3 + mx$ where 4 does not divide m and m is free of fourth powers. Here we considered positive $m \le 8000000$, of which 24.9% of the even twists have vanishing central value. This is less than the 27.4% for $m \le 10^6$, and 29.8% for $m \le 10^5$. Similar percentages occur for the negative m.

3 Computational techniques

The computations were carried out on a network of about 10 SPARC machines (mostly SPARC-V) over a 6-month period at the beginning of 2001. Our bound of $m \le 10^7$ was chosen as we were mainly interested in the question of extra vanishing, and 10^7 seemed sufficient to answer the question posed by [ZK] on whether the rate remained constant. With today's technology, extending the experiment to $m \le 10^8$ should be feasible, as should a similar experiment looking at cubic twists with odd functional equation.

As stated in [ZK], the computation of the $a_m(n)$ takes time $O(\log n)$ if n is prime and $O(1)$ time otherwise (using the multiplicativity relations, viewing the values for the primes dividing n as taking negligible time as they are already computed). We computed the values of $a_1(p)$ for $p \le 10^9$ once-and-for-all ahead of time, and then read these from disk as needed. Additionally, tricks such as fast modular exponentiation were used to speed up the computation of $m^{(p-1)/3} \bmod p$. Similarly, the computing of $e^{-2\pi n/\sqrt{N_m}}$ was faciliated by the fact that the exponential function is a homomorphism; for a given N, we computed various powers of $e^{-2\pi/\sqrt{N_m}}$ and then for each n multiplied these together as needed to get the desired value. For the computation of $L(E_m, 1)$, and the question of how far the infinite sum need be computed, we followed a method similar to that of [ZK], calculating the $|\text{III}|$-value $S_m = \frac{T^2}{c\Omega} L(E_m, 1)$ where T is the size of the torsion group, c is the global Tamagawa number, and Ω is the real period (see pages 54–56 of [ZK] or Section 4 for these). We then stop the calculation when S_m is sufficiently close to an integer (possibly zero). As a check, we expect all the S_m values to be squares, which indeed does turn out to be the case.

4 Random matrix models

In this section we make some comments about random matrix theory and the expected number of even cubic twists of $X_0(27)$ which have vanishing central

L-value. We follow the ideas of [CKRS] and [DFK]. In our case of cubic twists, we expect, similar to the case in [CKRS], to have symmetry type O^+, that is, orthogonal with positive determinant. This is because the sign of our functional equation is always $+1$. Note that [DFK] have unitary symmetry in their type of cubic twist, due to the fact that the functional equation has an essentially arbitrary complex number (related to a Gauss sum) appearing in it.

We write $E = X_0(27)$ and E_d for the dth cubic twist of E. As given in equations (20), (22), and (16) of [CKRS], the assumption of O^+ symmetry implies that $P_E(N, x) = c_E N^{3/8}/\sqrt{x}$ should approximate (for small x) the probability density function for values of $L(E_d, 1)$, where $N \sim \log d$ and we integrate $\int_0^X P_E(N, x)\, dx$ to get an expected probability that $L(E_d, 1)$ is less than X. The idea is that we know that the actual values of $L(E_d, 1)$ are discretised (due to the Birch–Swinnerton-Dyer formula), and thus we declare (in a somewhat arbitrary manner) sufficiently small values of $L(E_d, 1)$ to indicate that in fact we have $L(E_d, 1) = 0$. We recall that the BSD conjecture implies we have

$$\frac{L(E_d, 1)}{\Omega_d} = \prod_{p|3d} c_p \cdot \frac{|\text{III}_d|}{|T_d|^2}$$

where Ω_d is the real period of E_d, the c_p are Tamagawa numbers, III_d is the Shafarevitch–Tate group, and T_d is the torsion group of E_d. We are thus thinking of $|\text{III}_d|$ (which is a square) as our discretised variable, with everything else being computable. When $d > 2$ the torsion group is trivial. For cubefree d we have that $\Omega_d = \Omega_1/d^{1/3}$, except when $9|d$ in which case we have $\Omega_d = 3\Omega_1/d^{1/3}$. Note that in definition (8) of [CKRS], quadratic twists that are not relatively prime to the conductor are excluded; we will similarly exclude twists that are divisible by 3, though one could deal with them via making appropriate corrections. For the Tamagawa product we have that $c_3 = 3$ when $d \equiv \pm 1$ (mod 9), $c_3 = 2$ when $d \equiv \pm 2$ (mod 9), and $c_3 = 1$ otherwise, while $c_p = 3$ for primes $p \equiv 1$ (mod 3) and $c_p = 1$ for primes $p \equiv 2$ (mod 3). Given this divergent behaviour based upon prime divisibility, as in Conjecture 1 of [CKRS] we decided to restrict to prime twists, and additionally split the primes into congruence classes modulo 9. Indeed, it is calculable that the sign of the functional equation is odd when our twisting prime d is congruent to $4, 7, 8$ (mod 9), and by 3-descent we can verify that the rank is zero when d is 2 or 5 (mod 9). Moreover, again by 3-descent, we know that the rank is at most 2 (and the functional equation is even) when d is 1 mod 9. Computing as with equation (23) in [CKRS] we are led to:

Question 4.1. *Let V_T be the set of primes d less than T congruent to 1 modulo 9 with $L(E_d, 1) = 0$. Is there some constant $c \neq 0$ such that*

$$\sum_{d \in V_T} 1 \sim cT^{5/6}(\log T)^{-5/8} \quad \text{as} \quad T \to \infty \quad ?$$

Assuming an affirmative answer, our data give a constant of approximately $c = 1/6$. The argument is similar for quartic twists of $X_0(32)$ or sextic twists of $X_0(27)$, and we can expect respective asymptotics for prime twists of order $T^{7/8}(\log T)^{-5/8}$ and $T^{11/12}(\log T)^{-5/8}$, and upon restricting to various congruence classes we should get appropriate constants in front of these. Via techniques from prime number theory and considerations from Tamagawa numbers, one should be able to argue as in [CKRS] to get an asymptotic for all cubefree twists.

Finally we derive a version of Conjecture 2 of [CKRS] suitable for cubic, quartic, and sextic twists. For cubic twists, for a given prime $p \equiv 1 \pmod 3$ there are 3 solutions to $a^2 + 3b^2 = 4p$ with $a \equiv 2 \pmod 3$, which correspond to the three possibilities for the Frobenius trace a_p. The argument given from (27)-(31) in [CKRS] does not differ (see below), and so we are led to:

Question 4.2. *Let $p \geq 5$ be prime, and for $1 \leq q \leq p - 1$ let $F_p^q(T)$ be the set of cubefree positive integers $d \equiv q \pmod p$ that are less than T such that $x^3 + y^3 = d$ has even functional equation and positive rank. Letting $a_d(p)$ be the pth trace of Frobenius for $x^3 + y^3 = d$ (where d need not be cubefree), do we have*

$$\lim_{T \to \infty} \left(\sum_{d \in F_p^Y(T)} 1 \Big/ \sum_{d \in F_p^Z(T)} 1 \right) = \sqrt{\frac{p + 1 - a_Y(p)}{p + 1 - a_Z(p)}} \quad ?$$

In Tables 2-5 below we list vanishing frequencies in support of an affirmative answer to the above question; the c-column represents which congruence class is used. For $p = 7$, the classes $c = 3, 4$ have $a(p) = 5$, while $c = 1, 6$ have $a(p) = -1$ and $c = 2, 5$ have $a(p) = -4$. The experimental data from Table 3 here show ratios of $[0.100, 0.184, 0.211]$, while theory predicts $[\sqrt{3} : \sqrt{9} : \sqrt{12}]$. For $p = 13$ the theory predicts $[\sqrt{9} : \sqrt{12} : \sqrt{21}]$. We also have some data (see Tables 6-9) for the vanishing frequencies for positive quartic twists of $X_0^\flat(32)$. For $p = 5$ the ratios should be $[\sqrt{2} : \sqrt{4} : \sqrt{8} : \sqrt{10}]$; for $p = 13$ they should be $[\sqrt{8} : \sqrt{10} : \sqrt{18} : \sqrt{20}]$. We could also make a similar calculation for sextic twists of $X_0(27)$, but did not do so.

The heuristic for Conjecture 2 in [CKRS] is based upon supposed cancellation from a quadratic character, whereas in our cubic twist case the source of cancellation is perhaps not so transparent. Therefore we go through the details. We have that

$$\sum_{d \in F_p^q(T)} L(E_d, 1/2)^k = \sum_{d \in F_p^q(T)} \left(\sum_{n=1}^{\infty} \frac{a_d(n)}{n} \right)^k = \sum_{d \in F_p^q(T)} \sum_{n=1}^{\infty} \frac{b_d(n)}{n},$$

where $b_m(n) = \sum_{n=n_1 \cdots n_k} a_m(n_1) \cdots a_m(n_k)$ with the sum being over all ways of writing n as a product of k positive factors. If we invert the order of summation in this last expression, the sum over d should typically have much cancellation since the $b_d(n)$ are essentially randomly distributed. This, however, is not the case for n that are a power of p, as here the value of $a_d(p^r)$ is fixed since d is fixed modulo p. Thus we should get a main contribution in the above by restricting to values of n that are powers of p (indeed, if we did this argument with no congruence restriction we would expect $n = 1$ to give the main term). As in (31) of [CKRS] we thus get that

$$\sum_{d \in F_p^q(T)} L(E_d, 1/2)^k \sim \sum_{d \in F_p^q(T)} \sum_{p^r} \frac{b_d(p^r)}{p^r} = \sum_{d \in F_p^q(T)} \left(\sum_{p^r} \frac{a_d(p^r)}{p^r} \right)^k =$$

$$= \left(\frac{p}{p + 1 - a_d(p)} \right)^k \sum_{d \in F_p^q(T)} 1.$$

We complete our heuristic by first noting that the sets $F_p^q(T)$ have asymptotically equal sizes and then taking $k = -1/2$ as is suggested by the random matrix theory of [CKRS]. Note that a similar heuristic can be given for moments of higher derivatives, but the combinatorics become more difficult due to the presence of logarithms. In this context, the data of Elkies [E2] distinctly show a congruence-class phenomenon for rank 3 quadratic twists of $X_0(32)$.

5   Acknowledgments

The author was partially funded by an NSF VIGRE Postdoctoral Fellowship at The Pennsylvania State University for part of the time this work was done. He also thanks an anonymous referee for useful comments.

References

[BS]   B. J. Birch, N. M. Stephens, *The parity of the rank of the Mordell-Weil group.* Topology **5** (1966), 295–299.

[C]   J. W. S. Cassels, *Arithmetic on curves of genus 1. I. On a conjecture of Selmer.* J. Reine Angew. Math. **202** (1959), 52–99.

[CKRS]   J. B. Conrey, J. P. Keating, M. O. Rubinstein, N. C. Snaith, *On the frequency of vanishing of quadratic twists of modular L-functions.* In *Number theory for the millennium, I* (Urbana, IL, 2000), A K Peters, Natick, MA (2002), 301–315. Available online at `arxiv.org/math.NT/0012043`

[DFK]   C. David, J. Fearnley, H. Kisilevsky, *On the vanishing of twisted L-functions of elliptic curves.* Experiment. Math. **13** (2004), no. 2, 185–198. Available online at `arxiv.org/math.NT/0406012`

[De]   C. Delaunay, *Heuristics on Tate-Shafarevitch Groups of Elliptic Curves Defined over* **Q**. Experiment. Math. **10** (2001), no. 2, 191–196.

[E1]   N. D. Elkies, *Heegner point computations.* In *Algorithmic Number Theory*, Proceedings of the First International Symposium (ANTS-I) held at Cornell University, Ithaca, New York, May 6–9, 1994. Edited by L. M.

Adleman and M.-D. Huang. Lecture Notes in Computer Science, **877**. Springer-Verlag, Berlin (1994), 122–133.

[E2] N. D. Elkies, *Curves $Dy^2 = x^3 - x$ of odd analytic rank*. In *Algorithmic number theory*, Proceedings of the 5th International Symposium (ANTS-V) held at the University of Sydney, Sydney, July 7–12, 2002. Lecture Notes in Computer Science, **2369**. Springer-Verlag, Berlin (2002), 244–251. `arxiv.org/math.NT/0208056`

[ER] N. D. Elkies, N. F. Rogers, *Elliptic curves $x^3 + y^3 = k$ of high rank*. In *Algorithmic number theory*, Proceedings of the 6th International Symposium (ANTS-VI) held at the University of Vermont, Burlington, VT, June 13–18, 2004. Lecture Notes in Computer Science, **3076**. Springer-Verlag, Berlin (2004), 184–193. `arxiv.org/math.NT/0403116`

[L] D. B. Lieman, *Nonvanishing of L-series associated to cubic twists of elliptic curves*. Ann. of Math. (2) **140** (1994), no. 1, 81–108.

[M] L. Mai, *The analytic rank of a family of elliptic curves*. Canad. J. Math. **45** (1993), no. 4, 847–862. Also: *The average analytic rank of a family of elliptic curves*. J. Number Theory **45** (1993), no. 1, 45–60.

[N] T. Nagell, *L'analyse indéterminée de degré supérieur*. Mémorial des Sciences Mathématiques, Fascicule XXXIX, Paris, 1929.

[RVZ] F. Rodriguez Villegas, D. Zagier, *Which primes are sums of two cubes?* In *Number theory*, Proceedings of the Fourth Conference of the Canadian Number Theory Association held at Dalhousie University, Halifax, Nova Scotia, July 2–8, 1994. Edited by K. Dilcher. CMS Conference Proceedings, **15**. Published by the AMS, Providence, RI; (1995), 295–306.

[Sel] E. S. Selmer, *The Diophantine equation $ax^3 + by^3 + cz^3 = 0$*. Acta Math. **85** (1951), 203–362, Acta Math. **92** (1954), 191–197.

[Ste1] N. M. Stephens, *Conjectures concerning elliptic curves*. Bull. Amer. Math. Soc. **73** (1967), 160–163.

[Ste2] N. M. Stephens, *The diophantine equation $X^3 + Y^3 = DZ^3$ and the conjectures of Birch and Swinnerton-Dyer*. J. Reine Angew. Math. **231** (1968), 121–162.

[ST] C. L. Stewart, J. Top, *On Ranks of Twists of Elliptic Curves and Power-Free Values of Binary Forms*. JAMS, **8** (1995), no. 4, 943–973.

[Syl] J. J. Sylvester, *On Certain Ternary Cubic-Form Equations*. American Journal of Mathematics, **2** (1879), no. 3, 280–285, no. 4, 357–393, **3** (1880), no. 1, 58–88, no. 2, 179–189.

[ZK] D. Zagier, G. Kramarz, *Numerical Investigations Related to the L-series of Certain Elliptic Curves*. J. Indian Math. Soc. **52** (1987), 51–69.

School of Mathematics
University of Bristol
Bristol BS8 1TW
UK

Table 2: $p = 5$, $X_0(27)$

c	$\#r > 0$	$\#$curves	
1	140463	838612	0.167
2	140549	838570	0.168
3	140613	838575	0.168
4	140750	838637	0.168

Table 3: $p = 7$, $X_0(27)$

c	$\#r > 0$	$\#$curves	
1	109569	595982	0.184
2	125728	595952	0.211
3	59440	595912	0.100
4	58759	595903	0.099
5	125714	595963	0.211
6	110125	595937	0.185

Table 4: $p = 11$, $X_0(27)$

c	$\#r > 0$	$\#$curves	
1	64989	378410	0.172
2	65211	378408	0.172
3	65001	378430	0.172
4	65008	378444	0.172
5	64956	378423	0.172
6	65208	378426	0.172
7	65054	378411	0.172
8	64773	378422	0.171
9	65164	378396	0.172
10	65338	378401	0.173

Table 5: $p = 13$, $X_0(27)$

c	$\#r > 0$	$\#$curves	
1	44504	320075	0.139
2	52214	320099	0.163
3	51754	320124	0.162
4	67352	320151	0.210
5	43064	320116	0.135
6	68325	320090	0.213
7	68702	320124	0.215
8	43215	320104	0.135
9	67584	320107	0.211
10	51465	320072	0.161
11	51827	320135	0.162
12	44858	320042	0.140

Table 6: $p = 5$, $X_0(32)$

c	$\#r > 0$	$\#$curves	
1	156097	749089	0.208
2	104136	749107	0.139
3	236861	749125	0.316
4	215944	749182	0.288

Table 7: $p = 7$, $X_0(32)$

c	$\#r > 0$	$\#$curves	
1	128846	538523	0.239
2	128491	538505	0.239
3	128553	538517	0.239
4	128597	538505	0.239
5	128053	538495	0.238
6	128335	538512	0.238

Table 8: $p = 11$, $X_0(32)$

c	$\#r > 0$	$\#$curves	
1	82653	341092	0.242
2	82782	341070	0.243
3	82581	341069	0.242
4	82392	341072	0.242
5	82806	341113	0.243
6	82448	341061	0.242
7	82661	341108	0.242
8	82388	341045	0.242
9	82720	341091	0.243
10	82948	341083	0.243

Table 9: $p = 13$, $X_0(32)$

c	$\#r > 0$	$\#$curves	
1	85079	287669	0.296
2	60843	287670	0.212
3	85408	287673	0.297
4	53551	287693	0.186
5	60788	287689	0.211
6	60926	287684	0.212
7	81716	287656	0.284
8	81500	287704	0.283
9	85480	287661	0.297
10	53852	287654	0.187
11	81525	287683	0.283
12	53688	287668	0.187

Vanishing of L-functions of elliptic curves over number fields

Chantal David, Jack Fearnley and Hershy Kisilevsky *

Abstract

Let E be an elliptic curve over $\mathbb{Q}$, with L-function $L_E(s)$. For any primitive Dirichlet character χ, let $L_E(s,\chi)$ be the L-function of E twisted by χ. In this paper, we use random matrix theory to study vanishing of the twisted L-functions $L_E(s,\chi)$ at the central value $s=1$. In particular, random matrix theory predicts that there are infinitely many characters of order 3 and 5 such that $L_E(1,\chi)=0$, but that for any fixed prime $k \geqslant 7$, there are only finitely many character of order k such that $L_E(1,\chi)$ vanishes. With the Birch and Swinnerton-Dyer Conjecture, those conjectures can be restated to predict the number of cyclic extensions $K/\mathbb{Q}$ of prime degree such that E acquires new rank over K.

1 Introduction

Let E be an elliptic curve defined over $\mathbb{Q}$ with conductor N_E. For any number field $K/\mathbb{Q}$, let $E(K)$ be the group of points of E defined over K. By the Mordell-Weil Theorem, $E(K)$ is a finitely generated abelian group. Let $L_E(s,K)$ be the L-function of E over the field K.

Conjecture 1.1 (Birch and Swinnerton-Dyer conjecture over number fields). $L_E(s,K)$ *has analytic continuation to the whole complex plane, and*

$$ord_{s=1}L_E(s,K) \;\; = \;\; r_K(E)$$

where $r_K(E)$ is the rank of $E(K)$.

In this paper, we fix E an elliptic curve over $\mathbb{Q}$, and we study how the rank varies over abelian fields $K/\mathbb{Q}$ of fixed prime degree. For example, are there infinitely many such number fields where E acquires new rank over K (i.e. $r_K(E) > r_{\mathbb{Q}}(E)$)? With the Birch and Swinnerton-Dyer conjecture, one can rephrase the question in terms of vanishing of the L-function $L_E(s,K)$ at $s=1$. Let K be an abelian extension of $\mathbb{Q}$ with Galois group G and conductor

---

*The first and third authors are partially supported by grants from NSERC and FCAR

m. Let $\hat{G}$ be the group of characters of G which can be identified with a set of Dirichlet characters

$$\chi : (\mathbb{Z}/m\mathbb{Z})^* \to \mathbb{C}^*.$$

Let

$$L_E(s) = L_E(s, \mathbb{Q}) = \sum_{n \geqslant 1} \frac{a_n}{n^s}$$

be the L-function of E over $\mathbb{Q}$. For each primitive Dirichlet character χ, let

$$L_E(s, \chi) = \sum_{n \geqslant 1} \frac{\chi(n)\, a_n}{n^s}$$

be the L-function of E over $\mathbb{Q}$ twisted by the character χ. By the work of [16, 15, 1], $L_E(s)$ and $L_E(s, \chi)$ have analytic continuation to the whole complex plane. It also follows from properties of number fields that

$$L_E(s, K) = \prod_{\chi \in \hat{G}} L_E(s, \chi), \tag{1.1}$$

and the vanishing of the twisted L-functions $L_E(s, \chi)$ at $s = 1$ is equivalent, via the Birch and Swinnerton-Dyer conjecture, to the existence of rational points of infinite order on $E(K)$.

In this paper, we use random matrix theory to study the vanishing of the twisted L-functions $L_E(s, \chi)$ at $s = 1$. It has been known since the work of Montgomery [13] that certain statistics on probability spaces of random matrices (such as pair correlation between the eigenangles of the matrices) are similar to the same statistics on the zeroes of the Riemann zeta function. This intuition is supported by the extensive computations of Odlyzko [14] on the critical zeroes of the Riemann zeta function.

This was explored further in the work of Katz and Sarnak [8, 9], who extend the analogy between other probability spaces of matrices, and families of L-functions. Katz and Sarnak also studied the case of function fields, where they can actually prove some of those mysterious connections between random matrices and families of L-functions.

In order to study different statistics of number theoretic objects, Keating and Snaith [10, 11] introduced a new random variable on spaces of random matrices, the characteristic polynomial of the matrix evaluated at a given point. They computed the probability distribution of this new variable, which led to striking conjectures for the asymptotic behavior of the moments of the Riemann zeta function on the critical line. The ideas of Keating and Snaith have been applied to study vanishing of L-functions in families [3, 5, 17]. Families of quadratic twists are studied in [3], where a conjectural asymptotic for the number of quadratic twists with even non-zero rank is presented. This refines a conjecture of Goldfeld [6] which predicts that quadratic twists with rank greater than one have density zero. In [5], the authors used the ideas of Keating and Snaith to obtain a conjectural asymptotic for the number of cubic Dirichlet characters χ such that $L_E(1, \chi)$ vanishes. We present in this paper the case of characters of order k, for k any odd prime. More precisely, the conjectures that we obtain from the random matrix model are (the case $k = 3$ of [5] is included for completeness):

Conjecture 1.2. *Let k be an odd prime, let E be an elliptic curve defined over $\mathbb{Q}$, and let*

$$N_{E,k}(X) \;=\; \#\{\chi \text{ of order } k \; : \; \mathrm{cond}(\chi) \leqslant X \text{ and } L_E(1, \chi) = 0\}.$$

If $k = 3$, then

$$\log N_{E,k}(X) \;\sim\; \frac{1}{2}\log X \quad \text{as } X \to \infty.$$

If $k = 5$, then $N_{E,k}(X)$ is unbounded, but $N_{E,k}(X) \ll X^\varepsilon$ for any $\varepsilon > 0$ as $X \to \infty$.

If $k \geqslant 7$, then $N_{E,k}(X)$ is bounded.

In the light of (1.1), and under the Birch and Swinnerton-Dyer conjecture, one can rewrite $N_{E,k}(X)$ as

$$N_{E,k}(X) \;=\; (k-1)\,\#\{K/\mathbb{Q} \text{ cyclic of degree } k \; :$$
$$\mathrm{cond}(K) \leqslant X \text{ and } r_K(E) > r_{\mathbb{Q}}(E)\}.$$

The structure of the paper is as follows. In the second section, we use modular symbols to rewrite the special values $L_E(1, \chi)$ as a product of terms depending only on E, and some algebraic integer $n_E(\chi)$ depending on the character. In the third section, we use the embedding of number fields as lattices in $\mathbb{C}$ to give a discretisation of the algebraic integers $n_E(\chi)$. In the fourth section, we use this discretisation and the work of Keating and Snaith to obtain conjectures on the asymptotic behavior of $N_{E,k}(X)$. Finally, the last section presents some experimental results.

2   Special values and modular symbols

The notation of this section follows the introduction of [12]. Let E be an elliptic curve over $\mathbb{Q}$. By the work of [16, 15, 1], E is modular and let $f(z) = \sum_{n \geqslant 1} a_n\, e^{2\pi i n z}$ be the Fourier expansion of the modular form associated to E. Then, the L-function $L_E(s)$ is the Mellin transform

$$L_E(s) = \frac{(2\pi)^s}{\Gamma(s)} \int_0^\infty f(it)t^{s-1}\, dt. \tag{2.1}$$

$L_E(s)$ has analytic continuation to the whole complex plane, and satisfies the functional equation

$$\Lambda_E(s) = \left(\frac{\sqrt{N_E}}{2\pi}\right)^s \Gamma(s) L_E(s) = w_E \Lambda_E(2-s) \tag{2.2}$$

where $w_E = \pm 1$ is called the root number. From (2.1), we have that $L_E(1) = 2\pi \int_0^\infty f(it)\, dt$. For $a, m \in \mathbb{Q}, m > 0$, one defines the modular symbols

$$\lambda(a, m; E) \;=\; \lambda(a, m; f) = 2\pi \int_0^\infty f\left(it - \frac{a}{m}\right) dt. \tag{2.3}$$

Let χ be a primitive character of modulus m with Gauss sum

$$\tau(\chi) = \sum_{a \bmod m} \chi(a) e^{2\pi i a / m}.$$

The twisted L-function $L_E(s, \chi)$ satisfies the functional equation

$$\begin{aligned}
\Lambda_E(s, \chi) \;&=\; \left(\frac{m\sqrt{N_E}}{2\pi}\right)^s \Gamma(s) L_E(s, \chi) \\
&=\; \frac{w_E \chi(N_E) \tau(\chi)^2}{m} \Lambda_E(2 - s, \overline{\chi}).
\end{aligned}$$

From the identity

$$\chi(n) = \frac{1}{\tau(\overline{\chi})} \sum_{a \bmod m} \overline{\chi}(a) e^{2\pi i a n / m},$$

we have

$$\begin{aligned}
f_\chi(z) \;&=\; \sum_{n \geqslant 1} \chi(n) a_n e^{2\pi i n z} \\
&=\; \frac{1}{\tau(\overline{\chi})} \sum_{a \bmod m} \overline{\chi}(a) f\left(z + a/m\right)
\end{aligned}$$

by rearranging the sums (Birch's lemma). It then follows that

$$L_E(1, \chi) \;=\; \frac{1}{\tau(\overline{\chi})} \sum_{a \bmod m} \overline{\chi}(a) \lambda(a, m; E). \tag{2.4}$$

We define

$$\begin{aligned}
\lambda^+(a, m; E) \;&=\; \lambda(a, m; E) + \lambda(-a, m; E) \\
&=\; 2\pi \int_0^\infty \sum_{n \geqslant 1} a_n e^{-2\pi n t} \left(e^{2\pi i a n / m} + e^{-2\pi i a n / m}\right) dt \\
&=\; 4\pi \sum_{n \geqslant 1} a_n \left(\operatorname{Re}\left\{e^{2\pi i a n / m}\right\}\right) \int_0^\infty e^{-2\pi n t}\, dt \\
&=\; 2 \sum_{n \geqslant 1} \frac{a_n}{n} \operatorname{Re}\left\{e^{2\pi i a n / m}\right\}
\end{aligned}$$

which is a real number as E is defined over $\mathbb{Q}$. As in [12], there is a rational multiple Ω_E of the real period such that all $\lambda^+(a, m; E)$ satisfy

$$\Lambda(a, m; E) = \frac{\lambda^+(a, m; E)}{\Omega_E} \in \mathbb{Z}.$$

In this paper, χ has prime order $k \geqslant 3$. Then, $\chi(-1) = 1$, and we can rewrite (2.4) as

$$L_E(1, \chi) = \frac{\Omega_E}{2\tau(\overline{\chi})} \sum_{a \bmod m} \overline{\chi}(a)\Lambda(a, m; E) \qquad (2.5)$$

where the integers $\Lambda(a, m; E)$ do not depend on the character χ, but only on the conductor m. Then, we see from (2.5) that

$$\frac{2\tau(\overline{\chi})L_E(1, \chi)}{\Omega_E}$$

is an algebraic integer in the field obtained by adding a kth root of unity. In fact, one can prove the stronger result, in the theorem below, which is critical to the discretisation of the next section. The particular case $k = 3$ was proven in [5]. For any odd prime k, let $\mathbb{Q}(\xi_k)$ be the cyclotomic field obtained by adding a primitive kth root of unity ξ_k, and let $\mathbb{Q}(\xi_k)^+$ be the maximal real extension $\mathbb{Q} \subseteq \mathbb{Q}(\xi_k)^+ \subseteq \mathbb{Q}(\xi_k)$ of degree $(k-1)/2$ over $\mathbb{Q}$. The ring of integers of $\mathbb{Q}(\xi_k)^+$ will be denoted by $\mathbb{Z}[\xi_k]^+$.

Theorem 2.1. *Let k be an odd prime, and let χ be a primitive character of order k. Then,*

$$\frac{2\tau(\overline{\chi})L_E(1, \chi)}{\Omega_E} = \begin{cases} \chi(N_E)^{(k+1)/2} \, n_E(\chi) & \text{when } w_E = 1 \\[2ex] \left(\xi_k^{-1} - \xi_k\right)^{-1} \chi(N_E)^{(k+1)/2} \, n_E(\chi) & \text{when } w_E = -1 \end{cases}$$

where $n_E(\chi) \in \mathbb{Z}[\xi_k] \cap \mathbb{R} = \mathbb{Z}[\xi_k]^+$.

Proof: From the functional equation, we have

$$L_E^{alg}(1, \chi) = \frac{2\tau(\overline{\chi})L_E(1, \chi)}{\Omega_E} = \frac{2\tau(\overline{\chi})w_E\chi(N_E)\tau(\chi)^2}{m\Omega_E}L_E(1, \overline{\chi})$$

$$= w_E\chi(N_E)\overline{\frac{2\tau(\overline{\chi})L_E(1, \chi)}{\Omega_E}} = w_E\chi(N_E)\overline{L_E^{alg}(1, \chi)}$$

Let $z \in \mathbb{C}^*$ satisfying $z = w_E\chi(N_E)\overline{z}$. Then, $L_E^{alg}(1, \chi) = \alpha\overline{z}^{-1}$ with α real. If $w_E = 1$, we take $z = \chi(N_E)^{(k+1)/2}$, and $L_E^{alg}(1, \chi) = \alpha\overline{z}^{-1}$ with $\alpha \in \mathbb{R} \cap \mathbb{Z}[\xi_k] = \mathbb{Z}[\xi_k]^+$, which gives the result.
If $w_E = -1$, we take $z = \left(\xi_k - \xi_k^{-1}\right)\chi(N_E)^{(k+1)/2}$, and $L_E^{alg}(1, \chi) = \alpha\overline{z}^{-1}$ with $\alpha \in \mathbb{R} \cap \mathbb{Z}[\xi_k] = \mathbb{Z}[\xi_k]^+$, which gives the result. $\qquad\square$

3 Discretisation

By Theorem 2.1, $n_E(\chi)$ is an algebraic integer in $\mathbb{Z}[\xi_k]^+$, and there is then a natural discretisation on the algebraic integer $n_E(\chi)$ given by the geometry of numbers. Let ϕ be the map

$$\phi : \mathbb{Z}[\xi_k]^+ \;\longrightarrow\; \mathbb{R}^{(k-1)/2}$$
$$\alpha \;\longmapsto\; (\sigma_1(\alpha), \sigma_2(\alpha), \ldots, \sigma_{(k-1)/2}(\alpha))$$

where $\mathrm{Gal}(\mathbb{Q}(\xi_k)^+/\mathbb{Q}) = \{\sigma_1 = 1, \sigma_2, \ldots, \sigma_{(k-1)/2}\}$. Let $\alpha_1, \ldots, \alpha_{(k-1)/2}$ be an integral basis for $\mathbb{Z}[\xi_k]^+$. The image of $\mathbb{Z}[\xi_k]^+$ in $\mathbb{R}^{(k-1)/2}$ is the lattice generated by the linearly independent vectors

$$\omega_1 = \phi(\alpha_1), \;\ldots, \; \omega_{(k-1)/2} = \phi(\alpha_{(k-1)/2}).$$

Let $R \subseteq \mathbb{R}^{(k-1)/2}$ be the region

$$R \;=\; \big\{\, a_1\omega_1 + a_2\omega_2 + \cdots + a_{(k-1)/2}\omega_{(k-1)/2} :$$
$$-1 < a_i < 1 \text{ for } 1 \leqslant i \leqslant (k-1)/2 \,\big\}.$$

The discretisation given by the embedding of $\mathbb{Z}[\xi_k]^+$ in $\mathbb{R}^{(k-1)/2}$ is then

$$n_E(\chi) = 0 \;\Longleftrightarrow\; \phi(n_E(\chi)) \in R. \tag{3.1}$$

Let χ be any character of conductor m and order k. For any automorphism $\sigma \in \mathrm{Gal}(\mathbb{Q}(\xi_k)/\mathbb{Q})$, let χ^σ be the character

$$\chi^\sigma : (\mathbb{Z}/m\mathbb{Z})^* \;\longrightarrow\; \langle \xi_k \rangle \subseteq \mathbb{C}^*$$
$$a \;\longmapsto\; \sigma\,(\chi(a))$$

Then, χ^σ is also a character of conductor m and order k.

Lemma 3.1. *Let k be an odd prime, and χ a character of order k and conductor m. For any σ in $\mathrm{Gal}(\mathbb{Q}(\xi_k)/\mathbb{Q})$, we have*

$$|L_E(1, \chi^\sigma)| \;=\; \frac{c_{E,k}}{m^{1/2}} |n_E(\chi)^\sigma|$$

where $c_{E,k}$ is an explicit constant depending only on E and k.

Proof: Using (2.5), we have

$$L_E^{alg}(1, \chi)^\sigma \;=\; \left(\frac{2\tau(\overline{\chi}) L_E(1, \chi)}{\Omega_E} \right)^\sigma$$
$$=\; \sum_{a \bmod m} \overline{\chi}^\sigma(a) \Lambda(a, m; E) = L_E^{alg}(1, \chi^\sigma).$$

Suppose first that $\omega_E = 1$. Then,

$$n_E(\chi)^\sigma \;=\; L_E^{alg}(1, \chi)^\sigma \left(\chi(N_E)^{(k+1)/2} \right)^\sigma$$
$$=\; \frac{2\tau(\overline{\chi}^\sigma) L_E(1, \chi^\sigma)}{\Omega_E} \left(\chi(N_E)^{(k+1)/2} \right)^\sigma,$$

and taking absolute values we get the result with $c_{E,k} = \Omega_E/2$. The proof for $\omega_E = -1$ is similar, with a different explicit constant $c_{E,k}$. $\qquad\square$

We first consider the case $k = 5$. We have that $\mathbb{Z}[\xi_5]^+ = \mathbb{Z}[\alpha]$ with $\alpha = (1 + \sqrt{5})/2$, and $G_5 = \langle 1, \tau \rangle$, where the non-trivial automorphism τ sends $\sqrt{5}$ to $-\sqrt{5}$. Then, the lattice of $\mathbb{Z}[\alpha]$ in $\mathbb{R}^2$ is generated by $\omega_1 = (\alpha, \alpha^\tau)$ and $\omega_2 = (\alpha^\tau, \alpha)$. Let R be the region

$$R = \{a\omega_1 + b\omega_2 : -1 < a < 1, -1 < b < 1\}.$$

By (3.1), $n_E(\chi) = 0$ if and only if $\phi(n_E(\chi)) = (n_E(\chi), n_E(\chi)^\tau) \in R$. As the region R is not symmetric with respect to the absolute value, and we have a probability model for $|L_E(1, \chi)|$, we also consider the two regions of $\mathbb{R}^2$

$$
\begin{aligned}
R_1 &= \{(x, y) \ : \ -1 < x, y < 1\} \\
R_2 &= \left\{(x, y) \ : \ -\sqrt{5} < x, y < \sqrt{5}\right\}
\end{aligned}
$$

with the property that $R_1 \subseteq R \subseteq R_2$, and

$$(n_E(\chi), n_E(\chi)^\tau) \in R_i \quad\Longleftrightarrow\quad (|n_E(\chi)|, |n_E(\chi)^\tau|) \in |R_i| \tag{3.2}$$

where

$$|R_i| = \{(x, y) \in R_i \ : \ x, y \geqslant 0\}.$$

The following lemma is now immediate from (3.2) and Lemma 3.1

Lemma 3.2. *Let $\sigma \in Gal(\mathbb{Q}(\xi_5)/\mathbb{Q})$ be an automorphism which restricts to the non-trivial automorphism of $\mathbb{Q}(\sqrt{5})$. For $i = 1, 2$, we have*

$$(n_E(\chi), n_E(\chi)^\sigma) \in R_i \quad\Longleftrightarrow\quad |L_E(1, \chi)|, |L_E(1, \chi^\sigma)| \leqslant \frac{c_i}{\sqrt{m}}$$

where c_1, c_2 are explicit constants depending only on E.

We now suppose that $k \geqslant 7$. Let B be the non-zero constant

$$B = \max_{1 \leqslant i \leqslant (k-1)/2} \sum_{j=1}^{(k-1)/2} |\sigma_i(\alpha_j)|,$$

and let $R \subseteq R' \subseteq \mathbb{R}^{(k-1)/2}$ be the region

$$R' = \left\{(x_1, \ldots, x_{(k-1)/2}) \ : \ -B \leqslant x_i \leqslant B \text{ for } 1 \leqslant i \leqslant (k-1)/2.\right\}$$

Then, $n_E(\chi) = 0 \Rightarrow \phi(n_E(\chi)) \in R'$ by (3.1). The following lemma is now immediate from Lemma 3.1

Lemma 3.3. *Let $\sigma_1, \ldots, \sigma_{(k-1)/2} \in Gal(\mathbb{Q}(\xi_k)/\mathbb{Q})$ be a set of representatives for $G_k = Gal(\mathbb{Q}(\xi_k)^+/\mathbb{Q})$. Then, $\phi(n_E(\chi)) \in R'$ if and only if*

$$|L_E(1, \chi^{\sigma_i})| \leqslant \frac{c_k}{m^{1/2}} \quad \text{for } 1 \leqslant i \leqslant (k-1)/2$$

where c_k is an explicit constant depending only E and k.

4   Unitary Random Matrices

Let $U(N)$ be the set of unitary $N \times N$ matrices with complex coefficients which forms a probability space with respect to the Haar measure. For each $A \in U(N)$, let

$$P_A(\lambda) = \det(A - \lambda I)$$

be the characteristic polynomial of A. For any $s \in \mathbb{C}$, let

$$M_U(s, N) = \int_{U(N)} |P_A(1)|^s \, d\text{Haar}$$

be the moments for the distribution of $|P_A(1)|$ in $U(N)$ with respect to the Haar measure. In [10], Keating and Snaith proved that

$$M_U(s, N) \;\; = \;\; \prod_{j=1}^{N} \frac{\Gamma(j)\Gamma(j+s)}{\Gamma^2(j+s/2)}, \tag{4.1}$$

and then $M_U(s, N)$ is analytic for $\text{Re}(s) > -1$, and has meromorphic continuation to the whole complex plane. By Fourier inversion, the probability density of $|P_A(1)|$ is

$$p(x) = \frac{1}{2\pi i} \int_{(c)} M_U(s, N) x^{-s-1} ds$$

for some $c > -1$. Then, for any $I \subseteq \mathbb{R}$,

$$\text{Prob}\left(|P_A(1)| \in I\right) = \int_I p(x) \, dx.$$

In our application to the vanishing of twisted L-functions, we will be interested only in small values of x where the value of $p(x)$ is determined by the first pole of $M_U(s, N)$ at $s = -1$. More precisely, for

$$x \leqslant N^{-1/2},$$

one can show that

$$p(x) \sim G^2(1/2) N^{1/4} \quad \text{as } N \to \infty,$$

where $G(z)$ is the Barnes G-function, with special value

$$G(1/2) = \exp\left(\frac{3}{2}\zeta'(-1) - \frac{1}{4}\log \pi + \frac{1}{24}\log 2\right)$$

(see [10, p. 81] or [7, p. 58] for more details).

We now consider the moments for the special values of L-functions in families of twists. Fix $k \geqslant 3$, and let

$$S_k(X) \;\; = \;\; \{\chi \text{ of order } k \text{ and conductor} \leqslant X\}$$
$$N_k(X) \;\; = \;\; \#S_k(X) \sim b_k X$$

with an explicit constant b_k (see for example [2]). We then define for any $s \in \mathbb{C}$

$$M_E(s, X) \;=\; \frac{1}{N_k(X)} \sum_{\chi \in S_k(X)} |L_E(1, \chi)|^s \tag{4.2}$$

The family of twists of order k has unitary symmetry, as do the values $|\zeta(1/2 + it)|$ on the critical line. Then

Conjecture 4.1 (Keating and Snaith Conjecture for twists of order k).

$$M_E(s, X) \;\sim\; a_E(s/2) M_U(s, N) \quad \text{as } N = 2 \log X \to \infty,$$

where $a_E(s/2)$ is an arithmetic factor depending only on the curve E.

In the conjecture, the relation between N and X is obtained by equating the mean density of eigenangles of matrices in the unitary group, and the mean density of non-trivial zeroes of the twisted L-functions $L_E(s, \chi)$ at a fixed height (see [5]). The arithmetic factor $a_E(s)$ can not be obtained from the random matrix theory, and has to be determined separately for each family from its arithmetic. This was done for the family of cubic twists in [5], and could be done for the family of twists of order k for each k. The arithmetic factor $a_E(s)$ would then be a meromorphic function for all $s \in \mathbb{C}$. As it will be seen below, the only influence of the arithmetic factor $a_E(s)$ in our application is that the special value $a_E(-1/2)$ will be part of the constant of the conjectural asymptotic of $N_{E,k}(X)$. This would not provide any further information to the cases $k \geqslant 5$ considered in this paper in view of Conjecture 1.2.

From Conjecture 4.1, the probability density $p_E(x)$ for the distribution of the special values $|L_E(1, \chi)|$ for characters of order k is

$$p_E(x) \;=\; \frac{1}{2\pi i} \int_{(c)} M_E(s, X)\, x^{-s-1}\, ds$$

$$\sim\; \frac{1}{2\pi i} \int_{(c)} a_E(s/2)\, M_U(s, N)\, x^{-s-1}\, ds \tag{4.3}$$

as $N = 2 \log X \to \infty$. As above, when $x \leqslant N^{-1/2}$, the value of $p_E(x)$ is determined by the residue of $M_U(s, N)$ at $s = -1$, and it follows from (4.3) that

$$p_E(x) \sim C_E \log^{1/4} X \tag{4.4}$$

for $x \leqslant (2 \log X)^{-1/2}$, $X \to \infty$, and $C_E = 2^{1/4} a_E(-1/2) G^2(1/2)$.

Let χ be a character of order $k \geqslant 3$ and conductor m. We apply the above model to find the probability that $|L_E(1, \chi)| < cm^{-1/2}$, for some constant $c > 0$. For $x < cm^{-1/2} < (2 \log m)^{-1/2}$ (for m large enough), we have $p_E(x) \sim C_E \log^{1/4} m$, and then

$$\text{Prob}\left(|L_E(1, \chi)| < cm^{-1/2}\right) \;\sim\; \int_0^{cm^{-1/2}} C_E \log^{1/4} m \; dx$$

$$=\; c C_E \frac{\log^{1/4} m}{m^{1/2}}. \tag{4.5}$$

We now use the probability density of the random matrix model with the discretisation of Section 3 to obtain conjectures for the vanishing of the L-values $L_E(1, \chi)$. We first suppose that $k = 5$, and as in the previous section, let $\sigma \in \mathrm{Gal}(\mathbb{Q}(\xi_5)/\mathbb{Q})$ which restricts to the non-trivial automorphism of $\mathbb{Q}(\sqrt{5})$. We saw in the previous two sections that

$$L_E(1, \chi) = 0 \quad \Longleftrightarrow \quad \phi(n_E(\chi)) = (n_E(\chi), n_E(\chi)^\sigma) \in R.$$

As $R_1 \subseteq R \subseteq R_2$, and using Lemma 3.2, the probability that $L_E(1, \chi)$ is zero is bounded below by

$$\mathrm{Prob}\left(|L_E(1, \chi)| < \frac{c_1}{\sqrt{m}}\right) \mathrm{Prob}\left(|L_E(1, \chi^\sigma)| < \frac{c_1}{\sqrt{m}}\right)$$

and bounded above by

$$\mathrm{Prob}\left(|L_E(1, \chi)| < \frac{c_2}{\sqrt{m}}\right) \mathrm{Prob}\left(|L_E(1, \chi^\sigma)| < \frac{c_2}{\sqrt{m}}\right).$$

Assuming that $|L_E(1, \chi)|$ and $|L_E(1, \chi^\sigma)|$ are independent identically distributed random variables, and using (4.5), we get that the probability that $L_E(1, \chi)$ is zero is about

$$\frac{\log^{1/2} m}{m},$$

neglecting all constants which are not significant here. The sum of the probabilities is

$$\sum_{\chi \in S_3(X)} \frac{\log^{1/2} m}{m} \quad \sim \quad \frac{2b_3}{3} \log^{3/2} X. \tag{4.6}$$

As discussed in [5], the exact power of $\log X$ that is obtained with the random matrix approach depends subtly on the discretisation, and is difficult to predict. For example, rational torsion of order three on the elliptic curve seemed to cause extra vanishing of the twisted L-values $L_E(1, \chi)$ for cubic characters, and changed the power of logarithm in the conjectural asymptotic for $N_{E,k}(X)$ of [5]. For $k = 5$, the sum of the probabilities is just on the border between convergence and divergence, and the random matrix model seems to indicate that the number of quintic twists such that $L_E(1, \chi)$ vanishes is infinite, but that $N_{E,k}(X) \ll X^\varepsilon$ for any $\varepsilon > 0$. This agrees with the empirical evidence of Section 5.

We now suppose that $k \geqslant 7$. Let $\sigma_1 = 1, \ldots, \sigma_{(k-1)/2}$ be elements of the Galois group of $\mathbb{Q}(\xi_k)/\mathbb{Q}$ which form a set of representatives for the Galois group of $\mathbb{Q}(\xi_k)^+/\mathbb{Q}$. As we saw in the two previous sections,

$$L_E(1, \chi) = 0 \quad \Longleftrightarrow \quad \phi(n_E(\chi)) = (n_E(\chi)^{\sigma_1}, \ldots, n_E(\chi)^{\sigma_{(k-1)/2}}) \in R.$$

As $R \subseteq R'$, and using Lemma 3.3, the probability that $L_E(1, \chi)$ is zero is bounded by

$$\mathrm{Prob}\left(|L_E(1, \chi^{\sigma_1})| < \frac{c_k}{\sqrt{m}}\right), \ldots, \mathrm{Prob}\left(|L_E(1, \chi^{\sigma_{(k-1)/2}})| < \frac{c_k}{\sqrt{m}}\right).$$

Assuming that $|L_E(1, \chi^{\sigma_i})|$ are independent identically distributed random variables for $1 \leqslant i \leqslant (k-1)/2$, and using (4.5), we get that the probability that $L_E(1, \chi)$ is zero is

$$\frac{\log^{(k-1)/8} m}{m^{(k-1)/4}},$$

neglecting all constants which are not significant here. Summing the probabilities, this gives for $k \geqslant 7$

$$\sum_{\chi \in S_3(X)} \frac{\log^{(k-1)/8} m}{m^{(k-1)/4}} = O(1). \tag{4.7}$$

From the random matrix model, we then conjecture that the number of twists of order $k \geqslant 7$ such that $L_E(1, \chi)$ vanishes is bounded. This also agrees with the empirical evidence of Section 5.

5 Numerical Evidence

The following table shows the observed number of vanishing twists $L_E(1, \chi)$ for characters of orders three, five and seven, and for the first three elliptic curves in the Cremona catalogue [4]. For each elliptic curve E, the characters with conductor prime to N_E and less than two million were considered. Any two characters of conductor m and order k generating the same cyclic subgroup of the character group are conjugate, and hence the special values $L_E(1, \chi)$ vanish simultaneously by Lemma 3.1. The number in the table records one of each class of conjugate characters for which the special value vanishes, which is $1/(k-1)$ of the number of characters with vanishing special value. The twists of order eleven in the same range for the curve $E14$ were also computed, and no vanishing were found.

Curve	Cubic vanishing	Quintic vanishing	Septic vanishing
E11	1152	15	2
E14	4347	10	0
E15	2050	11	0

The results for cubic twists have been analyzed in [5] and support Conjecture 1.2. The results for quintic and septic twists are too sparse to either support or refute Conjecture 1.2, but they nevertheless illustrate the extreme scarcity of vanishing in higher order twists which is predicted by the conjecture.

Acknowledgments This paper was first presented at the Clay Mathematics Institute "Special Week on Ranks of Elliptic Curves and Random Matrix Theory" held at the Isaac Newton Institute in February 2004. The first and second authors would like to thank the organizers of the workshop and the Isaac Newton Institute for their hospitality and financial support. The first author would also like to thank B. Birch, B. Conrey and C. Hughes for helpful discussions.

References

[1] C. Breuil, B. Conrad, F. Diamond, and R. Taylor, On the modularity of elliptic curves over $\mathbb{Q}$: wild 3-adic exercises, *J. Amer. Math. Soc.* **14** (2001), 843–939 (electronic).

[2] H. Cohen, F. Diaz y Diaz and M. Olivier, A survey of discriminant counting, Algorithmic number theory (Sydney, 2002), 80–94, Lecture Notes in Comput. Sci., **2369**, Springer, Berlin, 2002.

[3] J.B. Conrey, J.P. Keating, M. Rubinstein, N.C. Snaith, On the frequency of vanishing of quadratic twists of modular L-functions, Number theory for the Milennium I, 301–315, Editors: M. A. Bennett, B. C. Berndt, N. Boston NH. G. , Diamond, A. J. Hildebrand, W. Philipp, A. K. Peters Ltd, Natick, 2002.

[4] J. Cremona, Algorithms for modular elliptic curves, Cambridge University Press, Cambridge, UK, 1992.

[5] C. David, J. Fearnley and H. Kisilevsky, On the Vanishing of Twisted L-Functions of Elliptic Curves, *Experimental Mathematics*, 13 (2004), 185–198.

[6] D. Goldfeld, Conjectures on elliptic curves over quadratic fields, Number theory, Carbondale 1979 (Proc. Southern Illinois Conf., Southern Illinois Univ., Carbondale, Ill., 1979), pp. 108–118, Lecture Notes in Math. **751**, Springer, Berlin, 1979.

[7] C. P. Hughes, On the characteristic polynomial of random unitary matrix and the Riemann zeta function, Ph.D. thesis, University of Bristol, England, 2001.

[8] N. M. Katz and P. Sarnak, Zeroes of zeta functions and symmetry, *Bull. Amer. Math. Soc. (N.S.)* **36** (1999), 1–26.

[9] N. M. Katz and P. Sarnak, Random matrices, Frobenius eigenvalues, and monodromy, American Mathematical Society Colloquium Publications **45**, American Mathematical Society, Providence, RI, 1999.

[10] J. P. Keating and N. C. Snaith, Random matrix theory and $\zeta(1/2 + it)$, *Comm. Math. Phys.* **214** (2000), 57–89.

[11] J. P. Keating and N. C. Snaith, Random matrix theory and *L*-functions at $s = 1/2$, *Comm. Math. Phys.* **214** (2000), 91–110.

[12] B. Mazur, J. Tate and J. Teitelbaum, On *p*-adic analogues of the conjectures of Birch and Swinnerton-Dyer, *Invent. Math.* **84** (1986), 1–48.

[13] H. Montgomery, The pair correlation of zeroes of the zeta function, *Proc. Sym. Pure Math.* **24** (1973), 181–193.

[14] A. Odlyzko, The 10^{20}-th zero of the Riemann zeta function and 70 millions of his neighbors, preprint, A.T.T., 1989.

[15] R. Taylor and A. Wiles, Ring-theoretic properties of certain Hecke algebras, *Ann. of Math. (2)* **141**, 553-572.

[16] A. Wiles, Modular elliptic curves and Fermat's last theorem, *Ann. of Math. (2)* **141**, 443-551.

[17] M. Watkins, Rank distribution in a family of cubic twists, in this volume.

Department of Mathematics and Statistics,
Concordia University,
1455 de Maisonneuve Blvd. West,
Montréal, Québec,
Canada H3G 1M8

cdavid@mathstat.concordia.ca

jack@mathstat.concordia.ca

kisilev@mathstat.concordia.ca

Computing central values of L-functions

*Fernando Rodriguez-Villegas**

1

How fast can we compute the value of an L-function at the center of the critical strip?

We will divide this question into two separate questions while also making it more precise. Fix an elliptic curve E defined over $\mathbb{Q}$ and let $L(E, s)$ be its L-series. For each fundamental discriminant D let $L(E, D, s)$ be the L-series of the twist E_D of E by the corresponding quadratic character; note that $L(E, 1, s) = L(E, s)$.

A. How fast can we compute the central value $L(E, 1)$?

B. How fast can we compute $L(E, D, 1)$ for D in some interval say $a \leqslant D \leqslant b$?

These questions are obviously related but, as we will argue below, are not identical.

We should perhaps clarify what *to compute* means. First of all, we know, thanks to the work of Wiles and others, that $L(E, s) = L(f, s)$ for some modular form f of weight 2; hence, $L(E, s)$, first defined on the half-plane $\Re(s) > 3/2$, extends to an analytic function on the whole s-plane which satisfies a functional equation as s goes to $2 - s$. In particular, it makes sense to talk about the value $L(E, 1)$ of our L-function at the center of symmetry $s = 1$. The same reasoning applies to $L(E, D, s)$.

As a first approximation to our question we may simply want to know the real number $L(E, D, 1)$ to some precision given in advance; but we can expect something better. The Birch–Swinnerton-Dyer conjectures predict a formula of type

$$L(E, D, 1) = \kappa_D \, m_D^2, \tag{1.1}$$

for some integer m_D and κ_D an explicit easily computable positive constant. (Up to the usual fudge factors the conjectures predict that m_D^2, if non-zero, should be the order of the Tate–Shafarevich group of E_D.) To compute $L(E, D, 1)$ would then mean to calculate m_D *exactly*.

---

*Support for this work was provided in part by a grant of the NSF

In fact, formulas à la Waldspurger have the form (1.1) with m_D the $|D|$-th coefficient of a modular form g of weight $3/2$ which is in Shimura correspondance with f. The main point of this note is to discuss informally how explicit versions of such formulas can be used for problem **B** above.

Let us also note the interesting fact that m_D, being related to the coefficient of a modular form, typically does not have a constant sign. The significance of the extra information provided by $\mathrm{sgn}(m_D)$ remains a tantalizing mystery.

2

There is a standard analytic method to compute $L(E, 1)$, which we now recall. If E has conductor N then the associated modular form f has level N and

$$f|_{w_N} = -\varepsilon f,$$

where w_N is the Fricke involution and ε is the sign of the functional equation for $L(E, s)$. Concretely, we have

$$f\left(\frac{i}{\sqrt{N}t}\right) = \varepsilon t^2 f\left(\frac{it}{\sqrt{N}}\right), \qquad t \in \mathbb{R}.$$

It follows that

$$\left(\frac{2\pi}{\sqrt{N}}\right)^{-s} \Gamma(s)L(f, s) = \int_0^\infty f\left(\frac{it}{\sqrt{N}}\right) t^s \frac{dt}{t}.$$

Now break the integral as $\int_0^1 + \int_1^\infty$, make the substitution $t \mapsto 1/t$ in the first and use the Fricke involution to obtain

$$\left(\frac{2\pi}{\sqrt{N}}\right)^{-s} \Gamma(s)L(f, s) = \int_1^\infty f\left(\frac{it}{\sqrt{N}}\right) t^s \frac{dt}{t} + \varepsilon \int_1^\infty f\left(\frac{it}{\sqrt{N}}\right) t^{2-s} \frac{dt}{t}.$$

(This is the classical argument to prove the functional equation of $L(E, s)$ and goes back to Riemann who used it for his zeta function.)

Now plug in $s = 1$ to get

$$\frac{\sqrt{N}}{2\pi} L(E, 1) = (1 + \varepsilon) \sum_{n \geqslant 1} \frac{a_n}{n} e^{-2\pi n/\sqrt{N}} \tag{2.1}$$

where f has Fourier expansion

$$f = \sum_{n \geqslant 1} a_n q^n, \qquad q = e^{2\pi i z}.$$

Assume for simplicity that $\gcd(N, D) = 1$. Then the conductor of E_D is ND^2 and (2.1) applied to E_D yields, more generally,

$$\frac{|D|\sqrt{N}}{2\pi} L(E, D, 1) = (1 + \varepsilon_D) \sum_{n \geqslant 1} \left(\frac{D}{n}\right) \frac{a_n}{n} e^{-2\pi n/|D|\sqrt{N}}, \tag{2.2}$$

with ε_D the sign in the functional equation of $L(E, D, s)$.

We know that $|a_n|$ grows no more than polynomially with n (a straightforward argument gives $|a_n| = O(n)$). It follows that for a fixed E and varying D we will need to take, very roughly, of the order of $O(|D|)$ terms in the sum to obtain a decent approximation to $L(E, D, 1)$. Assuming the Birch–Swinnerton-Dyer conjectures we may use (2.2) to compute m_D^2 in (1.1) exactly. However, if we know that m_D is the D-th coefficient of some specific modular form (i.e. we have a formula à la Waldspurger) we would get $|m_D|$ but would not be able to recover $\mathrm{sgn}(m_D)$.

Using this method to compute, say, $L(E, D, 1)$ for $|D| \leqslant X$ would take time of the order of $O(X^2)$. We will see below that using formulas of type (1.1) we can reduce this to $O(X^{3/2})$ for at least some fraction of such D's.

3

Before tackling $L(E, D, 1)$ let us consider the case of the special value of an Eisenstein series of weight 2 (as opposed to a cusp form as we have for $L(E, D, 1)$). What follows is meant only as an illustration of the general case.

Let the L-function be $L\left(\left(\frac{D}{\cdot}\right), s-1\right)L\left(\left(\frac{D}{\cdot}\right), s\right)$ with $D < 0$ the discriminant of an imaginary quadratic field K. Its value at $s = 1$ is essentially $h(D)^2$, where $h(D)$ is the class number of K, and we find an analogue of (1.1) with $h(D)$ playing the role of m_D. There are many excellent algorithms for computing the class number $h(D)$ (see for example [1] chap. 5). Unfortunately, these do not obviously generalize to the calculation of m_D. The main reason for this is that the class group of K is easy to describe (both its elements and the group operation) in terms of binary quadratic forms, whereas its elliptic analogue, the Tate-Shafarevich group of E_D, is notoriously intractable.

The standard analytic method of the previous section yields the following formula (which was known to Lerch, see [2] vol. III, p. 171)

$$h(D)^2 = \frac{w_D^2 \sqrt{|D|}}{2\pi} \sum_{n \geqslant 1} \left(\frac{D}{n}\right) \frac{\sigma(n)}{n} e^{-2\pi n/|D|}, \tag{3.1}$$

where w_D is the number of units in K and $\sigma(n) := \sum_{d|n} d$ is the divisor sum function. Again, we need to take, roughly, $O(|D|)$ number of terms in the sum to obtain a reasonable approximation of the left hand side. In this case, we in fact have an *exact* formula requiring D terms, namely, Dirichlet's class number formula

$$h(D) = -\frac{w_D}{2|D|} \sum_{n=1}^{|D|-1} n \left(\frac{D}{n}\right). \tag{3.2}$$

Neither one of these formulas is, however, particularly useful for computing $h(D)$ in practice. On the other hand, it may be worth pointing out that similar arguments yield the formula [2] vol. III, p. 153.

$$h(D) = w_D \sum_{n \geqslant 1} \left(\frac{D}{n}\right) \frac{1}{1 - (-1)^n e^{\pi n/\sqrt{|D|}}}, \qquad D \equiv 5 \bmod 8,$$

with the number of necessary steps now reduced to the order of $O(\sqrt{|D|})$. (Analogous formulas can be given for D in other congruence classes modulo 8.)

To make the connection to the general case of computing $L(E, D, 1)$ that we are considering we mention two other possible approaches to computing $h(D)$ that do generalize.

(I) The first is to follow Gauss and realize ideal classes of K as classes of primitive, positive definite binary quadratic forms of discriminant D. Each class has a unique representative $Q = (a, b, c)$ in the standard fundamental domain (what is known as a *reduced form*) and we can simply enumerate these. A straightforward algorithm is as follows: run over values of b with $b \equiv D \bmod 2$ and $0 \leqslant b \leqslant \sqrt{|D|/3}$; for each b decompose $(b^2 - D)/4$ as ac with $0 < a \leqslant c$. Add one or two to the total count as the case may be if $\gcd(a, b, c) = 1$.

Though this algorithm also takes time $O(|D|)$ the constant of proportionality is very small making the algorithm quite practical. An important point to notice for our purpose, however, is that if we wanted to compute $h(D)$ for $0 \leqslant |D| \leqslant X$ we may simply run over all triples a, b, c of size at most $\sqrt{X/3}$ checking the necessary conditions on (a, b, c) for it to be a reduced form. In this way we obtain an algorithm which will run in time $O(X^{3/2})$.

(II) The second approach is again to follow Gauss but in a different direction. He proved that $h(D)$ is related to the number of representations of $|D|$ as a sum of three squares. One precise form of this relation is the following identity (see [3] p.177)

$$\tfrac{1}{2} \sum_{x \equiv y \equiv z \bmod 2} q^{x^2 + y^2 + z^2} = \tfrac{1}{2} + 12 \sum_D H_2(D) \, q^{|D|} \tag{3.3}$$

where D runs through all negative discriminants (i.e. $D < 0$ and $D \equiv 0, 1 \bmod 4$), and H_2 is a variant of the Hurwitz class number (see [3], page 120). (For us it suffices to know that it is related to $h(D)$; for example for $D \equiv 5 \bmod 8$ a fundamental discriminant we have $H_2(D) = h(D)$.)

There are sophisticated techniques for computing the coefficients of the left hand side, such as *convolution* which uses the fast Fourier transform to compute products of q-series. But even a simple enumeration of the lattice points $x^2 + y^2 + z^2 \leqslant X, x \equiv y \equiv z \bmod 2$ would again take time $O(X^{3/2})$.

The two approaches (I) and (II) are of course related; they amount to *counting* (in an appropriate sense) the number of representations of D by a certain ternary quadratic form. In case (I) we count the number of solutions to $b^2 - 4ac = D$ up to $SL_2(\mathbb{Z})$-equivalence; in (II), the number of solutions to $|D| = x^2 + y^2 + z^2$ with $x \equiv y \equiv z \bmod 2$. Note the crucial difference that the ternary quadratic form involved is indefinite in case (I) and positive definite in case (II).

A more geometrical point of view is to think that we are dealing with *Heegner points*. In case (I) we may associate to a primitive positive definite binary quadratic form $Q = (a, b, c)$ the point $z_Q = (-b + \sqrt{D})/2a$ in the upper half plane $\mathcal{H}$. The respective actions of $SL_2(\mathbb{Z})$ on forms and $\mathcal{H}$ are compatible; hence, the class of Q determines a unique (Heegner) point in $SL_2(\mathbb{Z})\backslash\mathcal{H}$ of discriminant D.

It is a bit less intuitive how to think of Heegner points in case (II) but this was worked out by Gross [3]. The main ingredient is a positive definite quaternion algebra B over $\mathbb{Q}$ ramified, say, at ∞ and a prime N. Pick a maximal order R of B and let $I_1, \ldots, I_n$ be representatives for the (left) ideal classes of R. Let R_i be the right order of I_i for $i = 1, \cdots, n$.

Fix an imaginary quadratic field K of discriminant D. Then we can think of a Heegner point of discriminant D (what Gross calls a *special point*) as an (optimal) embedding of the ring of integers $\mathcal{O}_K$ into some R_i. Eichler has proved that the total number of such points, each counted up to conjugation by $R_i^\star$, is $(1 - (\frac{D}{N}))h(D)$. (In fact, the situation is quite analogous to that of case (I) if we take the *indefinite* algebra $B = M_2(\mathbb{Q})$ and $R = M_2(\mathbb{Z})$.)

For example, if $N = 2$ then the algebra B is the usual Hamilton quaternions and we may pick R to be the order discovered by Hurwitz (in standard notation)

$$R = \mathbb{Z} + \mathbb{Z}i + \mathbb{Z}j + \mathbb{Z}\tfrac{1}{2}(1 + i + j + k).$$

In this case there is only one class of left R-ideals represented by R itself. Hence a Heegner point is an embedding $\phi : \mathcal{O}_K \to R$.

How do we find such embeddings? The main thing we need is a $w \in R$ with $w^2 = D$. Such a quaternion, because D is a scalar, necessarily has trace $t(w) = 0$ and norm $n(w) = -D$ and conversely. Elements of trace 0 in R form a rank 3 lattice and hence $n(w) = -D$ is a representation of D by a certain ternary quadratic form associated to R. A few congruence conditions are needed to actually produce an optimal embedding ϕ out of w but the upshot is that the problem becomes one about representations of $-D$ by ternary quadratic forms. For example, in the case $N = 2$ Eichler's count of embeddings can be completely encoded in the identity (3.3); the presence of the factor 12 in that formula is due to the fact that this is the order of $R^\star/\pm 1$. More details on this setup are given below in §4 (II).

4

We now return to the main case of computing $L(E, D, 1)$ and describe analogues of cases (I) and (II) of the previous section. These analogues are the remarkable results of Gross and Zagier.

(I) Let us assume for simplicity that E has conductor a prime N, sign of the functional equation equal to -1, $L'(E,1) \neq 0$, and $E(\mathbb{Q}) = \langle P_0 \rangle$. If f is the weight 2 eigenform associated to E then we get a map

$$\Phi: \quad \begin{array}{ccc} X_0(N) & \longrightarrow & \mathbb{C}/L \\ z & \mapsto & 2\pi i \int_{i\infty}^{z} f(u)\,du \end{array} \tag{4.1}$$

where $X_0(N)$ is the modular curve of level N and $L \subset \mathbb{C}$ is a certain lattice of periods of f. It is known that $\mathbb{C}/L = E'(\mathbb{C})$ for some elliptic curve $E'/\mathbb{Q}$ isogenous to E. Since the L-function is unchanged by isogenies we may assume without loss of generality that $E' = E$.

Let K be an imaginary quadratic field of discriminant $D < -4$ in which N splits. Choose $b_* \in \mathbb{Z}$ such that $b_*^2 \equiv D \bmod N$; this is possible by the assumption that N splits in K. Note also that N does not divide b_*. We want to consider Heegner points on $X_0(N)$ of discriminant D. To define them concretely choose representatives $Q = (a, b, c)$ of the $h(D)$ classes of binary quadratic forms with $N \mid a$ and $b \equiv b_* \bmod N$. (For example, start with representatives (a, b, c) with $\gcd(a, N) = 1$ and compose them with the fixed form $(N, b_*, (b_*^2 - D)/2N)$.)

Then $z_Q := (-b + \sqrt{D})/2/a) \in X_0(N)$ is well defined and $P_D := \sum_Q \Phi(z_Q) \in E(K)$. Moreover, complex conjugation fixes P_D, by the assumption on the sign of the functional equation. Hence P_D actually is in $E(\mathbb{Q})$ (and is independent of the choice of b_*).

One consequence of the results of Gross–Zagier is the following [15],[4], [5]. By our assumption on $E(\mathbb{Q})$ we have $P_D = m_D P_0$ for some $m_D \in \mathbb{Z}$ and hence

$$L(E, D, 1) = \kappa_D m_D^2; \tag{4.2}$$

where κ_D is an explicit easily computable positive constant; i.e. we have a formula of type (1.1).

Usually one regards the Gross-Zagier formula as a way to compute a rational point P_D on E whose height is given in terms of $L(E, D, 1)L'(E, 1)$ and hence obtaining, when this value does not vanish, a confirmation of the predictions of the Birch–Swinnerton-Dyer conjecture. Here, instead, we are taking the point of view that the points of $E(\mathbb{Q})$ are known and use the Gross–Zagier formula as a means to computing $L(E, D, 1)$.

To calculate m_D in practice it is better to work on the $E(\mathbb{C}) = \mathbb{C}/L$ model of E rather than, say, a Weierstrass equation. Let $z_0 \in \mathbb{C}$ represent, modulo L, the point $P_0 \in E(\mathbb{Q})$. We first compute an approximation to

$$z_D := \sum_Q \sum_{n \geq 1} \frac{a_n}{n} e^{2\pi i n z_Q}.$$

Then we solve the linear equation below for integers n_1 and n_2

$$z_Q = m_D z_0 + n_1 \omega_1 + n_2 \omega_2,$$

where ω_1, ω_2 are a basis for L. (In fact, multiplying by 2 if necessary, we may assume that $\omega_1 \in \mathbb{R}$ and $\omega_1 \in i\mathbb{R}$ and hence by taking real parts solve only a three term equation instead.)

The result is a practical and reasonably efficient algorithm for computing m_D. The number m_D is the D-th Fourier coefficient of a weight $3/2$ modular form g of level $4N$ which is in Shimura correspondence with f. It is interesting that we can compute the Fourier coefficients of g directly without any knowledge of the whole vector space of modular forms in which g lies; though we do, of course, start by knowing f itself. (We have only described the calculation for certain D's but there is analogous way to get all coefficients.)

Together with my student Ariel Pacetti we implemented the above algorithm in GP. The corresponding routines can be found at

```
http://www.ma.utexas.edu/users/villegas/cnt/
```

under Heegner points.

Here is a sample example. Let E be the curve $y^2 + y = x^3 - x$ of conductor $N = 37$ (this is the elliptic curve over $\mathbb{Q}$ of positive rank with smallest conductor). This case was described in detail in [15]. It is known that $E(\mathbb{Q}) = \langle (0,0) \rangle$.

```
? e=ellinit([0,0,1,-1,0]);   anvec=ellan(e,5000); ? for(d=5,100,
if(isfundamental(-d) && kronecker(-d,37)==1,
        print(-d," ",ellheegnermult(e,-d,[0,0],0,anvec)[1])))


-7   -11 -40 -47 -67   -71 -83   -84 -95
 1    -1  -2   1  -6    -1   1     1   0
```

The first row is D, the second m_D (for typographical reasons we transposed the actual GP output). These values agree, fortunately, with Zagier's [15] formula (28) up to a global negative sign.

In our implementation at least the algorithm is not that well suited for computing $L(E, D, 1)$ for all $D < 0$ and $|D| < X$ for very large X; for this, it would be better to adapt (see §5) the ideas of (II) below but these have not been fully implemented as yet.

(II) Let B over $\mathbb{Q}$ be the (unique up to isomorphism) positive definite quaternion algebra ramified at ∞ and a prime N. Pick a maximal order R of B and let $I_1, \ldots, I_n$ be representatives for the (left) ideal classes of R. Let R_i be the right order of I_i for $i = 1, \cdots, n$. The class number n of R, in contrast with $h(D)$, has a simple formula and is roughly of size $N/12$.

For example, if $N \equiv 3 \bmod 4$ we can describe B as the algebra over $\mathbb{Q}$ with generators i, j such that $i^2 = -1, j^2 = -N$ and $ij = -ji$. Also in this case we can take $R = \mathbb{Z} + \mathbb{Z}i + \mathbb{Z}\frac{1}{2}(1 + j) + \mathbb{Z}i\frac{1}{2}(1 + j)$.

There are various ways to compute representatives $I_1, \cdots, I_n$ of the ideal classes (for algorithms for quaternion algebras see [12]). If $N \equiv 3 \bmod 4$ there is an algorithm which is completely analogous to that of Gauss §3 (I) for binary quadratic forms. It exploits the fact that our choice of R has an embedding of $\mathbb{Z}[i]$ and hence allows us to view R-left ideals as rank 2 modules over $\mathbb{Z}[i]$; then classes of R-ideals correspond to classes of positive definite binary Hermitian forms over $\mathbb{Z}[i]$ of discriminant $-N$. Instead of $\mathcal{H}$ we now need to work on hyperbolic 3-space where, as it turns out, the action of $SL_2(\mathbb{Z}[i])$ has a very simple fundamental domain. This yields an algorithm which is almost verbatim that of Gauss for binary forms over $\mathbb{Z}$. Details can be found in [13].

For example, if $N = 11$ then there are two classes of positive definite binary Hermitian forms of discriminant -11 over $\mathbb{Z}[i]$; namely, $(1, 1, 3)$ and $(2, 1+2i, 2)$ corresponding to the two ideals

$$I_0 := R = \mathbb{Z} + \mathbb{Z}i + \mathbb{Z}\tfrac{1}{2}(1 + j) + \mathbb{Z}i\tfrac{1}{2}(1 + j)$$

and

$$I_1 := 2\mathbb{Z} + \mathbb{Z}2i + \mathbb{Z}\tfrac{1}{2}(1 + 2i + j) + \mathbb{Z}i\tfrac{1}{2}(1 + 2i + j)$$

representing the $n = 2$ classes of left R-ideals.

Let $V_{\mathbb{Q}}$ be the $\mathbb{Q}$ vector space of functions on the set $\{I_0, \ldots, I_n\}$. For each $m \in \mathbb{Z}_{\geqslant 0}$ there is an operator $B(m)$ acting on $V_{\mathbb{Q}}$, the Brandt matrix of order m, which encodes the number of representations of m by certain quaternary quadratic forms (see [3] (1.4)). Let $\mathbb{B}$ be the algebra generated over $\mathbb{Z}$ by all the $B(m)$; it is commutative and $\mathbb{B} \otimes_{\mathbb{Z}} \mathbb{Q}$ is semisimple.

On the other hand, we have the space $M_{\mathbb{C}}$ of modular form of weight 2 on $\Gamma_0(N)$ (known to be of dimension n) and the Hecke operators T_m acting on $M_{\mathbb{C}}$. Let $\mathbb{T}$ be the algebra spanned by the T_m over $\mathbb{Z}$; like $\mathbb{B}$ it is commutative and $\mathbb{T} \otimes_{\mathbb{Z}} \mathbb{Q}$ is semisimple. This algebra preserves the $\mathbb{Q}$ vector space $M_{\mathbb{Q}} \subset M_{\mathbb{C}}$ of dimension n consisting of those modular forms in $M_{\mathbb{C}}$ with Fourier coefficients in $\mathbb{Q}$.

These two setups are closely related and indeed we have a special case of the Jacquet–Langlands correspondence. Eichler proved that T_m and $B(m)$ have the same trace for all $m \in \mathbb{N}$. Hence, by semisimplicity of the algebras the map $T_m \mapsto B(m)$ induces a ring isomorphism $\mathbb{T} \simeq \mathbb{B}$. It follows that eigenspaces of $V_{\mathbb{Q}}$ and $M_{\mathbb{Q}}$, under the action of $\mathbb{B}$ and $\mathbb{T}$ respectively, correspond to each other. Since we also have multiplicity one these eigenspaces are one-dimensional.

In conclusion, given $f = \sum_{n \geqslant 0} a_n q^n \in M_{\mathbb{C}}$ an eigenform for all Hecke operators T_m (so that $T_m f = a_m f$) there is an $e_f \in V_{\mathbb{Q}} \otimes_{\mathbb{Z}} K$ unique up to scalars such that $B(m)e_f = a_m e_f$. (Here K denotes the field $\mathbb{Q}(a_0, a_1, \ldots)$ generated by the Fourier coefficients of f.)

In fact, this correspondence gives an efficient way to compute Fourier coefficients of eigenforms in $M_{\mathbb{C}}$ (see [12]). An implementation of the corresponding algorithms can be found in the above mentioned website (under `qalgmodforms`).

Here is a sample GP session.

```
? R=qsetprime(11);

? brandt(R,2)~

[1 3]

[2 0]

? brandt(R,3)~

[2 3]

[2 1]
```

The first line defines R as a maximal order in the algebra ramified at 11 and ∞; the others compute the corresponding Brandt matrices. We find that these matrices have two eigenvectors: $e_E = (1/2, 1/3)$ and $e_f = (-1, 1)$ corresponding to an Eisenstein series and a cusp form, respectively.

The above implementation is intended for small to medium scale computations. For large scale computations one should use the *graph method* ideas of Mestre and Oesterlé [8], which exploit the sparse nature of the Brandt matrices.

Now following Gross we show how to associate a modular form of weight $3/2$ to an eigenvector e_f. Let R_i be the right order of I_i and let $L_i \subset R_i$ be the ternary lattice defined by

$$L_i: \qquad w \in R_i, \qquad t(w) = 0, \qquad w \in \mathbb{Z} \bmod 2R_i.$$

Let g_i be the corresponding theta series

$$g_i(\tau) := \tfrac{1}{2} \sum_{w \in L_i} q^{n(w)}, \qquad q = e^{2\pi i \tau}.$$

Gross [3] prop. 12.9 describes precisely how the D-th coefficient $a_i(D)$ of g_i relates to the optimal embeddings of imaginary quadratic orders of $Q(\sqrt{D})$ into R_i.

These theta series are modular forms of weight $3/2$ and level $4N$ and, in fact, belong to a certain subspace U defined by Kohnen. This subspace is determined by the condition that the coefficient of q^d of a form should be zero unless $D := -d$ is a discriminant, i.e., $D \equiv 0, 1 \bmod 4$, and $\left(\frac{D}{N}\right) \neq 1$. The weight $3/2$ Hecke operators T_{m^2} preserve U.

Define

$$g := \sum_i e_f(i)\, g_i = \sum_D m_D\, q^{|D|} \in U.$$

This form is identically zero if the sign in the functional equation of f is -1. If g is non-zero it is a modular form in Shimura correspondence with f; i.e., $T_{m^2}g = a_m g$, where $T_m f = a_m f$. Moreover, we have the Waldspurger formula [3] 13.5

$$L(f,1)L(f \otimes \chi_D, 1) = \kappa_f \frac{\delta_D}{\sqrt{|D|}}\, m_D^2, \tag{4.3}$$

where D is a fundamental discriminant with $\left(\frac{D}{N}\right) \neq 1$, χ_D is the associated quadratic character, $\kappa_f > 0$ is a constant depending only on f and $\delta_D := 2$ if $N \mid D$ and $\delta_D := 1$ otherwise.

Finally, let $E/\mathbb{Q}$ be an elliptic curve of prime conductor N and sign $+1$ in its functional equation. Let f and g be the corresponding modular forms of weight 2 and 3/2 respectively as above. Then if $L(f,1) \neq 0$ we obtain from (4.3) a formula of type (1.1) with m_D the Fourier coefficient of g. As in §3 (II) to compute m_D for $|D| < X$ we could run through all $w \in L_i$ with $n(w) \leqslant X$ whose total number is $O(X^{3/2})$. Again various computational techniques could also be used to speed up the calculation of m_D. Note that in any case all computations are done with integer arithmetic.

Tables of m_D's for several curves and the routines to compute them can be found at G. Tornaría's website

`http://www.ma.utexas.edu/users/tornaria/cnt/`

among other goodies (an interactive version of Cremona's tables of elliptic curves and an interactive table of ternary quadratic forms).

5

We conclude with some remarks about the general situation.

1. It follows from (4.3) that if $L(f,1) = 0$ then the form g vanishes identically. In this case we naturally need to do something else.

In [7] we work out an extension of Gross's work introducing an auxiliary prime l; the theta series g_i, for example, are modified by introducing an appropriate weight function. The complexity of algorithms only increase by a factor essentially proportional to l.

2. If the level N is not prime but square-free the situation is not too different from the one described above. The downside is that $L(E,D,1)$ can be computed this way only for a certain fraction of D's (determined by local conditions). One needs to consider a quaternion algebra B ramified at ∞ and at primes $l \mid N$ for which the Atkin-Lehner involution acts as $f|_{w_l} = -f$ and an Eichler order in B of level the product of the remaining primes factors of N.

3. If the level is not square-free things become quite a bit more complicated; for example, the algebra $\mathbb{B}$ of Brandt matrices typically does not act with multiplicity one and some modular forms are simply missing. The arithmetic of the corresponding orders, which are no longer Eichler orders in general, also becomes more involved and, moreover, one needs to consider two types of orders: one for the weight 2 side and another for the weight 3/2 side; see [9], [10], [11] some work on this case.

4. To compute twists $L(f \otimes \chi_l, 1)$ by *real* quadratic fields $\mathbb{Q}(\sqrt{l})$ one may consider a twist $f_D := f \otimes \chi_D$ by an auxiliary imaginary quadratic field $Q(\sqrt{D})$ and find a formula of type (1.1) for $L(f_D \otimes \chi_{Dl}, 1)$. The form f_D typically does not have square-free level so several corresponding difficulties ensue, see [11].

5. Forms of higher weight can also be handled using quaternion algebras by introducing harmonic polynomials as weight functions for the theta functions (both for the ideals I_i corresponding to forms of weight $2 + 2r$ and for the ternary lattices L_i corresponding to forms of weight $3/2 + r$) see [6], [14].

References

[1] H. Cohen, *A course in computational algebraic number theory*, GTM **138**, Springer-Verlag, Berlin, (1993).

[2] L. Dickson, *History of the theory of numbers*, Chelsea Pub. Co., New York (1966).

[3] B. Gross, *Heights and the special values of L-series*, Number theory (Montreal, Que., 1985), 115–187, CMS Conf. Proc., **7**, Amer. Math. Soc., Providence, RI, (1987).

[4] B. Gross and D. Zagier, *Heegner points and derivatives of L-series*, Invent. Math. **84** (1986), 225–320.

[5] B. Gross, W. Kohnen and D. Zagier, *Heegner points and derivatives of L-series. II*, Math. Ann. **278** (1987), 497–562.

[6] R. Hatcher, *Heights and L-series*, Canad. J. Math. **42**,(1990), 533–560.

[7] Z. Mao, F. Rodriguez-Villegas, G. Tornaría, *Computation of central value of quadratic twists of modular L-functions*, in this volume.

[8] J.-F., Mestre, *La méthode des graphes. Exemples et applications*, Proceedings of the international conference on class numbers and fundamental units of algebraic number fields (Katata, 1986), 217–242, Nagoya Univ., Nagoya, 1986.

[9] A. Pacetti and F. Rodriguez-Villegas (appendix by B. Gross), *Computing Weight 2 Modular forms of level p^2* Math. Comp. **251**, (2004), 1545–1557.

[10] A. Pacetti, G. Tornaría, *Examples of Shimura correspondence for level p^2 and real quadratic twists*, in this volume.

[11] A. Pacetti, G. Tornaría, *Shimura correspondence for level p^2 and the central values of L-series*, preprint (2005).

[12] A. Pizer, *An Algorithm for Computing Modular Forms on $\Gamma_0(N)$*, Journal of Algebra **64**, (1980), 340-390.

[13] F. Rodriguez-Villegas, *Explicit models of genus 2 curves with split CM* Algorithmic number theory (Leiden, 2000), 505–513, Lecture Notes in Comput. Sci., **1838**, Springer, Berlin, (2000).

[14] H. Rosson, G. Tornaría, *Central values of quadratic twists for a modular form of weight 4*, in this volume.

[15] D. Zagier, *Modular points, modular curves, modular surfaces and modular forms*, Workshop Bonn 1984 (Bonn, 1984), 225–248, Lecture Notes in Math., **1111**, Springer, Berlin, (1985).

Department of Mathematics
University of Texas at Austin,
TX 78712 USA

villegas@math.utexas.edu

Computation of central value of quadratic twists of modular $L-$functions

Z. Mao, F. Rodriguez-Villegas and G. Tornaría

1 Introduction

Let $f \in S_2(p)$ be a newform of weight two, prime level p. If $f(z) = \sum_{m=1}^{\infty} a(m)q^m$, where $q = e^{2\pi i z}$, and D is a fundamental discriminant, we define the twisted L-function

$$L(f, D, s) = \sum_{m=1}^{\infty} \frac{a(m)}{m^s} \left(\frac{D}{m}\right).$$

It will be convenient to also allow $D = 1$ as a fundamental discriminant, in which case we write simply $L(f, s)$ for $L(f, 1, s)$.

In this paper we consider the question of computing the twisted central values $\{L(f, D, 1) : |D| \leqslant x\}$ for some x.

It is well known that the fact that f is an eigenform for the Fricke involution yields a rapidly convergent series for $L(f, D, 1)$. Computing $L(f, D, 1)$ by means of this series, which we call the *standard method*, takes time very roughly proportional to $|D|$ and therefore time very roughly proportional to x^2 to compute $L(f, D, 1)$ for $|D| \leqslant x$. We will see that this can be improved to $x^{3/2}$ by using an explicit version of Waldspurger's theorem [W]; this theorem relates the central values $L(f, D, 1)$ to the $|D|$-th Fourier coefficient of weight $3/2$ modular forms in Shimura correspondence with f.

Concretely, the formulas we use have the basic form

$$L(f, D, 1) = \star \kappa_{\mp} \frac{|c_{\mp}(|D|)|^2}{\sqrt{|D|}}, \qquad \text{sign}(D) = \mp, \tag{1.1}$$

where $\star = 1$ if $p \nmid D$, $\star = 2$ if $p \mid D$, κ_- and κ_+ are positive constants independent of D, and $c_-(|D|)$ (resp. $c_+(|D|)$) is the $|D|$-th Fourier coefficient of a certain modular form g_- (resp. g_+) of weight $3/2$.

Gross [G] proves such a formula, and gives an explicit construction of the corresponding form g_-, in the case that $L(f, 1) \neq 0$. The purpose of this paper is to extend Gross's work to all cases. Specifically, we give an explicit construction of both g_- and g_+, regardless of the value of $L(f, 1)$, together with the corresponding values of κ_- and κ_+ in (1.1). The proof of the validity of this construction will be given in a later publication and relies partly in the results of [BM].

The construction gives g_- and g_+ as linear combinations of (generalized) theta series associated to positive definite ternary quadratic forms. Computing the Fourier coefficients of these theta series up to x is tantamount to running over all lattice points in an ellipsoid of volume proportional to $x^{3/2}$. Doing this takes time roughly proportional to $x^{3/2}$ which yields our claim above.

This approach to computing $L(f, D, 1)$ has several other advantages over the standard method. First, the numbers $c(|D|)$ are algebraic integers and are computed with exact arithmetic. Once $c(|D|)$ is know it is trivial to compute $L(f, D, 1)$ to any desired precision. Second, the $c(|D|)$'s have extra information; if f has coefficients in $\mathbb{Z}$, for example, (1.1) gives a specific square root of $L(f, D, 1)$ (if non-zero), whose sign remains a mystery.

Moreover, the actual running time of our method vs. the standard method is, in practice, significantly better even for small x.

2   Quaternion algebras and Brandt matrices

A quaternion algebra B over a field K is a central simple algebra of dimension 4 over K. When $2 \neq 0$ in K we can give B concretely by specifying a K-basis $\{1, i, j, k\}$ such that

$$i^2 = \alpha, \qquad j^2 = \beta, \qquad \text{and} \qquad k = ij = -ji,$$

for some non-zero $\alpha, \beta \in K$. If $K = \mathbb{Q}$ we typically rescale and assume that $\alpha, \beta \in \mathbb{Z}$. A general element of B then has the form $b = b_0 + b_1 i + b_2 j + b_3 k$, with $b_i \in K$ and multiplication in B is determined by the above defining relations and K-linearity.

The *conjugate* of b is defined as

$$\bar{b} = b_0 - b_1 i - b_2 j - b_3 k.$$

We define the (reduced) norm and trace of b by

$$\mathcal{N} b := b\bar{b} = b_0^2 - \alpha b_1^2 - \beta b_2^2 + \alpha\beta b_3^2, \qquad \mathsf{Tr}\, b := b + \bar{b} = 2b_0.$$

Let B be a quaternion algebra over $K = \mathbb{Q}$. For ν a rational prime we let $\mathbb{Q}_\nu$ be the field of ν-adic numbers and for $\nu = \infty$ we let $\mathbb{Q}_\nu = \mathbb{R}$. We call ν, a rational prime or ∞, a *place* of $\mathbb{Q}$.

The localization $B_\nu := B \otimes \mathbb{Q}_\nu$ is a quaternion algebra over $\mathbb{Q}_\nu$. It is a fundamental fact of Number Theory that B_ν is either isomorphic to the algebra $M_2(\mathbb{Q}_\nu)$ of 2×2 matrices, or a division algebra, which is unique up to isomorphism. (A *division algebra* is an algebra in which every non-zero element has a multiplicative inverse.) The two options are encoded in the *Hilbert symbol* $(\alpha, \beta)_\nu$, defined as $+1$ if B_ν is a matrix algebra, -1 if it is a division algebra. In the first case we say that B is *split* at ν, in the second, that B is *ramified* at ν.

For example, if $\nu = \infty$ so $\mathbb{Q}_\nu = \mathbb{R}$ then $(\alpha, \beta)_\infty = -1$ if and only if $\alpha < 0, \beta < 0$ in which case B_∞ is isomorphic to the usual Hamilton quaternions. A quaternion algebra B is *definite* if it ramifies at ∞ otherwise it is *indefinite* (this notation is consistent with the nature of the quadratic form on B_∞ determined by the norm $\mathcal{N}$).

A quaternion algebra B is ramified at a finite number of places and the total number of ramified places must be even (e.g. the *Hilbert reciprocity law* says that $\prod_\nu (\alpha, \beta)_\nu = 1$). The set of ramified places determines B up to isomorphism (the *local-global principle*). For any finite set S with an even number of places there is a (unique up to isomorphism) B which ramifies exactly at places in S.

Let B be a quaternion algebra over $\mathbb{Q}$. An *order* in B is a (full rank) lattice $R \subseteq B$ which is also a ring with $1 \in R$. As for number fields, an element of an order must be integral over $\mathbb{Z}$, i.e., must satisfy a monic equation with coefficients in $\mathbb{Z}$ (or even more concretely must have integral trace and norm). Unlike in the commutative case, however, the set of all integral elements of B is *not* a ring. The best next thing is to consider maximal orders (which always exist), i.e., orders not properly contained in another order. But maximal orders are not unique. In fact, if B is definite, a maximal order is in general not even unique up to isomorphism though there always is only a finite number of isomorphism classes of maximal orders in B.

As an illustration consider the classical case $\alpha = \beta = -1$ of the Hamilton quaternions. The algebra is definite and hence ramifies at $\nu = \infty$. It must ramify at least one other prime, which turns out to be only $\nu = 2$. To see this note that

$$\mathcal{N}(b_0 + b_1 i + b_2 j + b_3 k) = b_0^2 + b_1^2 + b_2^2 + b_3^2.$$

There always is a non-trivial solution to the congruence $b_0^2 + b_1^2 + b_2^2 + b_3^2 \equiv 0 \bmod p$ for p prime. If p is odd we can lift this solution to a solution in $\mathbb{Z}_p$ by Hensel's lemma obtaining a non-zero quaternion in B_p of zero norm. This implies that B_p cannot be a divison algebra and hence $(-1, -1)_p = 1$ for p odd. We must necessarily have then that $(-1, -1)_2 = -1$.

If we want to study the representation of numbers as sum of four squares it is natural to consider, as Lipschitz did, the arithmetic of the quaternions with $b_i \in \mathbb{Z}$. These quaternions form an order R', but, as it turns out, it is not maximal. Indeed, as Hurwitz noted, $\rho := \frac{1}{2}(1 + i + j + k)$ is integral ($\mathcal{N}\rho = 1$ and $\mathsf{Tr}\,\rho = 1$) and $R := R' + \mathbb{Z}\rho$ is also an order of B strictly containing R'.

Moreover, R is maximal and hence its arithmetic is significantly simpler than that of R'. Hurwitz showed, for example, that there is a left and right division algorithm in R, from which it follows that every positive integer is a sum of four squares.

Fix a prime p and let B be the quaternion algebra over $\mathbb{Q}$ ramified precisely at ∞ and p. Let R be a fixed maximal order in B. A right ideal I of R is a lattice in B that is stable under right multiplication by R. Two right ideals I and J are in the same *class* if $J = bI$ with $b \in B^\times$. The set of right ideal classes is finite; let n be its number. Chose a set of representatives $\{I_1, \ldots, I_n\}$ of the classes. (We should emphasize here that contrary to the commutative setting there is *no* natural group structure on the set of classes.)

Consider the vector space V of formal linear combinations

$$\sum_{i=1}^{n} a_i \, [I_i], \qquad a_i \in \mathbb{C}$$

(here $[I]$ denotes the class of I).

For each integer m there is an $n \times n$ matrix B_m acting on V. Pizer [P] gives an efficient algorithm for computing these *Brandt matrices*: its coefficients are given by the representation numbers of the norm form for certain quaternary lattices in B.

The Brandt matrices commute with each other and are self-adjoint with respect to the *height pairing* on V (see §1 and §2 of [G] for an account of this.) From this it follows that there is basis of V consisting of simultaneous eigenvectors of all B_m.

It follows from Eichler's trace formula that there is a one to one correspondence between Hecke eigenforms of weight 2 and level p (cf. [G, §5]) and eigenvectors in V of all Brandt matrices (up to a constant multiple).

If f is the Hecke eigenform we let e_f be the corresponding eigenvector (well defined up to a constant). Then $B_m e_f = a_m e_f$ where $T_m f = a_m f$ and T_m is the m-th Hecke operator.

3   Construction of g_- and g_+

Let e_f be the eigenvector for the Brandt matrices for R corresponding to f as in the last section. One can use linear algebra to find its coefficients

$$e_f = \sum_{i=1}^{n} a_i [I_i],$$

by computing the Brandt matrices, and from the knowledge of a few eigenvalues (i.e. Fourier coefficients) of f.

We will describe below the construction of certain generalized theta series $\Theta_{l^*}([I_i])$ corresponding to each ideal class $[I_i]$, and then define

$$\Theta_{l^*}(e_f) := \sum_{i=1}^{n} a_i \, \Theta_{l^*}([I_i]) = \sum_{n=1}^{\infty} c_{l^*}(n) \, q^n.$$

Here l^* is a fundamental discriminant for which we will consider three cases: $l^* = 1$, which is Gross's construction of g_-; $l^* = l$ for an odd prime $l \neq p$

such that $l \equiv 1 \pmod 4$, which will generalize Gross's construction of g_-; and $l^* = -l$ for an odd prime $l \neq p$ such that $l \equiv 3 \pmod 4$, which will give a construction of g_+.

Furthermore, for any fundamental discriminant D such that $Dl^* < 0$, the following formula holds

$$L(f, l^*, 1)\, L(f, D, 1) = \star \kappa_f \frac{|c_{l^*}(|D|)|^2}{\sqrt{|Dl^*|}}, \tag{3.1}$$

where $\star = 1$ if $p \nmid D$, $\star = 2$ if $p \mid D$, and $\kappa_f := \frac{\langle f, f \rangle}{\langle e_f, e_f \rangle}$ is a positive constant independent of D or l^*. Here $\langle e_f, e_f \rangle$ is the height of e_f, and $\langle f, f \rangle$ is the Petersson norm of f (cf. §4 and §7 of [G].) For $l = 1$, this formula was proved by Gross in [G, Proposition 13.5]. The proof of this formula for the case $l \neq 1$ will be given in a later publication.

Note that, as a corollary, we have $\Theta_{l^*}(e_f) \neq 0$ if and only if $L(f, l^*, 1) \neq 0$, and this happens for infinitely many $l^* > 0$ and for infinitely many $l^* < 0$, as follows from [BFH].

3.1 Gross's construction of Θ_1

Let $R_i := \{b \in B : bI_i \subset I_i\}$ be the left order of I_i. The R_i are maximal orders in B, and each conjugacy class of maximal orders has a representative R_i for some i.

We let $S_i^0 := \{b \in \mathbb{Z} + 2R_i : \operatorname{Tr} b = 0\}$, a ternary lattice, and define

$$\Theta_1([I_i]) := \frac{1}{2} \sum_{b \in S_i^0} q^{\mathcal{N} b}.$$

Then $\Theta_1([I_i])$ is a weight $\frac{3}{2}$ modular form of level $4p$ and trivial character.

3.2 Weight functions and Θ_l

Fix an odd prime $l \neq p$. In order to generalize Gross's method, we need to construct certain weight functions $\omega_l(I_i, \cdot)$ on S_i^0 with values in $\{0, \pm 1\}$. There is a choice of sign in the construction, and some care is needed to ensure that the choice is consistent from one ideal to another. It will be the case that $\omega_l(I_i, b) = 0$ unless $l \mid \mathcal{N} b$, and thus we define a *generalized theta series*

$$\Theta_l([I_i]) := \frac{1}{2} \sum_{b \in S_i^0} \omega_l(I_i, b) q^{\mathcal{N} b/l},$$

a modular form of weight $\frac{3}{2}$ and level $4p$ with trivial character. In addition, $\Theta_l([I_i])$ is already a cusp form whenever $l \neq 1$, although it might be zero.

Definition 3.1. *Given a pair (L, v), where L is an integral $\mathbb{Z}_l$-lattice of rank 3 with $l \nmid \det L$, and $v \in L$ is such that $l \mid \mathcal{N} v$ but $v \notin l\,L$, we define its* weight function $\omega_{l,v} : L \to \{0, \pm 1\}$ *to be*

$$
\omega_{l,v}(v') := \begin{cases} 0 & \text{if } l \nmid \mathcal{N} v', \\ \chi_l(\langle v, v' \rangle) & \text{if } l \nmid \langle v, v' \rangle, \\ \chi_l(k) & \text{if } v' - k\,v \in l\,L. \end{cases}
$$

Here χ_l is the quadratic character of conductor l, and $\mathcal{N} v := \frac{1}{2}\langle v, v \rangle$.

This is well defined, because if $v, v' \in L$ are such that $\mathcal{N} v \equiv \mathcal{N} v' \equiv \langle v, v' \rangle \equiv 0 \pmod{l}$, then v and v' must be collinear modulo l, since L is unimodular. This means that, assuming $v \notin l\,L$, there is indeed a well defined $k \in \mathbb{Z}/l\mathbb{Z}$ such that $v' - k\,v \in l\,L$.

Note that there are, for different choices of v, two different weight functions for each L, opposite to each other; the definition above singles out the one for which $\omega_{l,v}(v) = +1$.

We will apply the above definition to the ternary lattices $S_i^0(\mathbb{Z}_l) := S_i^0 \otimes \mathbb{Z}_l$. Fix a quaternion $b_0 \in S^0 := \{b \in \mathbb{Z} + 2R \ : \ \mathsf{Tr}\, b = 0\}$, and such that $l \mid \mathcal{N} b_0$ but $b_0 \notin lS^0$. For each I_i, find $x_i \in I_i$ such that $l \nmid n_i := \mathcal{N} x_i / \mathcal{N} I_i$. Then x_i is a local generator of I_i, and $b_i := x_i b_0 x_i^{-1} \in S_i^0(\mathbb{Z}_l)$. We finally set

$$
\omega_l(I_i, b) := \chi_l(n_i)\,\omega_{l,b_i}(b),
$$

where ω_{l,b_i} is the weight function of the pair $(S_i^0(\mathbb{Z}_l), b_i)$.

3.3   Odd weight functions and Θ_{-l}

When $l \equiv 3 \pmod 4$ the weight functions $\omega_l(I_i, \cdot)$ are odd, since χ_l is odd. Therefore, we will have $\Theta_l = 0$. To address this problem, we will construct a different kind of weight function $\omega_p(I_i, \cdot)$, and then define

$$
\Theta_{-l}([I_i]) := \frac{1}{2} \sum_{b \in S_i^0} \omega_p(I_i, b)\,\omega_l(I_i, b)\,q^{\mathcal{N} b/l},
$$

which will be a modular form of weight $3/2$, this time of level $4p^2$. Again, $\Theta_{-l}([I_i])$ is a cusp form, which might be zero. Note that we could have used the product of two odd weight functions ω_{l_1} and ω_{l_2}, but this construction would only lead us to the same g_-. By using the weight functions ω_p we get a construction of g_+ instead.

Definition 3.2. *Given a triple (L, v, ψ) where L is an integral $\mathbb{Z}_p$-lattice of rank 3 with level p and determinant p^2 (i.e. L is $\mathbb{Z}_p$-equivalent to $S^0(\mathbb{Z}_p)$,) $v \in L$ is such that $p \nmid \mathcal{N} v$, and $\psi : \mathbb{Z}/p\mathbb{Z} \to \mathbb{C}$ is a periodic function modulo p, the weight function $\omega_{\psi,v} : L \to \mathbb{C}$ is defined by*

$$\omega_{\psi,v}(v') := \psi(\langle v, v' \rangle).$$

Clearly, the weight function $\omega_{\psi,v}$ will be odd if and only if ψ itself is odd. From now on we assume that ψ is a fixed odd periodic function such that

$$|\psi(t)| = 1 \qquad \text{for } t \not\equiv 0 \pmod{p}. \tag{3.2}$$

Now fix $b_0 \in S^0$ such that $p \nmid \mathcal{N} b_0$. Find $x_i \in I_i$ such that $p \nmid \mathcal{N} x_i / \mathcal{N} I_i$; since B is ramified at p, the maximal order at p is unique, and so $b_i := x_i b_0 x_i^{-1} \in S^0(\mathbb{Z}_l) = S_i^0(\mathbb{Z}_p)$. We define

$$\omega_p(I_i, b) := \omega_{\psi, b_i}(b),$$

to be the weight function for the triple $(S_i^0(\mathbb{Z}_p), b_i, \psi)$. In practice, one can use the same b_0 and b_i for the definitions of both $\omega_l(I_i, \cdot)$ and $\omega_p(I_i, \cdot)$.

Note that different choices of ψ will, in general, yield different forms $\Theta_{-l}([I_i])$, but as long as (3.2) holds their coefficients will be the same up to a constant of absolute value 1; thus formula (3.1) will not be affected. Moreover, given two such odd periodic functions it is not difficult to produce another *periodic* function χ with the property that the ratio of the m-th coefficients of the respective theta series will be $\chi(m)$.

The case when ψ is actually a character of conductor p is of particular interest, since the generalized theta series $\Theta_{-l}([I_i])$ will be a modular form of level $4p^2$ and character ψ_1, where $\psi_1(m) = \left(\frac{-1}{m}\right)\psi(m)$. From a computational point of view, however, it will always be preferable to choose a real ψ, whose values will be 0 or ± 1, and so that the coefficients of $\Theta_{-l}([I_i])$ will be rational integers. Only in case $p \equiv 3 \pmod{4}$ it is possible to satisfy both requirements at the same time, by taking for ψ the quadratic character of conductor p.

4   Examples

4.1   11A

Let $f = f_{11A}$, the modular form of level 11, and consider $B = B(-1, -11)$, the quaternion algebra ramified precisely at ∞ and 11. A maximal order, and representatives for its right ideals classes, are given by

$$R = I_1 = \left\langle 1, i, \frac{1+j}{2}, \frac{i+k}{2} \right\rangle \qquad \text{with } \mathcal{N}\, I_1 = 1,$$

$$I_2 = \left\langle 2, 2i, \frac{1+2i+j}{2}, \frac{2+3i+k}{2} \right\rangle \qquad \text{with } \mathcal{N}\, I_2 = 2.$$

By computing the Brandt matrices (see Rodrigues-Villegas paper in this volume or §6 of [G] for this example), we find a vector

$$e_f = [I_2] - [I_1]$$

of height $\langle e_f, e_f \rangle = 5$ corresponding to f. Since $L(f, 1) \approx 0.25384186$, Gross's method works, and it's easy to compute

$$\Theta_1(e_f) = \Theta_1([I_2]) - \Theta_1([I_1]) = q^3 - q^4 - q^{11} - q^{12} + q^{15} + 2q^{16} + O(q^{20}),$$

as the difference of two regular theta series corresponding to the ternary quadratic forms (4.1) and (4.2).

4.1.1   Real twists in a case of rank 0

Let $l = 3$. One can compute $L(f, -3, 1) \approx 1.6844963$, and thus expect $\Theta_{-3}(e_f)$ to be nonzero. We can choose $b_0 = i + k \in S^0$ with norm 12, and let $\psi = \chi_{11}$ be the quadratic character of conductor 11.

Clearly we can take $x_1 = 1$ and $x_2 = 2$, so that $n_1 = 1$, $n_2 = 2$ and $b_1 = b_2 = i + k$. Bases for S_1^0 and S_2^0 are given by

$$S_1^0 = \langle 2i, j, i + k \rangle \qquad \text{with } b_1 = (0, 0, 1),$$

$$S_2^0 = \left\langle 4i, 2i + j, \frac{7i+k}{2} \right\rangle \qquad \text{with } b_2 = (-\tfrac{3}{2}, 0, 2).$$

The norm form in the given bases will be

$$\mathcal{N}_1(x_1, x_2, x_3) = 4x_1^2 + 11x_2^2 + 12x_3^2 + 4x_1x_3, \tag{4.1}$$

$$\mathcal{N}_2(x_1, x_2, x_3) = 16x_1^2 + 15x_2^2 + 15x_3^2 + 14x_2x_3 + 28x_1x_3 + 16x_1x_2. \tag{4.2}$$

This information is all that we need to compute Θ_{-3}. As an example, we show how to compute $\Theta_{-3}([I_1])$. A simple calculation shows that

$$\langle (x_1, x_2, x_3), b_1 \rangle = 4x_1 + 24x_3 \equiv x_1 \pmod{3},$$

d	$c_{-3}(d)$	$L(f,d,1)$	d	$c_{-3}(d)$	$L(f,d,1)$	d	$c_{-3}(d)$	$L(f,d,1)$
1	1	0.253842	92	-5	0.661621	141	-10	2.137734
5	-5	2.838038	93	5	0.658054	152	-10	2.058929
12	-5	1.831946	97	5	0.644343	157	-15	4.558227
37	5	1.043284	104	10	2.489124	168	10	1.958432
53	10	3.486786	113	-5	0.596986	177	5	0.476998
56	10	3.392105	124	-5	0.569892	181	-15	4.245281
60	-5	0.819271	133	10	2.201088	185	-5	0.466571
69	15	6.875768	136	10	2.176676	188	-10	1.851332
89	-5	0.672680	137	-5	0.542179			

Table 1: Coefficients of $\Theta_{-3}(e_f)$ and central values for $f = f_{11A}$

and thus $\omega_3(I_1, \cdot)$ can be computed by

$$\omega_3(I_1, (x_1, x_2, x_3)) = \begin{cases} 0 & \text{if } 3 \nmid \mathcal{N}_1(x_1, x_2, x_3), \\ \chi_3(x_1) & \text{if } x_1 \not\equiv 0 \pmod 3, \\ \chi_3(x_3) & \text{otherwise.} \end{cases}$$

Similarly, $\omega_{11}(I_1, \cdot)$ will be given by

$$\omega_{11}(I_1, (x_1, x_2, x_3)) = \chi_{11}(4x_1 + 2x_3).$$

Hence we compute

$$\Theta_{-3}([I_1]) = -2q^4 + 2q^5 + 2q^9 + 2q^{12} + 2q^{20} + 2q^{25} - 2q^{37} + O(q^{48}).$$

In a similar way one can easily get

$$\Theta_{-3}([I_2]) = q + q^4 - 3q^5 - 3q^{12} + 4q^{16} - 3q^{20} + 2q^{25} - 6q^{36} + 3q^{37} + O(q^{48}).$$

Table 1 shows the values of $c_{-3}(d)$ and $L(f,d,1)$, where $0 < d < 200$ is a fundamental discriminant such that $\left(\frac{d}{11}\right) = 1$. The formula

$$L(f,d,1) = k_{-3} \frac{c_{-3}(d)^2}{\sqrt{d}}$$

is satisfied, where

$$k_{-3} = \frac{1}{5} \cdot \frac{(f,f)}{L(f,-3,1)\sqrt{3}} = L(f,1) \approx 0.2538418608559106843377589233509\ldots$$

Note that when $\left(\frac{d}{11}\right) \neq 1$ it is trivial that $c_{-3}(d) = L(f,d,1) = 0$.

4.2   37 A

Let $f = f_{37A}$, the modular form of level 37 and rank 1, and consider $B = B(-2, -37)$, the quaternion algebra ramified precisely at ∞ and 37. A maximal order, and representatives for its right ideal classes, are given by

$$R = I_1 = \left\langle 1, i, \frac{1+i+j}{2}, \frac{2+3i+k}{4} \right\rangle \qquad \text{with } \mathcal{N}\, I_1 = 1,$$

$$I_2 = \left\langle 2, 2i, \frac{1+3i+j}{2}, \frac{6+3i+k}{4} \right\rangle \qquad \text{with } \mathcal{N}\, I_2 = 2,$$

$$I_3 = \left\langle 4, 2i, \frac{3+3i+j}{2}, \frac{6+i+k}{2} \right\rangle \qquad \text{with } \mathcal{N}\, I_3 = 4.$$

By computing the Brandt matrices, we find a vector

$$e_f = \frac{[I_3] - [I_2]}{2}$$

of height $\langle e_f, e_f \rangle = \tfrac{1}{2}$ corresponding to f. Since $L(f, 1) = 0$ we know that $2\,\Theta_1(e_f) = \Theta_1([I_3]) - \Theta_1([I_2]) = 0$. Indeed, one checks that R_2 and R_3 are conjugate, which explains the identity $\Theta_1([I_2]) = \Theta_1([I_3])$.

4.2.1   Imaginary twists in a case of rank 1

Let $l = 5$. One can compute $L(f, 5, 1) \approx 5.3548616$, and thus we expect $\Theta_5(e_f)$ to be nonzero. We note that, by the same reason that the orders are conjugate, we have $\Theta_5([I_3]) = -\Theta_5([I_2])$, except now there's an extra sign, ultimately coming from the fact that $\left(\frac{37}{5}\right) = -1$. Thus, $\Theta_5(e_f) = \Theta_5([I_3])$. A basis for S_3^0 is given by

$$S_3^0 = \left\langle 4i, 3i + j, \frac{3i + 2j + k}{4} \right\rangle,$$

with the norm in this basis

$$\mathcal{N}_3\,(x_1, x_2, x_3) = 32x_1^2 + 55x_2^2 + 15x_3^2 + 46x_2x_3 + 12x_1x_3 + 48x_1x_2.$$

Choose $b_3 = (0, 0, 1)$, with norm 15. Then

$$\langle (x_1, x_2, x_3), b_3 \rangle = 12x_1 + 46x_2 + 30x_3 \equiv 2x_1 + x_2 \pmod{5}, \qquad (4.3)$$

so that

$$\omega_5(I_3, (x_1, x_2, x_3)) = \begin{cases} 0 & \text{if } 5 \nmid \mathcal{N}_3\,(x_1, x_2, x_3), \\ \chi_5(2x_1 + x_2) & \text{if } 2x_1 + x_2 \not\equiv 0 \pmod{5}, \\ \chi_5(x_3) & \text{otherwise.} \end{cases}$$

$-d$	$c_5(d)$	$L(f,-d,1)$	$-d$	$c_5(d)$	$L(f,-d,1)$	$-d$	$c_5(d)$	$L(f,-d,1)$
-3	1	2.830621	-95	0	0.000000	-139	0	0.000000
-4	1	2.451389	-104	0	0.000000	-148	-3	7.254107
-7	-1	1.853076	-107	0	0.000000	-151	-2	1.595930
-11	1	1.478243	-111	1	0.930702	-152	-2	1.590671
-40	2	3.100790	-115	-6	16.458713	-155	2	1.575203
-47	-1	0.715144	-120	-2	1.790242	-159	1	0.388816
-67	6	21.562911	-123	3	3.978618	-164	-1	0.382843
-71	1	0.581853	-127	1	0.435051	-184	0	0.000000
-83	-1	0.538150	-132	3	3.840589	-195	2	1.404381
-84	-1	0.534937	-136	4	6.726557			

Table 2: Coefficients of $\Theta_5(e_f)$ and central values for $f = f_{37A}$

Table 2 shows the values of $c_5(d)$ and $L(f,-d,1)$, where $-200 < -d < 0$ is a fundamental discriminant such that $\left(\frac{-d}{37}\right) \neq -1$. The formula

$$L(f,-d,1) = k_5 \frac{c_5(d)^2}{\sqrt{d}} \cdot \begin{cases} 1 & \text{if } \left(\frac{-d}{37}\right) = +1, \\ 2 & \text{if } \left(\frac{-d}{37}\right) = 0, \\ 0 & \text{if } \left(\frac{-d}{37}\right) = -1, \end{cases}$$

is satisfied, where

$$k_5 = 2 \cdot \frac{(f,f)}{L(f,5,1)\sqrt{5}} \approx 4.9027787639735801217708449663733...$$

Note that in the case $\left(\frac{-d}{37}\right) = -1$ it is trivial that $c_5(d) = L(f,-d,1) = 0$.

4.2.2 Real twists in a case of rank 1

Let $l = 3$, since $L(f,-3,1) \approx 2.9934586$. Keep b_3 as above, and let ψ be the odd periodic function modulo 37 such that

$$\psi(x) = \begin{cases} +1 & \text{if } 1 \leqslant x \leqslant 18, \\ -1 & \text{if } 19 \leqslant x \leqslant 36. \end{cases}$$

Using again (4.3), we have that

$$\omega_3(I_1,(x_1,x_2,x_3)) = \begin{cases} 0 & \text{if } 3 \nmid \mathcal{N}_1(x_1,x_2,x_3), \\ \chi_3(x_2) & \text{if } x_2 \not\equiv 0 \pmod 3, \\ \chi_3(3) & \text{otherwise.} \end{cases}$$

and $\omega_{11}(I_1,\cdot)$ will be given by

$$\omega_{37}(I_1,(x_1,x_2,x_3)) = \psi(12x_1 + 9x_2 + 30x_3).$$

d	$c_{-3}(d)$	$L(f,d,1)$	d	$c_{-3}(d)$	$L(f,d,1)$	d	$c_{-3}(d)$	$L(f,d,1)$
5	1	5.354862	76	1	1.373493	133	-1	1.038263
8	1	4.233390	88	1	1.276415	140	-3	9.107764
13	-1	3.320944	89	-1	1.269224	156	-1	0.958674
17	1	2.904081	92	2	4.993434	161	-2	3.774681
24	-1	2.444149	93	2	4.966515	165	1	0.932162
29	2	8.893941	97	0	0.000000	168	-1	0.923801
56	-1	1.600071	105	1	1.168527	172	1	0.912996
57	1	1.585973	109	-1	1.146885	177	0	0.000000
60	-1	1.545815	113	0	0.000000	193	-1	0.861895
61	0	0.000000	124	0	0.000000			
69	0	0.000000	129	1	1.054237			

Table 3: Coefficients of $\Theta_{-3}(e_f)$ and central values for $f = f_{37A}$

Table 3 shows the values of $c_{-3}(d)$ and $L(f,d,1)$, where $0 < d < 200$ is a fundamental discriminant such that $\left(\frac{d}{37}\right) = -1$. The formula

$$L(f,d,1) = k_{-3}\,\frac{c_{-3}(d)^2}{\sqrt{d}}$$

is satisfied, where

$$k_{-3} = 2 \cdot \frac{(f,f)}{L(f,-3,1)\sqrt{3}} \approx 11.97383458492783851932803991781\ldots$$

Note that in the case $\left(\frac{d}{37}\right) \neq -1$ it is trivial that $c_{-3}(d) = L(f,d,1) = 0$.

4.3   43A

Let $f = f_{43A}$, the modular form of level 43 and rank 1. Let $B = B(-1,-43)$, the quaternion algebra ramified precisely at ∞ and 43. A maximal order, and representatives for its right ideals classes, are given by

$$R = I_1 = \left\langle 1, i, \frac{1+j}{2}, \frac{i+k}{2} \right\rangle \qquad \text{with } \mathcal{N}\, I_1 = 1,$$

$$I_2 = \left\langle 2, 2i, \frac{1+2i+j}{2}, \frac{2+3i+k}{2} \right\rangle \qquad \text{with } \mathcal{N}\, I_2 = 2,$$

$$I_3 = \left\langle 3, 3i, \frac{1+2i+j}{2}, \frac{2+5i+k}{2} \right\rangle \qquad \text{with } \mathcal{N}\, I_3 = 3,$$

$$I_4 = \left\langle 3, 3i, \frac{1+4i+j}{2}, \frac{4+5i+k}{2} \right\rangle \qquad \text{with } \mathcal{N}\, I_4 = 3.$$

By computing the Brandt matrices, we find a vector

$$e_f = \frac{[I_4] - [I_3]}{2}$$

$-d$	$c_5(d)$	$L(f,-d,1)$	$-d$	$c_5(d)$	$L(f,-d,1)$	$-d$	$c_5(d)$	$L(f,-d,1)$
-3	1	3.148135	-91	-1	0.571601	-151	-1	0.443737
-7	1	2.060938	-104	1	0.534684	-155	-1	0.437974
-8	-1	1.927831	-115	-3	4.576227	-159	1	0.432430
-19	2	5.003768	-116	-1	0.506273	-163	7	20.927447
-20	-1	1.219267	-119	-1	0.499851	-168	-2	1.682749
-39	-1	0.873136	-120	0	0.000000	-179	-1	0.407556
-43	2	6.652268	-123	-5	12.291402	-184	-3	3.617825
-51	1	0.763535	-131	0	0.000000	-191	0	0.000000
-55	1	0.735246	-132	3	4.271393	-199	0	0.000000
-71	0	0.000000	-136	-1	0.467568			
-88	3	5.231366	-148	-4	7.171386			

Table 4: Coefficients of $\Theta_5(e_f)$ and central values for $f = f_{43A}$

of height $\langle e_f, e_f \rangle = \frac{1}{2}$ corresponding to f.

4.3.1 Imaginary twists in a case of rank 1

We can use $l = 5$, since $L(f, 5, 1) \approx 4.8913446$ is nonzero; again, we find
$\Theta_5(e_f) = \Theta_5([I_4])$. Table 4 shows the values of $c_5(d)$ and $L(f, -d, 1)$, where
$-200 < -d < 0$ is a fundamental discriminant such that $\left(\frac{-d}{43}\right) \neq -1$. The
formula

$$L(f, -d, 1) = k_5 \frac{c_5(d)^2}{\sqrt{d}} \cdot \begin{cases} 1 & \text{if } \left(\frac{-d}{43}\right) = +1, \\ 2 & \text{if } \left(\frac{-d}{43}\right) = 0, \\ 0 & \text{if } \left(\frac{-d}{43}\right) = -1, \end{cases}$$

is satisfied, where

$$k_5 = 2 \cdot \frac{(f,f)}{L(f,5,1)\sqrt{5}} \approx 5.4527296726817343855570722785283\ldots$$

Note that in the case $\left(\frac{-d}{43}\right) = -1$ it is trivial that $c_5(d) = L(f, -d, 1) = 0$.

4.3.2 Real twists in a case of rank 1

We can use $l = 3$, since $L(f, -3, 1) \approx 3.1481349$, and let $\psi = \chi_{43}$ be the
quadratic character of conductor 43. Table 5 shows the values of $c_{-3}(d)$ and
$L(f, d, 1)$, where $0 < d < 200$ is a fundamental discriminant such that $\left(\frac{d}{43}\right) = -1$. The formula

$$L(f, d, 1) = k_{-3} \frac{c_{-3}(d)^2}{\sqrt{d}}$$

is satisfied, where

$$k_{-3} = 2 \cdot \frac{(f,f)}{L(f,-3,1)\sqrt{3}} \approx 10.93737905993516764875873543879\ldots$$

Note that in the case $\left(\frac{d}{43}\right) \neq -1$ it is trivial that $c_{-3}(d) = L(f, d, 1) = 0$.

d	$c_{-3}(d)$	$L(f,d,1)$	d	$c_{-3}(d)$	$L(f,d,1)$	d	$c_{-3}(d)$	$L(f,d,1)$
5	1	4.891345	76	0	0.000000	137	2	3.737773
8	-1	3.866947	77	-3	11.217870	141	-2	3.684374
12	1	3.157349	85	1	1.186325	149	0	0.000000
28	-1	2.066970	88	-1	1.165929	156	1	0.875691
29	-1	2.031020	89	1	1.159360	157	2	3.491592
33	-1	1.903953	93	3	10.207380	161	-1	0.861986
37	2	7.192376	104	1	1.072498	168	-2	3.375348
61	1	1.400388	105	0	0.000000	177	-2	3.288415
65	-1	1.356615	113	-2	4.115608	184	1	0.806314
69	-1	1.316706	120	0	0.000000			
73	1	1.280123	136	1	0.937873			

Table 5: Coefficients of $\Theta_{-3}(e_f)$ and central values for $f = f_{43A}$

4.4   389A

Let $f = f_{389A}$, the modular form of level 389 and rank 2. Let $B = B(-2, -389)$, the quaternion algebra ramified precisely at ∞ and 389. A maximal order, with 33 ideal classes, is given by

$$R = \left\langle 1, i, \frac{1+i+j}{2}, \frac{2+3i+k}{4} \right\rangle.$$

There is a vector e_f of height $\langle e_f, e_f \rangle = \frac{5}{2}$ corresponding to f.

4.4.1   Imaginary twists in a case of rank 2

We can use $l = 5$, since $L(f, 5, 1) \approx 8.9092552$. We have omitted the 33 ideal classes; however, the computation of $\Theta_l(e_f)$ involves only 14 distinct theta series. In table 6 we give the value of e_f and the coefficients of the norm form $\mathcal{N}_i$ and of b_i on chosen bases of S_i^0.

Each row in the table allows one to compute an individual theta series

$$h_i(z) := \frac{1}{2} \sum_{b \in \mathbb{Z}^3} w_5(I_i, b) q^{\mathcal{N}_i(b)/5}.$$

The ternary form corresponding to a sextuple $(A_1, A_2, A_3, A_{23}, A_{13}, A_{12})$ is

$$\mathcal{N}_i(x_1, x_2, x_3) = A_1 x_1^2 + A_2 x_2^2 + A_3 x_3^2 + A_{23} x_2 x_3 + A_{13} x_1 x_3 + A_{12} x_1 x_2,$$

and $\omega_5(I_i, \cdot)$ is the weight function of the pair $(\mathbb{Z}^3, b_i)$. As an example, we show how to compute $h_1(z)$. First, we have

$$\mathcal{N}_1(x_1, x_2, x_3) = 15 x_1^2 + 107 x_2^2 + 416 x_3^2 - 100 x_2 x_3 - 8 x_1 x_3 - 14 x_1 x_2.$$

A simple calculation shows that

$$\langle (x_1, x_2, x_3), (2, 4, 0) \rangle \equiv 4x_1 + 3x_2 + 4x_3 \pmod{5}.$$

i	a_i	$\mathcal{N}_i$	b_i
1	$1/2$	15, 107, 416, -100, -8, -14	2, 4, 0
2	$-1/2$	15, 104, 415, 104, 2, 4	0, 4, 1
3	$-1/2$	23, 136, 203, 68, 2, 8	2, 1, 4
4	$1/2$	23, 72, 407, 72, 10, 20	1, 1, 0
5	$-1/2$	31, 51, 407, -46, -26, -10	1, 2, 0
6	$1/2$	31, 103, 204, 56, 20, 18	2, 0, 3
7	$1/2$	39, 128, 160, -116, -8, -36	1, 1, 4
8	$-1/2$	39, 40, 399, 40, 2, 4	1, 0, 1
9	$1/2$	40, 47, 399, 18, 40, 36	4, 3, 0
10	$-1/2$	47, 107, 135, 42, 22, 38	4, 3, 1
11	$-1/2$	56, 84, 139, 56, 4, 12	3, 1, 4
12	$1/2$	56, 92, 151, 76, 52, 44	4, 2, 3
13	$1/2$	71, 83, 132, -16, -12, -70	2, 3, 4
14	$-1/2$	71, 103, 124, -36, -64, -66	4, 0, 2

Table 6: Coefficients of the ternary forms and of b_i

Thus, ω_5 can be computed as

$$\omega_5(I_1,(x_1,x_2,x_3)) = \begin{cases} 0 & \text{if } 5 \nmid \mathcal{N}_1(x_1,x_2,x_3), \\ \chi_5(4x_1+3x_2+4x_3) & \text{if } \not\equiv 0 \pmod 5, \\ \chi_5(x_2) & \text{otherwise,} \end{cases}$$

and we have

$$h_1(z) = q^3 - q^{12} - q^{27} + q^{39} + q^{40} + q^{48} - q^{83} - 2q^{92} + O(q^{100}).$$

Finally, we combine all of the theta series in

$$\Theta_5(e_f) = \sum_{i=1}^{14} a_i h_i(z)$$

Table 7 shows the values of $c_5(d)$ and $L(f,-d,1)$, where $0 < -d < 200$ is a fundamental discriminant such that $\left(\frac{-d}{389}\right) \neq +1$. The formula

$$L(f,-d,1) = k_5 \frac{c_5(d)^2}{\sqrt{d}} \cdot \begin{cases} 1 & \text{if } \left(\frac{-d}{389}\right) = -1, \\ 2 & \text{if } \left(\frac{-d}{389}\right) = 0, \\ 0 & \text{if } \left(\frac{-d}{389}\right) = +1, \end{cases}$$

is satisfied, where

$$k_5 = \frac{2}{5} \cdot \frac{(f,f)}{L(f,5,1)\sqrt 5} \approx 7.886950806206592817689630792605\ldots$$

Note that when $\left(\frac{-d}{389}\right) = +1$ it is trivial that $c_5(d) = L(f,-d,1) = 0$.

$-d$	$c_5(d)$	$L(f,-d,1)$	$-d$	$c_5(d)$	$L(f,-d,1)$	$-d$	$c_5(d)$	$L(f,-d,1)$
-3	1	4.553533	-83	-1	0.865705	-139	-1	0.668962
-8	-1	2.788458	-84	1	0.860537	-148	6	23.338921
-15	-1	2.036402	-88	-4	13.452028	-151	2	2.567324
-23	1	1.644543	-103	0	0.000000	-152	-1	0.639716
-31	1	1.416538	-104	-1	0.773379	-155	3	5.701456
-39	1	1.262923	-107	0	0.000000	-163	8	39.536232
-40	1	1.247036	-115	-1	0.735462	-167	-1	0.610311
-43	-3	10.824738	-116	-2	2.929140	-191	1	0.570680
-47	0	0.000000	-123	3	6.400282	-195	1	0.564796
-51	-2	4.417576	-131	1	0.689086	-199	-1	0.559091
-56	1	1.053938	-132	-2	2.745884			
-71	1	0.936009	-136	-2	2.705202			

Table 7: Coefficients of $\Theta_5(e_f)$ and central values for $f = f_{389A}$

References

[BM]   Baruch E.M., Mao Z., *Central values of automorphic L-functions*, preprint.

[BFH]   Bump, D., Friedberg, S., Hoffstein, J., *Nonvanishing theorems for L-functions of modular forms and their derivatives*, Invent. Math. 102 (1990), p. 543-618.

[G]   Gross, B., *Heights and the special values of $L-$series*, Canadian Math. Soc. Conf. Proceedings, volume 7, (1987) p. 115-187.

[P]   Pizer A., *An algorithm for computing modular forms on $\Gamma_0(N)$*, J. Algebra 64 (1980), p. 340-390.

[W]   Waldspurger J-L., *Sur les coefficients de Fourier des formes modulaires de poids demi-entier*, J. Math. pures Appl. 60 (1981), p. 375-484.

Zhengyu Mao
Department of Mathematics and Computer Science
Rutgers university
Newark, NJ 07102-1811
zmao@andromeda.rutgers.edu

Fernando Rodriguez-Villegas
Gonzalo Tornaría
Department of Mathematics,
University of Texas at Austin,
Austin, TX 78712 USA
villegas@math.utexas.edu
tornaria@math.utexas.edu

Examples of the Shimura correspondence for level p^2 and real quadratic twists

Ariel Pacetti and Gonzalo Tornaría

Abstract

We give examples of the Shimura correspondence for rational modular forms f of weight 2 and level p^2, for primes $p \leqslant 19$, computed as an application of a method we introduced in [5]. Furthermore, we verify in this examples a conjectural formula for the central values $L(f, -pd, 1)$ and, in case $p \equiv 3 \pmod 4$, a formula for the central values $L(f, d, 1)$ corresponding to the real quadratic twists of f.

1   Introduction

Let f be a newform of weight 2. We let $L(f, s)$ denote its Hecke L-series, and for D a fundamental discriminant we define its *twisted L-series* as

$$L(f, D, s) := L\left(f \otimes \left(\tfrac{D}{\cdot}\right), s\right),$$

where $f \otimes \left(\tfrac{D}{\cdot}\right)$ is (the newform corresponding to) the twist of f by the quadratic character $n \mapsto \left(\tfrac{D}{n}\right)$. Recall that $L(f, D, s)$ is an entire function of the complex plane with a functional equation relating the its values at s and $2 - s$, its central value being $L(f, D, 1)$.

In the case of prime level p, a method due to Gross [3] constructs, provided $L(f, 1) \neq 0$, a nonzero modular form Θ_f of weight $\sfrac{3}{2}$ and level $4p$ which maps to f under the Shimura correspondence [8]. By Waldspurger's formula [11] the Fourier coefficients of Θ_f are related to the central values $L(f, -d, 1)$ for imaginary fundamental discriminants $-d < 0$, and Gross gives an explicit formula for such central values.

The authors have extended Gross's method to the case of level p^2 (under a technical hypothesis, see §4, and cf. [7].) We show in [5] how to construct *two* modular forms Θ_f^+ and Θ_f^- of weight $\sfrac{3}{2}$ and level $4p^2$, with character $\varkappa_p(n) := \left(\tfrac{p}{n}\right)$, mapping to f under the Shimura correspondence.

In this paper we outline the main ideas of our method, and conjecture a formula relating the central values $L(f, -pd, 1)$ for imaginary fundamental discriminants $-pd < 0$, with the Fourier coefficients of Θ_f^+ and Θ_f^-. In particular, such formula would imply that $\Theta_f^+ = \Theta_f^- = 0$ if and only if $L(f, 1) = 0$.

When $p \equiv 3 \pmod 4$, we apply this method and the conjectured formula to the computation of the central values $L(f, d, 1)$ for *real* fundamental discriminants $d > 0$, giving an algorithm that is particularly well suited to the case where f has level p. The proviso here would be that $L(f, -p, 1) \neq 0$.

Finally, we give examples of this algorithm applied to the rational modular forms f_{49A}, f_{11A}, f_{121A}, f_{121B}, f_{121C}, f_{121D}, f_{17A}, f_{289A}, f_{19A}, f_{361A}, and f_{361B}. The routines used for these calculations, which will be made available in [1], were written by the authors for the PARI/GP system [6].

More examples can be found among the data presented at the "Special Week on Ranks of Elliptic Curves and Random Matrix Theory" held at the Isaac Newton Institute, which includes the application of the latter algorithm to the rational modular forms of level $p \equiv 3 \pmod 4$, with $p < 500$ [9].

For a different approach to computing the central values for real quadratic twists, which works only for prime level, see [4].

2   Quaternion algebras and Shimura correspondence

Let a, b be negative integers and let $\mathbb{H} = \mathbb{H}(a, b)$ be the definite quaternion algebra over $\mathbb{Q}$ with basis $\{1, i, j, k = ij\}$ where $i^2 = a$, $j^2 = b$, $ij = -ji$. For $x \in \mathbb{H}$, we denote by $\mathcal{N}x$ the *reduced norm* of x, i.e.

$$\mathcal{N}\, x_0 + x_1 i + x_2 j + x_3 k := x_0^2 - ax_1^2 - bx_2^2 + abx_3^2.$$

The norm of a lattice $\mathfrak{a} \subseteq \mathbb{H}$ is defined to be $\mathcal{N}\mathfrak{a} := \gcd\{\mathcal{N}x \; : \; x \in \mathfrak{a}\}$.

An *order* in $\mathbb{H}$ is a (full rank) lattice $R \subseteq \mathbb{H}$ which is also a ring with $1 \in R$. Since the determinant of the quadratic form $\mathcal{N}$ in the above basis is $16\,(ab)^2$, the determinant of R is a rational square. Its positive square root will be denoted by $\mathrm{D}(R)$.

We let $\widetilde{\mathfrak{I}}(R)$ be the set of *left R-ideals*, i.e. the lattices $\mathfrak{a} \subseteq \mathbb{H}$ such that $\mathfrak{a}_p = R_p x_p$ for every prime p, with $x_p \in \mathbb{H}_p^\times$. An equivalence relation is defined on $\widetilde{\mathfrak{I}}(R)$ where two left ideals $\mathfrak{a}, \mathfrak{b} \in \widetilde{\mathfrak{I}}(R)$ are in the same class if $\mathfrak{a} = \mathfrak{b}x$, for some $x \in \mathbb{H}^\times$; we write $[\mathfrak{a}]$ for the class of $\mathfrak{a}$. The set of all left R-ideal classes, which we denote by $\mathfrak{I}(R)$, is known to be finite.

Let $\mathcal{M}(R)$ be the $\mathbb{R}$-vector space with basis $\mathfrak{I}(R)$, with the *height pairing*

$$\langle [\mathfrak{a}], [\mathfrak{b}] \rangle := \begin{cases} \frac{1}{2} \# R_r(\mathfrak{a})^\times & \text{if } [\mathfrak{a}] = [\mathfrak{b}], \\ 0 & \text{otherwise,} \end{cases}$$

as an inner product. Note that $\mathfrak{I}(R)$ is an orthogonal basis of this space.

For each integer $m \geqslant 1$ we define *Hecke operators* $t_m : \mathcal{M}(R) \to \mathcal{M}(R)$ by

$$t_m[\mathfrak{a}] := \sum_{[\mathfrak{b}] \in \mathfrak{I}(R)} B_m\big([\mathfrak{b}], [\mathfrak{a}]\big) \cdot [\mathfrak{b}],$$

where B_m is the classical *Brandt matrix*

$$B_m([\mathfrak{a}], [\mathfrak{b}]) := \#\{\mathfrak{c} \in [\mathfrak{a}^{-1}\mathfrak{b}] \ : \ \mathcal{N}\mathfrak{c} = m, \ \mathfrak{c} \text{ integral}\}.$$

The Hecke operators t_m, with $\gcd(m, \mathrm{D}(R)) = 1$, generate a commutative ring $\mathbb{T}_0$ and thus, by the spectral theorem, $\mathcal{M}(R)$ has an orthogonal basis of eigenvectors for $\mathbb{T}_0$. When f is a newform of weight 2, say $f_{|T(m)} = \lambda_m f$, we set

$$\mathcal{M}(R)^f := \{\mathbf{v} \in \mathcal{M}(R) \ : \ t_m \mathbf{v} = \lambda_m \mathbf{v} \text{ for } (m, \mathrm{D}(R)) = 1\},$$

to be the f-isotypical component of $\mathcal{M}(R)$.

2.1 Modular forms of weight $^3/_2$

The *discriminant* of a quaternion $x \in \mathbb{H}$ is defined to be $\Delta x := (x - \bar{x})^2$. This is a quadratic form of rank 3 which we will use to construct modular forms of weight $^3/_2$.

Let R be an order in $\mathbb{H}$. We define $\Omega(R) := \gcd\{\Delta x \ : \ x \in R\}$, and note that $Q(x) := -\Delta x / \Omega(R)$, in the ternary lattice $R/\mathbb{Z}$, is a primitive, positive definite ternary quadratic form. Its theta series,

$$\Theta(R) := \frac{1}{2} \sum_{x \in R/\mathbb{Z}} q^{Q(x)},$$

depends only on the $\mathbb{Z}$-equivalence class of the ternary quadratic form Q; in the examples such a ternary quadratic form will be given by its coefficients a_1, a_2, a_3, a_{23}, a_{13}, a_{12}, meaning that in some basis of $R/\mathbb{Z}$,

$$Q(X_1, X_2, X_3) = a_1 X_1^2 + a_2 X_2^2 + a_3 X_3^2 + a_{23} X_2 X_3 + a_{13} X_1 X_3 + a_{12} X_1 X_2. \quad (2.1)$$

We will also write Q to stand for its theta series.

Now let $\mathfrak{a} \in \widetilde{\mathcal{I}}(R)$. We set $\Theta([\mathfrak{a}]) := \Theta(R_r(\mathfrak{a}))$, and extend by linearity to all of $\mathcal{M}(R)$. Note that the ternary forms corresponding to R and $R_r(\mathfrak{a})$ are in the same genus since $\mathfrak{a}$, being principal, induces local isometries by conjugation. In particular, $\Omega(R) = \Omega(R_r(\mathfrak{a}))$.

Note that $\Theta(\mathbf{v})$ is in the space $M_{3/2}(N, \varkappa)$ of modular forms of weight $^3/_2$, level $N = 4\frac{\mathrm{D}(R)}{\Omega(R)}$, and character $\varkappa = \left(\frac{\Omega(R)}{\cdot}\right)$. Moreover,

Proposition 2.1. *The map Θ is Hecke-linear, i.e.*

$$\Theta(\mathbf{v})_{|T(m^2)} = \Theta(t_m \mathbf{v}),$$

for any $m \geqslant 1$ such that $(m, 2\,\mathrm{D}(R)) = 1$.

This means that for a newform f of weight 2, any nonzero modular form in $\Theta(M(R)^f)$ will map to f under the Shimura correspondence.

3　Gross's formula for level p

Let f be a newform of weight 2 and prime level p, and let $\mathcal{O}$ be a maximal order in the quaternion algebra ramified at p and ∞. It follows from Eichler's trace formula [2] that $\dim \mathcal{M}(\mathcal{O})^f = 1$. Thus $\mathbf{e}_f \in \mathcal{M}(\mathcal{O})^f$ is well defined up to a constant; we write

$$\Theta_f := \Theta(\mathbf{e}_f) = \sum_{d \geqslant 1} c_f(d) q^d.$$

We also define the Peterson norm of f to be

$$\langle f, f \rangle := 8\pi^2 \int_{\Gamma_0(N)\backslash \mathfrak{h}} |f(z)|^2 dx\, dy$$

Theorem 3.1 (Gross [3, Proposition 13.5, p. 179]). *Let $-d < 0$ be a fundamental discriminant. Then*

$$L(f, -d, 1)\, L(f, 1) = \star \frac{\langle f, f \rangle}{\sqrt{d}} \frac{c_f(d)^2}{\langle \mathbf{e}_f, \mathbf{e}_f \rangle},$$

where $\star = 1$ if $p \nmid d$, $\star = 2$ if $p \mid d$.

4　On certain non-maximal orders and level p^2

The aim here is to give a formula like the one in Theorem 3.1 that applies also to modular forms of level p^2. Keep the notation of the previous section, except f is now a newform of weight 2 and level p or p^2. Let $\tilde{\mathcal{O}} \subseteq \mathcal{O}$ be the unique suborder of index p in $\mathcal{O}$, namely

$$\tilde{\mathcal{O}} := \{x \in \mathcal{O} \ : \ p \mid \Delta\, x\}.$$

We have the following result due to Pizer ([7, Theorem 8.2, p.223]) :

$$\dim \mathcal{M}(\tilde{\mathcal{O}})^f = \begin{cases} 2 & \text{if } f \text{ is not the twist of a level } p \text{ form,} \\ 1 & \text{if } f \text{ is a level } p \text{ form or the quadratic twist of a level } p \text{ form,} \\ 0 & \text{otherwise.} \end{cases}$$

In what follows we will assume that f is not in the last case, i.e. $\mathcal{M}(\tilde{\mathcal{O}})^f \neq 0$.

Clearly, $\mathrm{D}(\tilde{\mathcal{O}}) = p^2$, but $\Omega(\tilde{\mathcal{O}}) = p$, and we have $\Theta\left(\mathcal{M}(\tilde{\mathcal{O}})\right) \subseteq M_{3/2}(4p, \varkappa_p)$. Thus $\Theta\left(\mathcal{M}(\tilde{\mathcal{O}})^f\right) = 0$ unless f is a level p form.

We now investigate suborders of index p in $\tilde{\mathcal{O}}$. One can prove that any such order contains $\mathbb{Z} + p\mathcal{O}$; conversely, any of the $p + 1$ lattices $\mathcal{O}'$ such that $\mathbb{Z} + p\mathcal{O} \subsetneq \mathcal{O}' \subsetneq \tilde{\mathcal{O}}$ is an order. Let $x \in \mathcal{O}'$ such that $x \notin \mathbb{Z} + p\mathcal{O}$. Then

$$\sigma := \left(\frac{\Delta\, x/p}{p}\right)$$

is well defined and nonzero, and we call σ the *sign* of $\mathcal{O}'$. The orders $\mathcal{O}'$ split in two local conjugacy classes: $\frac{p+1}{2}$ of sign $+$, and $\frac{p+1}{2}$ of sign $-$.

The space $\mathcal{M}(\mathcal{O}')$ depends only on the sign of $\mathcal{O}'$. Thus, we fix $\mathcal{O}^+$ and $\mathcal{O}^-$ to be two such orders, with signs $+$ and $-$ respectively. In what follows σ will denote either $+$ or $-$. Note that $D(\mathcal{O}^\sigma) = p^3$, $\Omega(\mathcal{O}^\sigma) = p$ and so we have Hecke-linear maps

$$\Theta : \mathcal{M}(\mathcal{O}^\sigma) \to M_{3/2}(4p^2, \varkappa_p).$$

The space $\mathcal{M}(\mathcal{O}^\sigma)$ is too big for our purposes, since it represents weight 2 modular forms of level p^3; indeed $\dim \mathcal{M}(\mathcal{O}^\sigma) = O(p^3)$, compared to $\dim \mathcal{M}(\tilde{\mathcal{O}}) = O(p^2)$.

For $\mathfrak{a} \in \tilde{\mathcal{I}}(\tilde{\mathcal{O}})$ the $\mathcal{O}^\sigma$-*subideals* of $\mathfrak{a}$ are the $\mathfrak{b} \in \tilde{\mathcal{I}}(\mathcal{O}^\sigma)$ such that $\mathfrak{b} \subseteq \mathfrak{a}$ and $N\mathfrak{b} = N\mathfrak{a}$. It can be proved that the number of $\mathcal{O}^\sigma$-subideals of $\mathfrak{a}$ is exactly p and, moreover, that they all have the same right order. Thus, we can define Hecke-linear maps

$$\Theta^\sigma : \mathcal{M}(\tilde{\mathcal{O}}) \to M_{3/2}(4p^2, \varkappa_p),$$

given, for $[\mathfrak{a}] \in \mathcal{I}(\tilde{\mathcal{O}})$, by

$$\Theta^\sigma([\mathfrak{a}]) := \Theta([\mathfrak{b}]),$$

where $\mathfrak{b}$ is any $\mathcal{O}^\sigma$-subideal of $\mathfrak{a}$.

If $\Theta^\sigma(\mathcal{M}(\tilde{\mathcal{O}})^f) = 0$, we let $\mathbf{e}_f^\sigma$ to be any nonzero vector in $\mathcal{M}(\tilde{\mathcal{O}})^f$. Otherwise, it follows from the strong multiplicity one theorem of Ueda ([10, Theorem 3.11, p.181]) that $\dim \Theta^\sigma(\mathcal{M}(\tilde{\mathcal{O}})^f) = 1$, and thus there is, up to a constant, a unique $\mathbf{e}_f^\sigma \in \mathcal{M}(\tilde{\mathcal{O}})^f$ orthogonal to $\ker \Theta^\sigma$. We write

$$\Theta_f^\sigma := \Theta^\sigma(\mathbf{e}_f^\sigma) = \sum_{\left(\frac{d}{p}\right)=\sigma} c_f(d) q^d.$$

Let us also introduce the rational constant

$$\alpha_f := \frac{1}{2} \cdot \begin{cases} 1 & \text{if } f \text{ is not the twist of a level } p \text{ form,} \\ \frac{p}{p-1} & \text{if } f \text{ is the quadratic twist of a level } p \text{ form,} \\ p+1 & \text{if } f \text{ is a level } p \text{ form.} \end{cases}$$

Conjecture 1. *Let d be an integer such that $-pd < 0$ is a fundamental discriminant, and such that $\left(\frac{d}{p}\right) = \sigma$. Then*

$$L(f, -pd, 1)\, L(f, 1) = \alpha_f \frac{\langle f, f \rangle}{\sqrt{pd}} \frac{c_f(d)^2}{\langle \mathbf{e}_f^\sigma, \mathbf{e}_f^\sigma \rangle}.$$

5   An Algorithm for the Real Quadratic Twists

Assume now that $p \equiv 3 \pmod 4$, and let f be as before a newform of weight 2 and level p or p^2. Let f^* be the twist of f by the quadratic character of conductor p. For any positive fundamental discriminant d, we have

$$L(f, d, s) = L(f^*, -pd, s).$$

Thus, the formula of conjecture 1 would be able to compute the central values of $L(f, d, s)$ for positive fundamental discriminants d prime to p, provided that $L(f^*, 1) = L(f, -p, 1) \neq 0$.

The algorithm consists of computing the Brandt matrices for $\tilde{\mathcal{O}}$ and finding the eigenspace $\mathcal{M}(\tilde{\mathcal{O}})^{f^*}$. When f has level p there is a better algorithm for computing $\mathcal{M}(\tilde{\mathcal{O}})^{f^*}$, by exploiting the two linear maps

- $\psi : \mathcal{M}(\mathcal{O}) \to \mathcal{M}(\tilde{\mathcal{O}})$ given for $\mathfrak{a} \in \widetilde{\mathcal{I}}(\mathcal{O})$ by

$$\psi([\mathfrak{a}]) = \sum_{\mathfrak{b} \text{ subideal}} [\mathfrak{b}],$$

where the sum is over all $\tilde{\mathcal{O}}$-subideals of $\mathfrak{a}$, i.e. the ideals $\mathfrak{b} \in \widetilde{\mathcal{I}}(\tilde{\mathcal{O}})$ such that $\mathfrak{b} \subseteq \mathfrak{a}$ and $\mathcal{N}\mathfrak{b} = \mathcal{N}\mathfrak{a}$. This map commutes with the Hecke operators, and thus

$$\mathcal{M}(\tilde{\mathcal{O}})^f = \psi(\mathcal{M}(\mathcal{O})^f).$$

- $\varphi : \mathcal{M}(\tilde{\mathcal{O}}) \to \mathcal{M}(\tilde{\mathcal{O}})$, given, for $\mathfrak{b} \in \widetilde{\mathcal{I}}(\tilde{\mathcal{O}})$ by

$$\varphi([\mathfrak{b}]) = \chi(\mathfrak{b})[\mathfrak{b}] \text{ for } \mathfrak{b} \in \widetilde{\mathcal{I}}(\tilde{\mathcal{O}}),$$

where $\chi(\mathfrak{b})$ is the *sign* of $\mathfrak{b}$ (namely $\chi(\mathfrak{b}) := \left(\frac{\mathcal{N}(x)/\mathcal{N}(\mathfrak{b})}{p} \right)$ for $x \in \mathfrak{b}$ such that $p \nmid (\mathcal{N}(x)/\mathcal{N}(\mathfrak{b}))$, see [7, Proposition 5.1]). This map corresponds to twisting by the quadratic character of conductor p; hence

$$\mathcal{M}(\tilde{\mathcal{O}})^{f^*} = \varphi(\mathcal{M}(\tilde{\mathcal{O}})^f).$$

Thus, it will be enough to compute the Brandt matrices for $\mathcal{O}$ to find $\mathcal{M}(\mathcal{O})^f$, and $\mathcal{M}(\tilde{\mathcal{O}})^{f^*} = \varphi(\psi(\mathcal{M}(\mathcal{O})^f))$. This is a big improvement since $\dim \mathcal{M}(\mathcal{O}) = O(p)$, while $\dim \mathcal{M}(\tilde{\mathcal{O}}) = O(p^2)$.

6   Example: level 7^2

Let $\mathbb{H} = \mathbb{H}(-1, -7)$, the quaternion algebra ramified precisely at ∞ and 7. A maximal order, having a unique left ideal class, is given by

$$\mathcal{O} = \mathfrak{a}_1 = \left\langle 1, i, \frac{1+j}{2}, \frac{i+k}{2} \right\rangle.$$

Its index p suborder is given by $\tilde{\mathcal{O}} = \left\langle 1, 7i, \frac{1+j}{2}, \frac{7i+k}{2} \right\rangle$; inequivalent $\tilde{\mathcal{O}}$-subideals for the $\mathcal{O}$-ideal are show in Table 6.1.

We fix two index p suborders of $\tilde{\mathcal{O}}$

$$\mathcal{O}^+ = \left\langle 1, 7i, \frac{1+j}{2}, \frac{7i+7k}{2} \right\rangle,$$

$$\mathcal{O}^- = \left\langle 1, 7i, \frac{1+7j}{2}, \frac{1+7i+5j+k}{2} \right\rangle$$

$\mathcal{O}$-ideals	$\tilde{\mathcal{O}}$-subideals		χ	$+$ genus	$-$ genus
$\mathfrak{a}_1$	$\mathfrak{b}_{1,1} = \left\langle 1, 7i, \frac{1+j}{2}, \frac{7i+k}{2} \right\rangle$		$+$	Q_1^+	Q_1^-
	$\mathfrak{b}_{1,2} = \left\langle 7, 1+i, \frac{7+j}{2}, \frac{8+i+k}{2} \right\rangle$		$+$	Q_2^+	Q_2^-
	$\mathfrak{b}_{1,3} = \left\langle 7, 3+i, \frac{7+j}{2}, \frac{10+i+k}{2} \right\rangle$		$-$	Q_3^+	Q_2^-
	$\mathfrak{b}_{1,4} = \left\langle 7, 5+i, \frac{7+j}{2}, \frac{12+i+k}{2} \right\rangle$		$-$	Q_3^+	Q_1^-

Table 6.1: Maps $\Theta_{\mathcal{O}+}$ and $\Theta_{\mathcal{O}-}$ from the $\tilde{\mathcal{O}}$-Ideals to ternary quadratic forms in the $+$ and $-$ genus respectively, level 7^2.

	a_1	a_2	a_3	a_{23}	a_{13}	a_{12}
Q_1^+	1,	28,	56,	-28,	0,	0
Q_2^+	4,	8,	49,	0,	0,	-4
Q_3^+	8,	9,	25,	2,	4,	8
Q_1^-	12,	12,	13,	-8,	-8,	-4
Q_2^-	5,	17,	17,	6,	2,	2

Table 6.2: Coefficients of ternary quadratic forms, level 7^2.

in the $+$ and $-$ genus respectively. Table 6.1 shows the maps from $\tilde{\mathcal{O}}$-ideals to ternary quadratic forms of level 7^2 in the $+$ genus and in the $-$ genus, computed via $\mathcal{O}^+$- and $\mathcal{O}^-$-subideals respectively. The actual coefficients of the ternary quadratic forms are given in Table 6.2, with the notation as in (2.1).

6.1 f_{49A}

By computing the Brandt matrices for $\tilde{\mathcal{O}}$, we find the space $\mathcal{M}(\tilde{\mathcal{O}})^{f_{49A}}$ of dimension 2, spanned by $\mathbf{e}_{f_{49A}}^+ = [\mathfrak{b}_{1,1}] - [\mathfrak{b}_{1,2}]$ and

$$\mathbf{e}_{f_{49A}}^- = \frac{[\mathfrak{b}_{1,1}] - [\mathfrak{b}_{1,2}] - [\mathfrak{b}_{1,3}] + [\mathfrak{b}_{1,4}]}{2},$$

with heights $\left\langle \mathbf{e}_{f_{49A}}^+, \mathbf{e}_{f_{49A}}^+ \right\rangle = 2 \left\langle \mathbf{e}_{f_{49A}}^-, \mathbf{e}_{f_{49A}}^- \right\rangle = 2$. Using Table 6.1, we see that

$$\Theta_{f_{49A}}^+ = Q_1^+ - Q_2^+,$$

and

$$\Theta_{f_{49A}}^- = Q_1^- - Q_2^-.$$

Table 6.3 shows the values of $c_{f_{49A}}(d)$ and $L(f_{49A}, -7d, 1) = L(f_{49A}, d, 1)$, where $0 < d < 200$ is a fundamental discriminant such that $7 \nmid d$. The formula

$$L(f_{49A}, -7d, 1) = L(f_{49A}, d, 1) = k_{f_{49A}} \frac{c_{f_{49A}}(d)^2}{\sqrt{d}} \cdot \begin{cases} 1 & \text{if } \left(\frac{d}{7}\right) = +1 \\ 2 & \text{if } \left(\frac{d}{7}\right) = -1 \end{cases}$$

d	$c_{f_{49A}}(d)$	$L(f_{49A}, d, 1)$	d	$c_{f_{49A}}(d)$	$L(f_{49A}, d, 1)$
1	1	0.966656	109	2	0.370355
8	-2	1.367058	113	-4	1.454965
29	2	0.718014	120	4	1.411891
37	2	0.635669	137	2	0.330348
44	0	0.000000	141	-4	1.302514
53	0	0.000000	149	0	0.000000
57	0	0.000000	156	4	1.238311
60	4	1.996716	165	-4	1.204065
65	-2	0.479596	172	0	0.000000
85	-2	0.419394	177	-8	4.650131
88	-4	1.648734	184	-4	1.140205
92	4	1.612493	193	0	0.000000
93	-4	1.603801	197	0	0.000000
d	$c_{f_{49A}}(d)$	$L(f_{49A}, d, 1)$	d	$c_{f_{49A}}(d)$	$L(f_{49A}, d, 1)$
5	-1	0.864603	97	-1	0.196298
12	2	2.232396	101	-1	0.192372
13	1	0.536204	104	4	3.033229
17	1	0.468897	124	-4	2.777864
24	-2	1.578542	129	4	2.723498
33	-2	1.346185	136	2	0.663120
40	0	0.000000	145	2	0.642211
41	1	0.301933	152	-2	0.627249
61	1	0.247535	157	-3	1.388656
69	-2	0.930974	173	1	0.146987
73	-3	2.036493	181	1	0.143702
76	-2	0.887064	185	-4	2.274238
89	3	1.844376	188	0	0.000000

Table 6.3: Coefficients of $\Theta^{+}_{f_{49A}}$ (top), $\Theta^{-}_{f_{49A}}$ (bottom), and central values for f_{49A}

is satisfied, where

$$k_{f_{49A}} = \frac{1}{4} \cdot \frac{\langle f_{49A}, f_{49A} \rangle}{L(f_{49A}, 1)\sqrt{7}} = 0.96665585280840577336653384189 = L(f_{49A}, 1).$$

7   Example: level 11^2

Let $\mathbb{H} = \mathbb{H}(-1, -11)$, the quaternion algebra ramified precisely at ∞ and 11. A maximal order, and representatives for its left ideals classes, are given by

$$\mathcal{O} = \mathfrak{a}_1 = \left\langle 1, i, \frac{1+j}{2}, \frac{i+k}{2} \right\rangle,$$

$$\mathfrak{a}_2 = \left\langle 2, 2i, \frac{3+2i+j}{2}, \frac{2+3i+k}{2} \right\rangle.$$

Its index p suborder is given by

$$\tilde{\mathcal{O}} = \left\langle 1, 11i, \frac{1+j}{2}, \frac{11i+k}{2} \right\rangle;$$

inequivalent $\tilde{\mathcal{O}}$-subideals for each $\mathcal{O}$-ideal are show in Table 7.1.

We fix two index p suborders of $\tilde{\mathcal{O}}$

$$\mathcal{O}^+ = \left\langle 1, 11i, \frac{1+j}{2}, \frac{11i+11k}{2} \right\rangle,$$

$$\mathcal{O}^- = \left\langle 1, 11i, \frac{1+11j}{2}, \frac{1+11i+j+k}{2} \right\rangle$$

in the $+$ and $-$ genus respectively. Table 7.1 shows the maps from $\tilde{\mathcal{O}}$-ideals to ternary quadratic forms of level 11^2 in the $+$ genus and in the $-$ genus, computed via $\mathcal{O}^+$- and $\mathcal{O}^-$-subideals respectively. The actual coefficients of the ternary quadratic forms are given in Table 7.2, with the notation as in (2.1).

7.1 f_{11A}

By computing the Brandt matrices for $\mathcal{O}$, we find the space $\mathcal{M}(\tilde{\mathcal{O}})^{f_{11A}} = \psi_{\tilde{\mathcal{O}}}\left(\mathcal{M}(\mathcal{O})^{f_{11A}}\right)$ of dimension 1, spanned by

$$\mathbf{e}^+_{f_{11A}} = \mathbf{e}^-_{f_{11A}} = \psi_{\tilde{\mathcal{O}}}\left(\frac{[\mathfrak{a}_1] - [\mathfrak{a}_2]}{2}\right),$$

with heights $\langle \mathbf{e}^+_{f_{11A}}, \mathbf{e}^+_{f_{11A}} \rangle = \langle \mathbf{e}^-_{f_{11A}}, \mathbf{e}^-_{f_{11A}} \rangle = 15$. Using Table 7.1, we see that

$$\Theta^+_{f_{11A}} = Q^+_1 + 2Q^+_2 + 2Q^+_3 + Q^+_4 - 3Q^+_5 - 3Q^+_6,$$

$\mathcal{O}$-ideals	$\tilde{\mathcal{O}}$-subideals	χ	$+$ genus	$-$ genus
$\mathfrak{a}_1$	$\mathfrak{b}_{1,1} = \left\langle 1, 11i, \frac{1+j}{2}, \frac{11i+k}{2} \right\rangle$	$+$	Q_1^+	Q_1^-
	$\mathfrak{b}_{1,2} = \left\langle 11, 5+i, \frac{11+j}{2}, \frac{16+i+k}{2} \right\rangle$	$+$	Q_2^+	Q_2^-
	$\mathfrak{b}_{1,3} = \left\langle 11, 9+i, \frac{11+j}{2}, \frac{20+i+k}{2} \right\rangle$	$+$	Q_2^+	Q_3^-
	$\mathfrak{b}_{1,4} = \left\langle 11, 4+i, \frac{11+j}{2}, \frac{4+i+k}{2} \right\rangle$	$-$	Q_3^+	Q_3^-
	$\mathfrak{b}_{1,5} = \left\langle 11, 10+i, \frac{11+j}{2}, \frac{10+i+k}{2} \right\rangle$	$-$	Q_4^+	Q_1^-
	$\mathfrak{b}_{1,6} = \left\langle 11, 3+i, \frac{11+j}{2}, \frac{14+i+k}{2} \right\rangle$	$-$	Q_3^+	Q_2^-
$\mathfrak{a}_2$	$\mathfrak{b}_{2,1} = \left\langle 22, 14+2i, \frac{3+2i+j}{2}, \frac{10+3i+k}{2} \right\rangle$	$+$	Q_5^+	Q_4^-
	$\mathfrak{b}_{2,2} = \left\langle 22, 8+2i, \frac{19+2i+j}{2}, \frac{34+3i+k}{2} \right\rangle$	$+$	Q_5^+	Q_4^-
	$\mathfrak{b}_{2,3} = \left\langle 22, 12+2i, \frac{23+2i+j}{2}, \frac{18+3i+k}{2} \right\rangle$	$-$	Q_6^+	Q_4^-
	$\mathfrak{b}_{2,4} = \left\langle 2, 22i, \frac{3+22i+j}{2}, \frac{2+11i+k}{2} \right\rangle$	$-$	Q_6^+	Q_4^-

Table 7.1: Maps $\Theta_{\mathcal{O}^+}$ and $\Theta_{\mathcal{O}^-}$ from the $\tilde{\mathcal{O}}$-Ideals to ternary quadratic forms in the $+$ and $-$ genus respectively, level 11^2.

	a_1	a_2	a_3	a_{23}	a_{13}	a_{12}
Q_1^+	1,	44,	132,	$-44,$	0,	0
Q_2^+	16,	16,	25,	$-4,$	$-4,$	-12
Q_3^+	5,	36,	36,	28,	4,	4
Q_4^+	4,	12,	121,	0,	0,	-4
Q_5^+	5,	9,	124,	$-8,$	$-4,$	-2
Q_6^+	4,	33,	45,	$-22,$	$-4,$	0
Q_1^-	8,	13,	61,	2,	4,	8
Q_2^-	13,	21,	21,	$-2,$	$-6,$	-6
Q_3^-	17,	21,	21,	$-2,$	$-14,$	-14
Q_4^-	13,	21,	24,	$-16,$	$-4,$	-6

Table 7.2: Coefficients of ternary quadratic forms, level 11^2.

and

$$\Theta^-_{f_{11A}} = 2Q^-_1 + 2Q^-_2 + 2Q^-_3 - 6Q^-_4.$$

Table 7.3 shows the values of $c_{f_{11A}}(d)$ and $L(f_{11A}, -11d, 1) = L(f_{121D}, d, 1)$, where $0 < d < 200$ is a fundamental discriminant such that $11 \nmid d$. The formula

$$L(f_{11A}, -11d, 1) = L(f_{121D}, d, 1) = k_{f_{11A}} \frac{c_{f_{11A}}(d)^2}{\sqrt{d}}$$

is satisfied, where

$$k_{f_{11A}} = \frac{2}{5} \cdot \frac{\langle f_{11A}, f_{11A} \rangle}{L(f_{11A}, 1)\sqrt{11}} = 1.7593990386620401412515859974 = L(f_{121D}, 1).$$

7.2 f_{121A}

By computing the Brandt matrices for $\tilde{O}$, we find the space $\mathcal{M}(\tilde{O})^{f_{121A}}$ of dimension 2, spanned by

$$e^+_{f_{121A}} = \frac{2[b_{1,1}] - [b_{1,2}] - [b_{1,3}] + [b_{1,4}] - 2[b_{1,5}] + [b_{1,6}]}{2},$$

and

$$e^-_{f_{121A}} = \frac{[b_{1,2}] - [b_{1,3}] - [b_{1,4}] + [b_{1,6}]}{2},$$

with heights $\langle e^+_{f_{121A}}, e^+_{f_{121A}} \rangle = 3 \langle e^-_{f_{121A}}, e^-_{f_{121A}} \rangle = 3$. Using Table 7.1, we see that

$$\Theta^+_{f_{121A}} = Q^+_1 - Q^+_2 + Q^+_3 - Q^+_4,$$

and

$$\Theta^-_{f_{121A}} = Q^-_2 - Q^-_3.$$

Table 7.4 shows the values of $c_{f_{121A}}(d)$ and $L(f_{121A}, -11d, 1) = L(f_{121C}, d, 1)$, where $0 < d < 200$ is a fundamental discriminant such that $11 \nmid d$. The formula

$$L(f_{121A}, -11d, 1) = L(f_{121C}, d, 1) = k_{f_{121A}} \frac{c_{f_{121A}}(d)^2}{\sqrt{d}} \cdot \begin{cases} 1 & \text{if } \left(\frac{d}{11}\right) = +1 \\ 3 & \text{if } \left(\frac{d}{11}\right) = 1 \end{cases}$$

is satisfied, where

$$k_{f_{121A}} = \frac{1}{6} \cdot \frac{\langle f_{121A}, f_{121A} \rangle}{L(f_{121A}, 1)\sqrt{11}} = 1.6661569203942160899376922029 = L(f_{121C}, 1).$$

d	$c_{f_{121D}}(d)$	$L(f_{121D}, d, 1)$	d	$c_{f_{121D}}(d)$	$L(f_{121D}, d, 1)$
1	1	1.759399	113	-1	0.165510
5	-1	0.786827	124	3	1.421988
12	-1	0.507895	133	2	0.610237
37	1	0.289243	136	-2	0.603469
53	-2	0.966688	137	-1	0.150316
56	-2	0.940438	141	2	0.592673
60	3	2.044237	152	-2	0.570824
69	-1	0.211807	157	1	0.140415
89	3	1.678463	168	-2	0.542962
92	-1	0.183430	177	-7	6.479982
93	1	0.182441	181	1	0.130775
97	-3	1.607759	185	3	1.164182
104	2	0.690093	188	2	0.513269

d	$c_{f_{121D}}(d)$	$L(f_{121D}, d, 1)$	d	$c_{f_{121D}}(d)$	$L(f_{121D}, d, 1)$
8	2	2.488166	101	2	0.700267
13	0	0.000000	105	-2	0.686799
17	2	1.706868	109	-2	0.674079
21	2	1.535729	120	-2	0.642442
24	-2	1.436543	129	2	0.619626
28	-2	1.329981	140	-2	0.594785
29	0	0.000000	145	-4	2.337762
40	2	1.112742	149	2	0.576542
41	-4	4.396351	156	-4	2.253835
57	4	3.728610	161	-2	0.554640
61	0	0.000000	172	2	0.536612
65	4	3.491625	173	2	0.535059
73	0	0.000000	184	-2	0.518818
76	0	0.000000	193	4	2.026309
85	-2	0.763334	197	-2	0.501408

Table 7.3: Coefficients of $\Theta^{+}_{f_{11A}}$ (top), $\Theta^{-}_{f_{11A}}$ (bottom), and central values for f_{121D}

d	$c_{f_{121C}}(d)$	$L(f_{121C}, d, 1)$	d	$c_{f_{121C}}(d)$	$L(f_{121C}, d, 1)$
1	1	1.666157	113	1	0.156739
5	1	0.745128	124	-2	0.598501
12	-2	1.923912	133	-2	0.577897
37	-1	0.273915	136	4	2.285948
53	1	0.228864	137	-2	0.569398
56	-2	0.890598	141	2	0.561263
60	-2	0.860400	152	2	0.540573
69	0	0.000000	157	0	0.000000
89	1	0.176612	168	4	2.056749
92	-2	0.694835	177	2	0.500944
93	2	0.691090	181	-3	1.114600
97	-1	0.169173	185	1	0.122498
104	-2	0.653521	188	2	0.486068

d	$c_{f_{121C}}(d)$	$L(f_{121C}, d, 1)$	d	$c_{f_{121C}}(d)$	$L(f_{121C}, d, 1)$
8	0	0.000000	101	0	0.000000
13	1	1.386326	105	0	0.000000
17	-1	1.212307	109	1	0.478767
21	0	0.000000	120	2	1.825183
24	-2	4.081234	129	-2	1.760363
28	2	3.778489	140	-2	1.689792
29	-1	0.928193	145	-1	0.415100
40	2	3.161310	149	1	0.409491
41	1	0.780630	156	2	1.600792
57	2	2.648255	161	2	1.575739
61	-2	2.559954	172	0	0.000000
65	1	0.619984	173	0	0.000000
73	-2	2.340107	184	2	1.473969
76	-2	2.293456	193	1	0.359798
85	1	0.542160	197	-1	0.356126

Table 7.4: Coefficients of $\Theta^{+}_{f_{121A}}$ (top), $\Theta^{-}_{f_{121A}}$ (bottom), and central values for f_{121C}

7.3　f_{121B}

By computing the Brandt matrices for $\tilde{\mathcal{O}}$, we find the space $\mathcal{M}(\tilde{\mathcal{O}})^{f_{121B}}$ of dimension 2. Using Table 7.1, we can check that

$$\Theta_{\mathcal{O}+}\left(\mathcal{M}(\tilde{\mathcal{O}})^{f_{121B}}\right) = \Theta_{\mathcal{O}-}\left(\mathcal{M}(\tilde{\mathcal{O}})^{f_{121B}}\right) = 0,$$

which is expected since $L(f_{121B}, 1) = 0$.

7.4　f_{121C}

We readily find the space $\mathcal{M}(\tilde{\mathcal{O}})^{f_{121C}} = \varphi\left(\mathcal{M}(\tilde{\mathcal{O}})^{f_{121A}}\right)$ of dimension 2, spanned by

$$\mathbf{e}^+_{f_{121C}} = \mathbf{e}^-_{f_{121C}} = \frac{2[\mathfrak{b}_{1,1}] - [\mathfrak{b}_{1,2}] - [\mathfrak{b}_{1,3}] - [\mathfrak{b}_{1,4}] + 2[\mathfrak{b}_{1,5}] - [\mathfrak{b}_{1,6}]}{2},$$

and

$$[\mathfrak{b}_{1,2}] - [\mathfrak{b}_{1,3}] + [\mathfrak{b}_{1,4}] - [\mathfrak{b}_{1,6}],$$

with heights $\left\langle \mathbf{e}^+_{f_{121C}}, \mathbf{e}^+_{f_{121C}} \right\rangle = \left\langle \mathbf{e}^-_{f_{121C}}, \mathbf{e}^-_{f_{121C}} \right\rangle = 3$. Using Table 7.1, we see that

$$\Theta^+_{f_{121C}} = Q^+_1 - Q^+_2 - Q^+_3 + Q^+_4,$$

and

$$\Theta^-_{f_{121C}} = 2Q^-_1 - Q^-_2 - Q^-_3.$$

Table 7.5 shows the values of $c_{f_{121C}}(d)$ and $L(f_{121C}, -11d, 1) = L(f_{121A}, d, 1)$, where $0 < d < 200$ is a fundamental discriminant such that $11 \nmid d$. The formula

$$L(f_{121C}, -11d, 1) = L(f_{121A}, d, 1) = k_{f_{121C}} \frac{c_{f_{121C}}(d)^2}{\sqrt{d}}$$

is satisfied, where

$$k_{f_{121C}} = \frac{1}{6} \cdot \frac{\langle f_{121C}, f_{121C} \rangle}{L(f_{121C}, 1)\sqrt{11}} = 1.0197948617829165568371 17278 = L(f_{121A}, 1).$$

d	$c_{f_{121A}}(d)$	$L(f_{121A}, d, 1)$	d	$c_{f_{121A}}(d)$	$L(f_{121A}, d, 1)$
1	1	1.019795	113	-1	0.095934
5	-1	0.456066	124	-6	3.296890
12	2	1.177558	133	2	0.353710
37	-5	4.191331	136	-2	0.349787
53	1	0.140080	137	2	0.348508
56	-2	0.545103	141	2	0.343529
60	6	4.739578	152	-2	0.330865
69	-4	1.964302	157	4	1.302216
89	-3	0.972882	168	4	1.258862
92	2	0.425284	177	2	0.306610
93	-2	0.422991	181	1	0.075801
97	-3	0.931900	185	3	0.674791
104	-4	1.599986	188	2	0.297505

d	$c_{f_{121A}}(d)$	$L(f_{121A}, d, 1)$	d	$c_{f_{121A}}(d)$	$L(f_{121A}, d, 1)$
8	2	1.442208	101	2	0.405894
13	3	2.545562	105	4	1.592349
17	-1	0.247337	109	1	0.097679
21	-4	3.560600	120	-2	0.372376
24	-2	0.832659	129	2	0.359152
28	-2	0.770892	140	-2	0.344754
29	3	1.704340	145	5	2.117234
40	-4	2.579900	149	-1	0.083545
41	-1	0.159265	156	2	0.326596
57	-2	0.540301	161	-2	0.321484
61	0	0.000000	172	8	4.976552
65	1	0.126490	173	2	0.310134
73	0	0.000000	184	10	7.518027
76	6	4.211226	193	-5	1.835161
85	-5	2.765307	197	1	0.072657

Table 7.5: Coefficients of $\Theta^{+}_{f_{121C}}$ (top), $\Theta^{-}_{f_{121C}}$ (bottom), and central values for f_{121A}

d	$c_{f_{11A}}(d)$	$L(f_{11A}, d, 1)$	d	$c_{f_{11A}}(d)$	$L(f_{11A}, d, 1)$
1	1	0.253842	113	-5	0.596986
5	-5	2.838038	124	-5	0.569892
12	-5	1.831946	133	10	2.201088
37	5	1.043284	136	10	2.176676
53	10	3.486786	137	-5	0.542179
56	10	3.392105	141	-10	2.137734
60	-5	0.819271	152	-10	2.058929
69	15	6.875768	157	-15	4.558227
89	-5	0.672680	168	10	1.958432
92	-5	0.661621	177	5	0.476998
93	5	0.658054	181	-15	4.245281
97	5	0.644343	185	-5	0.466571
104	10	2.489124	188	-10	1.851332

Table 7.6: Coefficients of $\Theta^{+}_{f_{121D}}$ and central values for f_{11A}

7.5 f_{121D}

We readily find the space $\mathcal{M}(\tilde{\mathbb{O}})^{f_{121D}} = \varphi\left(\mathcal{M}(\tilde{\mathbb{O}})^{f_{11A}}\right)$ of dimension 1, spanned by

$$\mathbf{e}^{+}_{f_{121D}} = \varphi \circ \psi_{\tilde{\mathbb{O}}}\left(\frac{[\mathfrak{a}_1] - [\mathfrak{a}_2]}{2}\right),$$

with height $\langle \mathbf{e}^{+}_{f_{121D}}, \mathbf{e}^{+}_{f_{121D}} \rangle = 15$. Using Table 7.1, we see that

$$\Theta^{+}_{f_{121D}} = Q^{+}_1 + 2Q^{+}_2 - 2Q^{+}_3 - Q^{+}_4 - 3Q^{+}_5 + 3Q^{+}_6;$$

on the other hand $\Theta_{\mathbb{O}^-}\left(\mathcal{M}(\tilde{\mathbb{O}})^{f_{121D}}\right) = 0$, which is expected since $\varepsilon(f_{11A}) = +1$. Table 7.6 shows the values of $c_{f_{121D}}(d)$ and $L(f_{121D}, -11d, 1) = L(f_{11A}, d, 1)$, where $0 < d < 200$ is a fundamental discriminant such that $\left(\frac{d}{11}\right) = +1$. The formula

$$L(f_{121D}, -11d, 1) = L(f_{11A}, d, 1) = k_{f_{121D}} \frac{c_{f_{121D}}(d)^2}{\sqrt{d}}$$

is satisfied, where

$$k_{f_{121D}} = \frac{11}{300} \cdot \frac{\langle f_{121D}, f_{121D} \rangle}{L(f_{121D}, 1)\sqrt{11}} = 0.25384186085591068433775892331 = L(f_{11A}, 1).$$

8 Example: level 17^2

Let $\mathbb{H} = \mathbb{H}(-17, -3)$, the quaternion algebra ramified precisely at ∞ and 17. A maximal order, and representatives for its left ideals classes, are given by

$$\mathcal{O} = \mathfrak{a}_1 = \left\langle 1, i, \frac{1+j}{2}, \frac{3i + 2j + k}{6} \right\rangle,$$

$$\mathfrak{a}_2 = \left\langle 2, 2i, \frac{3 + 2i + j}{2}, \frac{3i + 2j + k}{6} \right\rangle.$$

Its index p suborder is given by $\tilde{\mathcal{O}} = \left\langle 1, i, \dfrac{1 + 17j}{2}, \dfrac{3 + 3i + 17j + k}{6} \right\rangle$; inequiv-

alent $\tilde{\mathcal{O}}$-subideals for each $\mathcal{O}$-ideal are show in Table 8.1.

We fix two index p suborders of $\tilde{\mathcal{O}}$

$$\mathcal{O}^+ = \left\langle 1, 17i, \frac{1 + 17j}{2}, \frac{3 + 33i + 17j + k}{6} \right\rangle,$$

$$\mathcal{O}^- = \left\langle 1, 17i, \frac{1 + 17j}{2}, \frac{3 + 99i + 17j + k}{6} \right\rangle$$

in the $+$ and $-$ genus respectively. Table 8.1 shows the maps from $\tilde{\mathcal{O}}$-ideals to ternary quadratic forms of level 17^2 in the $+$ genus and in the $-$ genus, computed via $\mathcal{O}^+$- and $\mathcal{O}^-$-subideals respectively. The actual coefficients of the ternary quadratic forms are given in Table 8.2, with the notation as in (2.1).

8.1 f_{17A}

By computing the Brandt matrices for $\mathcal{O}$, we find the space $\mathcal{M}(\tilde{\mathcal{O}})^{f_{17A}} = \psi_{\tilde{\mathcal{O}}}\left(\mathcal{M}(\mathcal{O})^{f_{17A}}\right)$ of dimension 1, spanned by

$$\mathbf{e}^+_{f_{17A}} = \mathbf{e}^-_{f_{17A}} = \psi_{\tilde{\mathcal{O}}}\left(\frac{[\mathfrak{a}_1] - [\mathfrak{a}_2]}{2}\right),$$

with heights $\left\langle \mathbf{e}^+_{f_{17A}}, \mathbf{e}^+_{f_{17A}} \right\rangle = \left\langle \mathbf{e}^-_{f_{17A}}, \mathbf{e}^-_{f_{17A}} \right\rangle = 18$. Using Table 8.1, we see that

$$\Theta^+_{f_{17A}} = 3Q_1^+ + 6Q_2^+ - Q_3^+ - 2Q_4^+ - 2Q_5^+ - 2Q_6^+ - 2Q_7^+,$$

and

$$\Theta^-_{f_{17A}} = 3Q_1^- + 6Q_2^- - 2Q_3^- - 2Q_4^- - 2Q_5^- - 2Q_6^- - Q_7^-.$$

Table 8.3 shows the values of $c_{f_{17A}}(d)$ and $L(f_{17A}, -17d, 1) = L(f_{289A}, -d, 1)$, where $0 > -d > -200$ is a fundamental discriminant such that $17 \nmid d$. The formula

$$L(f_{17A}, -17d, 1) = L(f_{289A}, -d, 1) = k_{f_{17A}} \frac{c_{f_{17A}}(d)^2}{\sqrt{d}}$$

A. Pacetti and G. Tornaría

$\mathcal{O}$-ideals	$\tilde{\mathcal{O}}$-subideals	χ	$+$ genus	$-$ genus
$\mathfrak{a}_1$	$\mathfrak{b}_{1,1} = \left\langle 17, i, \frac{1+j}{2}, \frac{36+3i+2j+k}{6} \right\rangle$	$+$	Q_1^+	Q_1^-
	$\mathfrak{b}_{1,2} = \left\langle 17, i, \frac{25+j}{2}, \frac{84+3i+2j+k}{6} \right\rangle$	$+$	Q_2^+	Q_2^-
	$\mathfrak{b}_{1,3} = \left\langle 17, i, \frac{21+j}{2}, \frac{42+3i+2j+k}{6} \right\rangle$	$+$	Q_2^+	Q_2^-
	$\mathfrak{b}_{1,4} = \left\langle 17, i, \frac{31+j}{2}, \frac{96+3i+2j+k}{6} \right\rangle$	$-$	Q_1^+	Q_1^-
	$\mathfrak{b}_{1,5} = \left\langle 17, i, \frac{23+j}{2}, \frac{12+3i+2j+k}{6} \right\rangle$	$-$	Q_2^+	Q_2^-
	$\mathfrak{b}_{1,6} = \left\langle 17, i, \frac{29+j}{2}, \frac{24+3i+2j+k}{6} \right\rangle$	$-$	Q_2^+	Q_2^-
$\mathfrak{a}_2$	$\mathfrak{b}_{2,1} = \left\langle 2, 2i, \frac{3+2i+17j}{2}, \frac{9+9i+17j+k}{6} \right\rangle$	$+$	Q_3^+	Q_3^-
	$\mathfrak{b}_{2,2} = \left\langle 34, 2i, \frac{55+2i+j}{2}, \frac{144+3i+2j+k}{6} \right\rangle$	$+$	Q_4^+	Q_4^-
	$\mathfrak{b}_{2,3} = \left\langle 34, 2i, \frac{59+2i+j}{2}, \frac{84+3i+2j+k}{6} \right\rangle$	$+$	Q_5^+	Q_4^-
	$\mathfrak{b}_{2,4} = \left\langle 34, 2i, \frac{35+2i+j}{2}, \frac{36+3i+2j+k}{6} \right\rangle$	$+$	Q_6^+	Q_3^-
	$\mathfrak{b}_{2,5} = \left\langle 34, 2i, \frac{7+2i+j}{2}, \frac{48+3i+2j+k}{6} \right\rangle$	$+$	Q_7^+	Q_5^-
	$\mathfrak{b}_{2,6} = \left\langle 34, 2i, \frac{27+2i+j}{2}, \frac{156+3i+2j+k}{6} \right\rangle$	$+$	Q_7^+	Q_6^-
	$\mathfrak{b}_{2,7} = \left\langle 34, 2i, \frac{67+2i+j}{2}, \frac{168+3i+2j+k}{6} \right\rangle$	$+$	Q_6^+	Q_7^-
	$\mathfrak{b}_{2,8} = \left\langle 34, 2i, \frac{43+2i+j}{2}, \frac{120+3i+2j+k}{6} \right\rangle$	$+$	Q_5^+	Q_6^-
	$\mathfrak{b}_{2,9} = \left\langle 34, 2i, \frac{47+2i+j}{2}, \frac{60+3i+2j+k}{6} \right\rangle$	$+$	Q_4^+	Q_5^-
	$\mathfrak{b}_{2,10} = \left\langle 34, 2i, \frac{51+2i+j}{2}, \frac{3i+2j+k}{6} \right\rangle$	$-$	Q_3^+	Q_3^-
	$\mathfrak{b}_{2,11} = \left\langle 34, 2i, \frac{63+2i+j}{2}, \frac{24+3i+2j+k}{6} \right\rangle$	$-$	Q_4^+	Q_4^-
	$\mathfrak{b}_{2,12} = \left\langle 34, 2i, \frac{23+2i+j}{2}, \frac{12+3i+2j+k}{6} \right\rangle$	$-$	Q_5^+	Q_4^-
	$\mathfrak{b}_{2,13} = \left\langle 34, 2i, \frac{31+2i+j}{2}, \frac{96+3i+2j+k}{6} \right\rangle$	$-$	Q_6^+	Q_3^-
	$\mathfrak{b}_{2,14} = \left\langle 34, 2i, \frac{19+2i+j}{2}, \frac{72+3i+2j+k}{6} \right\rangle$	$-$	Q_7^+	Q_5^-
	$\mathfrak{b}_{2,15} = \left\langle 34, 2i, \frac{15+2i+j}{2}, \frac{132+3i+2j+k}{6} \right\rangle$	$-$	Q_7^+	Q_6^-
	$\mathfrak{b}_{2,16} = \left\langle 34, 2i, \frac{3+2i+j}{2}, \frac{108+3i+2j+k}{6} \right\rangle$	$-$	Q_6^+	Q_7^-
	$\mathfrak{b}_{2,17} = \left\langle 34, 2i, \frac{11+2i+j}{2}, \frac{192+3i+2j+k}{6} \right\rangle$	$-$	Q_5^+	Q_6^-
	$\mathfrak{b}_{2,18} = \left\langle 34, 2i, \frac{39+2i+j}{2}, \frac{180+3i+2j+k}{6} \right\rangle$	$-$	Q_4^+	Q_5^-

Table 8.1: Maps $\Theta_{\mathcal{O}+}$ and $\Theta_{\mathcal{O}-}$ from the $\tilde{\mathcal{O}}$-Ideals to ternary quadratic forms in the $+$ and $-$ genus respectively, level 17^2.

	a_1	a_2	a_3	a_{23}	a_{13}	a_{12}
Q_1^+	4,	51,	103,	-34,	-4,	0
Q_2^+	15,	32,	47,	24,	10,	4
Q_3^+	4,	52,	103,	36,	4,	4
Q_4^+	15,	35,	47,	-18,	-10,	-14
Q_5^+	8,	43,	60,	-16,	-4,	-4
Q_6^+	16,	32,	47,	-24,	-4,	-12
Q_7^+	15,	16,	100,	-12,	-4,	-12
Q_1^-	7,	11,	292,	-8,	-4,	-6
Q_2^-	23,	28,	40,	20,	12,	20
Q_3^-	7,	39,	79,	30,	6,	2
Q_4^-	12,	23,	75,	-10,	-8,	-4
Q_5^-	7,	20,	147,	-8,	-6,	-4
Q_6^-	27,	28,	39,	4,	26,	24
Q_7^-	3,	23,	295,	-22,	-2,	-2

Table 8.2: Coefficients of ternary quadratic forms, level 17^2.

is satisfied, where

$$k_{f_{17A}} = \frac{1}{2} \cdot \frac{\langle f_{17A}, f_{17A} \rangle}{L(f_{17A}, 1)\sqrt{17}} = 1.3318791063852161592204774762.$$

8.2 f_{289A}

We readily find the space $\mathcal{M}(\tilde{\mathcal{O}})^{f_{289A}} = \varphi\left(\mathcal{M}(\tilde{\mathcal{O}})^{f_{17A}}\right)$ of dimension 1. Using Table 8.1, we can check that

$$\Theta_{\mathcal{O}^+}\left(\mathcal{M}(\tilde{\mathcal{O}})^{f_{289A}}\right) = \Theta_{\mathcal{O}^-}\left(\mathcal{M}(\tilde{\mathcal{O}})^{f_{289A}}\right) = 0,$$

which is expected since $L(f_{289A}, 1) = 0$.

9 Example: level 19^2

Let $\mathbb{H} = \mathbb{H}(-1, -19)$, the quaternion algebra ramified precisely at ∞ and 19. A maximal order, and representatives for its left ideals classes, are given by

$$\mathcal{O} = \mathfrak{a}_1 = \left\langle 1, i, \frac{1+j}{2}, \frac{i+k}{2} \right\rangle,$$

$$\mathfrak{a}_2 = \left\langle 2, 2i, \frac{3+2i+j}{2}, \frac{2+3i+k}{2} \right\rangle.$$

Its index p suborder is given by $\tilde{\mathcal{O}} = \left\langle 1, 19i, \frac{1+j}{2}, \frac{19i+k}{2} \right\rangle$; inequivalent $\tilde{\mathcal{O}}$-subideals for each $\mathcal{O}$-ideal are show in Table 9.1.

$-d$	$c_{f_{289A}}(d)$	$L(f_{289A}, -d, 1)$	$-d$	$c_{f_{289A}}(d)$	$L(f_{289A}, -d, 1)$
-4	2	2.663758	-104	4	2.089624
-8	-2	1.883561	-111	-2	0.505665
-15	2	1.375559	-115	2	0.496793
-19	-2	1.222216	-120	0	0.000000
-35	-2	0.900515	-123	-6	4.323294
-43	2	0.812439	-127	2	0.472741
-47	0	0.000000	-132	0	0.000000
-52	0	0.000000	-151	2	0.433547
-55	2	0.718362	-152	0	0.000000
-59	-2	0.693584	-155	2	0.427916
-67	0	0.000000	-168	-4	1.644107
-83	-2	0.584771	-179	2	0.398197
-84	4	2.325119	-183	-2	0.393821
-87	2	0.571170	-191	0	0.000000
-103	0	0.000000	-195	4	1.526046

$-d$	$c_{f_{289A}}(d)$	$L(f_{289A}, -d, 1)$	$-d$	$c_{f_{289A}}(d)$	$L(f_{289A}, -d, 1)$
-3	-1	0.768961	-95	-4	2.186367
-7	-1	0.503403	-107	3	1.158819
-11	3	3.614190	-116	2	0.494647
-20	-2	1.191269	-131	-3	1.047301
-23	1	0.277716	-139	-1	0.112969
-24	2	1.087475	-143	-2	0.445509
-31	-1	0.239213	-148	-2	0.437919
-39	-4	3.412341	-159	2	0.422500
-40	2	0.842354	-163	7	5.111720
-56	0	0.000000	-164	-4	1.664037
-71	3	1.422585	-167	-1	0.103064
-79	-1	0.149848	-184	-8	6.283996
-88	-2	0.567915	-199	1	0.094414
-91	-2	0.558475			

Table 8.3: Coefficients of $\Theta^+_{f_{17A}}$ (top), $\Theta^-_{f_{17A}}$ (bottom), and central values for f_{289A}

We fix two index p suborders of $\tilde{\mathcal{O}}$

$$\mathcal{O}^+ = \left\langle 1, 19i, \frac{1+j}{2}, \frac{19i+19k}{2} \right\rangle,$$

$$\mathcal{O}^- = \left\langle 1, 19i, \frac{1+19j}{2}, \frac{19i+18j+k}{2} \right\rangle$$

in the $+$ and $-$ genus respectively. Table 9.1 shows the maps from $\tilde{\mathcal{O}}$-ideals to ternary quadratic forms of level 19^2 in the $+$ genus and in the $-$ genus, computed via $\mathcal{O}^+$- and $\mathcal{O}^-$-subideals respectively. The actual coefficients of the ternary quadratic forms are given in Table 9.2, with the notation as in (2.1).

9.1 $\;f_{19A}$

By computing the Brandt matrices for $\mathcal{O}$, we find the space $\mathcal{M}(\tilde{\mathcal{O}})^{f_{19A}} = \psi_{\tilde{\mathcal{O}}}\left(\mathcal{M}(\mathcal{O})^{f_{19A}}\right)$ of dimension 1, spanned by

$$\mathbf{e}^+_{f_{19A}} = \mathbf{e}^-_{f_{19A}} = \psi_{\tilde{\mathcal{O}}}\left(\frac{[\mathfrak{a}_1] - [\mathfrak{a}_2]}{2}\right),$$

with heights $\left\langle \mathbf{e}^+_{f_{19A}}, \mathbf{e}^+_{f_{19A}} \right\rangle = \left\langle \mathbf{e}^-_{f_{19A}}, \mathbf{e}^-_{f_{19A}} \right\rangle = 15$. Using Table 9.1, we see that

$$\Theta^+_{f_{19A}} = Q_1^+ + 2Q_2^+ + 2Q_3^+ + 2Q_4^+ + 2Q_5^+ + Q_6^+$$
$$- Q_7^+ - 2Q_8^+ - 2Q_9^+ - 2Q_{10}^+ - 2Q_{11}^+ - Q_{12}^+,$$

and

$$\Theta^-_{f_{19A}} = 2Q_1^- + 2Q_2^- + 2Q_3^- + 2Q_4^- + 2Q_5^-$$
$$- 2Q_6^- - 2Q_7^- - 2Q_8^- - 2Q_9^- - 2Q_{10}^-.$$

Table 9.3 shows the values of $c_{f_{19A}}(d)$ and $L(f_{19A}, -19d, 1) = L(f_{361B}, d, 1)$, where $0 < d < 200$ is a fundamental discriminant such that $19 \nmid d$. The formula

$$L(f_{19A}, -19d, 1) = L(f_{361B}, d, 1) = k_{f_{19A}} \frac{c_{f_{19A}}(d)^2}{\sqrt{d}}$$

is satisfied, where

$$k_{f_{19A}} = \frac{2}{3} \cdot \frac{\langle f_{19A}, f_{19A} \rangle}{L(f_{19A}, 1)\sqrt{19}} = 1.893639859594845381072872862 = L(f_{361B}, 1).$$

9.2 $\;f_{361A}$

By computing the Brandt matrices for $\tilde{\mathcal{O}}$, we find the space $\mathcal{M}(\tilde{\mathcal{O}})^{f_{361A}}$ of dimension 2. Using Table 9.1, we can check that

$$\Theta_{\mathcal{O}^+}\left(\mathcal{M}(\tilde{\mathcal{O}})^{f_{361A}}\right) = \Theta_{\mathcal{O}^-}\left(\mathcal{M}(\tilde{\mathcal{O}})^{f_{361A}}\right) = 0,$$

which is expected since $L(f_{361A}, 1) = 0$.

$\mathcal{O}$-ideals	$\tilde{\mathcal{O}}$-subideals	χ	$+$ genus	$-$ genus
$\mathfrak{a}_1$	$\mathfrak{b}_{1,1} = \left\langle 1, 19i, \frac{1+j}{2}, \frac{19i+k}{2} \right\rangle$	$+$	Q_1^+	Q_1^-
	$\mathfrak{b}_{1,2} = \left\langle 19, 14 + i, \frac{19+j}{2}, \frac{14+i+k}{2} \right\rangle$	$+$	Q_2^+	Q_2^-
	$\mathfrak{b}_{1,3} = \left\langle 19, 9 + i, \frac{19+j}{2}, \frac{28+i+k}{2} \right\rangle$	$+$	Q_3^+	Q_3^-
	$\mathfrak{b}_{1,4} = \left\langle 19, 17 + i, \frac{19+j}{2}, \frac{36+i+k}{2} \right\rangle$	$+$	Q_3^+	Q_4^-
	$\mathfrak{b}_{1,5} = \left\langle 19, 15 + i, \frac{19+j}{2}, \frac{34+i+k}{2} \right\rangle$	$+$	Q_2^+	Q_5^-
	$\mathfrak{b}_{1,6} = \left\langle 19, 16 + i, \frac{19+j}{2}, \frac{16+i+k}{2} \right\rangle$	$-$	Q_4^+	Q_4^-
	$\mathfrak{b}_{1,7} = \left\langle 19, 11 + i, \frac{19+j}{2}, \frac{30+i+k}{2} \right\rangle$	$-$	Q_5^+	Q_5^-
	$\mathfrak{b}_{1,8} = \left\langle 19, 1 + i, \frac{19+j}{2}, \frac{20+i+k}{2} \right\rangle$	$-$	Q_6^+	Q_1^-
	$\mathfrak{b}_{1,9} = \left\langle 19, 7 + i, \frac{19+j}{2}, \frac{26+i+k}{2} \right\rangle$	$-$	Q_5^+	Q_2^-
	$\mathfrak{b}_{1,10} = \left\langle 19, 6 + i, \frac{19+j}{2}, \frac{6+i+k}{2} \right\rangle$	$-$	Q_4^+	Q_3^-
$\mathfrak{a}_2$	$\mathfrak{b}_{2,1} = \left\langle 38, 2 + 2i, \frac{59+2i+j}{2}, \frac{22+3i+k}{2} \right\rangle$	$+$	Q_7^+	Q_6^-
	$\mathfrak{b}_{2,2} = \left\langle 38, 14 + 2i, \frac{71+2i+j}{2}, \frac{2+3i+k}{2} \right\rangle$	$+$	Q_8^+	Q_7^-
	$\mathfrak{b}_{2,3} = \left\langle 38, 12 + 2i, \frac{31+2i+j}{2}, \frac{18+3i+k}{2} \right\rangle$	$+$	Q_9^+	Q_8^-
	$\mathfrak{b}_{2,4} = \left\langle 38, 26 + 2i, \frac{7+2i+j}{2}, \frac{58+3i+k}{2} \right\rangle$	$+$	Q_9^+	Q_9^-
	$\mathfrak{b}_{2,5} = \left\langle 38, 24 + 2i, \frac{43+2i+j}{2}, \frac{74+3i+k}{2} \right\rangle$	$+$	Q_8^+	Q_{10}^-
	$\mathfrak{b}_{2,6} = \left\langle 38, 36 + 2i, \frac{55+2i+j}{2}, \frac{54+3i+k}{2} \right\rangle$	$+$	Q_7^+	Q_6^-
	$\mathfrak{b}_{2,7} = \left\langle 38, 16 + 2i, \frac{35+2i+j}{2}, \frac{62+3i+k}{2} \right\rangle$	$+$	Q_8^+	Q_7^-
	$\mathfrak{b}_{2,8} = \left\langle 38, 6 + 2i, \frac{63+2i+j}{2}, \frac{66+3i+k}{2} \right\rangle$	$+$	Q_9^+	Q_8^-
	$\mathfrak{b}_{2,9} = \left\langle 38, 32 + 2i, \frac{51+2i+j}{2}, \frac{10+3i+k}{2} \right\rangle$	$+$	Q_9^+	Q_9^-
	$\mathfrak{b}_{2,10} = \left\langle 38, 22 + 2i, \frac{3+2i+j}{2}, \frac{14+3i+k}{2} \right\rangle$	$+$	Q_8^+	Q_{10}^-
	$\mathfrak{b}_{2,11} = \left\langle 38, 18 + 2i, \frac{75+2i+j}{2}, \frac{46+3i+k}{2} \right\rangle$	$-$	Q_{10}^+	Q_8^-
	$\mathfrak{b}_{2,12} = \left\langle 38, 34 + 2i, \frac{15+2i+j}{2}, \frac{70+3i+k}{2} \right\rangle$	$-$	Q_{10}^+	Q_9^-
	$\mathfrak{b}_{2,13} = \left\langle 38, 30 + 2i, \frac{11+2i+j}{2}, \frac{26+3i+k}{2} \right\rangle$	$-$	Q_{11}^+	Q_{10}^-
	$\mathfrak{b}_{2,14} = \left\langle 38, 2i, \frac{19+2i+j}{2}, \frac{38+3i+k}{2} \right\rangle$	$-$	Q_{12}^+	Q_6^-
	$\mathfrak{b}_{2,15} = \left\langle 38, 8 + 2i, \frac{27+2i+j}{2}, \frac{50+3i+k}{2} \right\rangle$	$-$	Q_{11}^+	Q_7^-
	$\mathfrak{b}_{2,16} = \left\langle 38, 4 + 2i, \frac{23+2i+j}{2}, \frac{6+3i+k}{2} \right\rangle$	$-$	Q_{10}^+	Q_8^-
	$\mathfrak{b}_{2,17} = \left\langle 38, 20 + 2i, \frac{39+2i+j}{2}, \frac{30+3i+k}{2} \right\rangle$	$-$	Q_{10}^+	Q_9^-
	$\mathfrak{b}_{2,18} = \left\langle 38, 10 + 2i, \frac{67+2i+j}{2}, \frac{34+3i+k}{2} \right\rangle$	$-$	Q_{11}^+	Q_{10}^-
	$\mathfrak{b}_{2,19} = \left\langle 2, 38i, \frac{3+38i+j}{2}, \frac{2+19i+k}{2} \right\rangle$	$-$	Q_{12}^+	Q_6^-
	$\mathfrak{b}_{2,20} = \left\langle 38, 28 + 2i, \frac{47+2i+j}{2}, \frac{42+3i+k}{2} \right\rangle$	$-$	Q_{11}^+	Q_7^-

Table 9.1: Maps $\Theta_{\mathcal{O}+}$ and $\Theta_{\mathcal{O}-}$ from the $\tilde{\mathcal{O}}$-Ideals to ternary quadratic forms in the $+$ and $-$ genus respectively, level 19^2.

	a_1	a_2	a_3	a_{23}	a_{13}	a_{12}
Q_1^+	1,	76,	380,	-76,	0,	0
Q_2^+	5,	76,	92,	-76,	-4,	0
Q_3^+	25,	36,	36,	-4,	-16,	-16
Q_4^+	17,	44,	44,	12,	16,	16
Q_5^+	20,	24,	73,	4,	8,	20
Q_6^+	4,	20,	361,	0,	0,	-4
Q_7^+	5,	16,	365,	16,	2,	4
Q_8^+	16,	24,	77,	20,	8,	4
Q_9^+	20,	36,	45,	20,	16,	12
Q_{10}^+	9,	44,	77,	28,	6,	8
Q_{11}^+	5,	61,	92,	16,	4,	2
Q_{12}^+	4,	77,	96,	40,	4,	4
Q_1^-	8,	21,	181,	2,	4,	8
Q_2^-	29,	29,	37,	-6,	-6,	-18
Q_3^-	13,	48,	48,	20,	8,	8
Q_4^-	29,	32,	32,	-12,	-8,	-8
Q_5^-	29,	29,	41,	-14,	-14,	-18
Q_6^-	21,	32,	53,	-20,	-14,	-16
Q_7^-	29,	32,	37,	-28,	-6,	-8
Q_8^-	12,	13,	184,	-12,	-4,	-4
Q_9^-	8,	29,	124,	20,	4,	4
Q_{10}^-	21,	29,	53,	26,	14,	2

Table 9.2: Coefficients of ternary quadratic forms, level 19^2.

d	$c_{f_{361B}}(d)$	$L(f_{361B}, d, 1)$	d	$c_{f_{361B}}(d)$	$L(f_{361B}, d, 1)$
1	1	1.893640	104	4	2.970987
5	-1	0.846861	120	-2	0.691460
17	1	0.459275	137	-1	0.161785
24	2	1.546150	140	-3	1.440376
28	1	0.357864	149	-1	0.155133
44	-1	0.285477	156	6	5.458051
61	-5	6.061393	157	-4	2.418063
73	1	0.221634	161	0	0.000000
77	-1	0.215800	168	-2	0.584390
85	-3	1.848547	172	3	1.299498
92	2	0.789702	177	-2	0.569339
93	2	0.785445	188	1	0.138108
101	2	0.753697	197	2	0.539665

d	$c_{f_{361B}}(d)$	$L(f_{361B}, d, 1)$	d	$c_{f_{361B}}(d)$	$L(f_{361B}, d, 1)$
8	0	0.000000	97	0	0.000000
12	-2	2.186587	105	2	0.739201
13	0	0.000000	109	0	0.000000
21	-2	1.652904	113	2	0.712555
29	2	1.406560	124	-8	10.883448
33	-2	1.318562	129	2	0.666903
37	2	1.245250	136	4	2.598052
40	4	4.790572	141	-2	0.637893
41	-2	1.182947	145	2	0.629033
53	0	0.000000	165	-2	0.589679
56	-4	4.048772	173	-2	0.575883
60	-2	0.977871	181	6	5.067113
65	0	0.000000	184	-4	2.233616
69	4	3.647479	185	2	0.556893
88	-4	3.229803	193	4	2.180915
89	0	0.000000			

Table 9.3: Coefficients of $\Theta^+_{f_{19A}}$ (top), $\Theta^-_{f_{19A}}$ (bottom), and central values for f_{361B}

d	$c_{f_{19A}}(d)$	$L(f_{19A}, d, 1)$	d	$c_{f_{19A}}(d)$	$L(f_{19A}, d, 1)$
1	1	0.453253	104	12	6.400100
5	3	1.824309	120	6	1.489542
17	-3	0.989371	137	3	0.348516
24	-6	3.330718	140	-3	0.344762
28	-3	0.770911	149	-9	3.007688
44	-9	5.534770	156	6	1.306415
61	3	0.522298	157	0	0.000000
73	-3	0.477444	161	-12	5.143876
77	3	0.464877	168	6	1.258893
85	-3	0.442460	172	3	0.311042
92	6	1.701177	177	-6	1.226470
93	-6	1.692006	188	9	2.677608
101	6	1.623614	197	6	1.162546

Table 9.4: Coefficients of $\Theta^{+}_{f_{361B}}$ and central values for f_{19A}

9.3 f_{361B}

We readily find the space $\mathcal{M}(\tilde{\mathcal{O}})^{f_{361B}} = \varphi\left(\mathcal{M}(\tilde{\mathcal{O}})^{f_{19A}}\right)$ of dimension 1, spanned by

$$\mathbf{e}^{+}_{f_{361B}} = \varphi \circ \psi_{\tilde{\mathcal{O}}}\left(\frac{[\mathfrak{a}_1] - [\mathfrak{a}_2]}{2}\right),$$

with height $\langle \mathbf{e}^{+}_{f_{361B}}, \mathbf{e}^{+}_{f_{361B}}\rangle = 15$. Using Table 9.1, we see that

$$\Theta^{+}_{f_{361B}} = Q^{+}_1 + 2Q^{+}_2 + 2Q^{+}_3 - 2Q^{+}_4 - 2Q^{+}_5 - Q^{+}_6$$
$$- Q^{+}_7 - 2Q^{+}_8 - 2Q^{+}_9 + 2Q^{+}_{10} + 2Q^{+}_{11} + Q^{+}_{12};$$

on the other hand $\Theta_{\mathcal{O}^{-}}\left(\mathcal{M}(\tilde{\mathcal{O}})^{f_{361B}}\right) = 0$, which is expected since $\varepsilon(f_{19A}) = +1$. Table 9.4 shows the values of $c_{f_{361B}}(d)$ and $L(f_{361B}, -19d, 1) = L(f_{19A}, d, 1)$, where $0 < d < 200$ is a fundamental discriminant such that $\left(\frac{d}{19}\right) = +1$. The formula

$$L(f_{361B}, -19d, 1) = L(f_{19A}, d, 1) = k_{f_{361B}}\frac{c_{f_{361B}}(d)^2}{\sqrt{d}}$$

is satisfied, where

$$k_{f_{361B}} = \frac{19}{540}\cdot\frac{\langle f_{361B}, f_{361B}\rangle}{L(f_{361B}, 1)\sqrt{19}} = 0.453253244496103603578839 1869 = L(f_{19A}, 1).$$

References

[1] *Computational number theory*, `http://www.ma.utexas.edu/users/villegas/cnt/`.

[2] Martin Eichler, *Zur Zahlentheorie der Quaternionen-Algebren*, J. Reine Angew. Math. **195** (1955), 127–151 (1956).

[3] Benedict H. Gross, *Heights and the special values of L-series*, Number theory (Montreal, Que., 1985), CMS Conf. Proc., vol. 7, Amer. Math. Soc., Providence, RI, 1987, pp. 115–187.

[4] Zhengyu Mao, Fernando Rodriguez-Villegas, and Gonzalo Tornaría, *Computation of central values of quadratic twists of modular L-functions*, in this volume.

[5] Ariel Pacetti and Gonzalo Tornaría, *Shimura correspondence for level p^2 and the central values of L-series*, preprint.

[6] *PARI/GP, version* 2.2.8, http://pari.math.u-bordeaux.fr/, 2004.

[7] Arnold Pizer, *Theta series and modular forms of level p^2M*, Compositio Math. **40** (1980), no. 2, 177–241.

[8] Goro Shimura, *On modular forms of half integral weight*, Ann. of Math. (2) **97** (1973), 440–481.

[9] Gonzalo Tornaría, *Data about the central values of the L-series of (imaginary and real) quadratic twists of elliptic curves*, http://www.ma.utexas.edu/users/tornaria/cnt/, 2004.

[10] Masaru Ueda, *On twisting operators and newforms of half-integral weight*, Nagoya Math. J. **131** (1993), 135–205.

[11] Jean-Loup Waldspurger, *Sur les coefficients de Fourier des formes modulaires de poids demi-entier*, J. Math. Pures Appl. (9) **60** (1981), no. 4, 375–484.

Ariel Pacetti
Departamento de Matemática,
Universidad de Buenos Aires,
Pabellón I, Ciudad Universitaria. C.P:1428,
Buenos Aires, Argentina
apacetti@dm.uba.ar

Gonzalo Tornaría
Department of Mathematics,
University of Texas at Austin,
Austin, TX 78712 USA
tornaria@math.utexas.edu

Central values of quadratic twists for a modular form of weight 4

Holly Rosson and Gonzalo Tornaría

1   Introduction

Let f be the unique cuspidal modular form of weight 4 and level 7, whose q-expansion begins $f = q - q^2 - 2q^3 - 7q^4 + 16q^5 + 2q^6 - 7q^7 + \ldots$. We consider here the problem of computing the central values for the family of twisted L-functions

$$L(f, D, s) = \sum_{m=1}^{\infty} \frac{a(m)}{m^s} \left(\frac{D}{m} \right),$$

where D is a fundamental discriminant and $a(m)$ is the m-th Fourier coefficient of f.

By Waldspurger's formula [W], these central values $L(f, D, 2)$ are related to the $|D|$-th Fourier coefficient of weight $5/2$ modular forms in Shimura correspondence with f. The question is, then, how to find the said modular forms, and specifically how to compute their Fourier coefficients.

A constructive version of Waldspurger's formula was proved by Gross in [G] for the case of weight 2 and prime level. This has been generalized in several ways (cf. [BSP1], [BSP2], [MRVT], [PT]). In all these constructions, the modular forms of half integral weight are obtained as linear combinations of (generalized) ternary theta series coming from the arithmetic of quaternion algebras.

For the case of higher weight as done in [BSP2] these theta series involve spherical polynomials of *even* degree, and thus apply only to the construction of modular forms of weight $\equiv 3/2 \pmod 2$ in correspondence with even weights $\equiv 2 \pmod 4$.

Indeed, to obtain modular forms of weight $5/2$ from ternary quadratic forms it is necessary to utilize spherical polynomials of degree 1. However, such a theta series vanishes trivially, since such polynomials are odd functions. We will show how to solve this problem by employing the weight functions defined in [MRVT]. Although we only show here the simplest example, it is clear that a combination of the techniques of [BSP2] with those of [MRVT] should be enough to completely solve the problem in question for any modular form of even weight and prime level.

Much of our work was done at the Isaac Newton Institute during the Special Week on Ranks of Elliptic Curves and Random Matrix Theory. We thank the institute and the organizers of this program for their support and hospitality, as well as Zhengyu Mao and Fernando Rodriguez-Villegas for many helpful conversations.

2   Spherical polynomials and f

Let $B = B(-1, -7)$ be the quaternion algebra ramified at 7 and ∞. A maximal order, with class number 1, is given by

$$R = \left\langle 1, i, \frac{1+j}{2}, \frac{i+k}{2} \right\rangle.$$

The norm in the given basis is the quaternary quadratic form

$$\mathcal{N}(a, b, c, d) = a^2 + b^2 + 2c^2 + 2d^2 + ac + bd,$$

of level 7. It follows from results of Eichler on the Basis Problem [E] that f can be obtained as a (generalized) theta series for $\mathcal{N}$ with some spherical polynomial of degree 2.

The group of automorphisms of $\mathcal{N}$ has order 32, and is generated by the involutions

$$\mathcal{N}(a, b, c, d) = \mathcal{N}(a, -b, c, -d) = \mathcal{N}(b, a, d, c) = \mathcal{N}(a+c, b, -c, d).$$

Let $P(a, b, c, d)$ be a spherical polynomial of degree 2. We can assume without loss of generality that P is *even* (i.e. invariant) with respect to these involutions: if P were *odd* with respect to any of the above involutions the theta series weighted by P would be zero.

We claim that such P is unique up to a constant. Indeed, P is a quadratic form in 4 variables. Since $P(a, b, c, d) = P(a, -b, c, -d)$, it follows that $P(a, b, c, d) = P_1(a, c) + P_2(b, d)$ for some quadratic forms P_1 and P_2 in 2 variables. From $P(a, b, c, d) = P(b, a, d, c)$ we conclude that $P_1 = P_2$. The last involution implies that $P_1(a, c) = P_1(a+c, -c)$, and it follows that P_1 is a linear combination of the polynomials $a^2 + ac$ and c^2. The last condition on P_1 comes from the fact that P satisfies the Laplace differential equation

$$\Delta_{\mathcal{N}}(P) = 0,$$

where $\Delta_{\mathcal{N}}$ is the Laplacian operator with respect to the quadratic form $\mathcal{N}$. Note that $\Delta_{\mathcal{N}}(P) = 2\Delta_{\mathcal{N}}(P_1)$, and since P_1 is a quadratic form we can compute

$$\Delta_{\mathcal{N}}(P_1) = \operatorname{Tr}(M(P_1) \cdot M(\mathcal{N})^{-1}).$$

Here $M(P_1)$ and $M(\mathcal{N})$ are the matrices of P_1 and $\mathcal{N}$ respectively. Hence

$$\Delta_{\mathcal{N}}\left[\alpha(a^2 + ac) + \beta(c^2)\right] = \frac{6\alpha + 4\beta}{7} = 0,$$

and it follows that, up to a constant, $P_1(a, c) = 2a^2 + 2ac - 3c^2$.

Therefore we can compute the Fourier expansion of f by

$$f := \frac{1}{4} \sum_{(a,b,c,d)\in\mathbb{Z}^4} \left(2a^2 + 2ac - 3c^2\right) q^{N(a,b,c,d)}$$

$$= \sum_{m=1}^{\infty} a(m)\, q^m$$

$$= q - q^2 - 2q^3 - 7q^4 + 16q^5 + 2q^6 - 7q^7 + O(q^8).$$

Note that we used $P_1(a, c)$ as the spherical polynomial instead of $P(a, b, c, d)$ to simplify the computations, but as explained above the resulting theta series is the same up to a constant.

Finally, the standard method to compute the central values uses the quickly convergent series

$$L(f, D, 2) = (1 + \varepsilon_D) \sum_{m=1}^{\infty} \left(1 + \frac{2\pi m}{\sqrt{N_D}}\right) \exp\left(-\frac{2\pi m}{\sqrt{N_D}}\right) \left(\frac{D}{m}\right) \frac{a(m)}{m^2}, \qquad (2.1)$$

where ε_D and N_D are the sign and the level of the functional equation for $L(f, D, s)$ and are easily seen to be

$$\varepsilon_D = \operatorname{sign}(D) \cdot \begin{cases} \left(\frac{D}{7}\right) & \text{if } 7 \nmid D, \\ 1 & \text{if } 7 \mid D; \end{cases} \qquad N_D = D^2 \cdot \begin{cases} 7 & \text{if } 7 \nmid D, \\ 1 & \text{if } 7 \mid D. \end{cases}$$

Although the convergence of this series is exponential, the number of terms that is required to achieve a given precision is $O(\sqrt{N_D}) = O(|D|)$. Therefore, computing $L(f, D, 2)$ for $|D| \leqslant x$ will take time roughly proportional to x^2, with a big constant. In the next section we will show how to compute the exact value for the ratios

$$\frac{L(f, D, 2)}{L(f, 1, 2)} \quad D > 0, \qquad \frac{L(f, D, 2)}{L(f, -4, 2)} \quad D < 0,$$

in time proportional to $x^{3/2}$, with a much smaller constant. Of course, the special values $L(f, 1, 2)$ and $L(f, -4, 2)$ can be computed very quickly by the series (2.1), since the respective levels 7 and $7 \cdot 16$ are very small.

For instance, using the first 1000 Fourier coefficients, the series (2.1) gives about 1000 decimal places for $L(f, 1, 2)$, and about 250 decimal places for $L(f, -4, 2)$. However, the same 1000 Fourier coefficients will only give about 4 decimal places for $L(f, -191, 2)$ or $L(f, 197, 2)$. Note that all the central values that appear in Tables 3.1 and 3.2 were computed using the first 1000 Fourier coefficients of f, and thus their accuracy is actually less than what is displayed for the last few entries.

3   Two modular forms of weight $5/2$

Consider the ternary lattice corresponding to R, namely

$$S^0 := \{b \in \mathbb{Z} + 2R \ : \ \operatorname{Tr} b = 0\} = \left\langle\, 2i, \ j, \ i+k \,\right\rangle.$$

In this section we only deal with quaternions in S^0, which we will write in that basis as triples of integers, so that (x, y, z) is the quaternion $x(2i) + y(j) + z(i+k)$, and S^0 corresponds to $\mathbb{Z}^3$. With this convention, the norm restricted to S^0 is the ternary quadratic form

$$Q(x, y, z) := 4x^2 + 7y^2 + 8z^2 + 4xz,$$

whose corresponding bilinear form is

$$\langle (x, y, z), (x', y', z') \rangle := 8xx' + 14yy' + 16zz' + 4xz' + 4zx'.$$

As explained in the introduction, in order to obtain modular forms of weight $5/2$ we need to compute a theta series of Q with spherical polynomials of degree 1. Since such polynomials are odd, we need to combine them with *odd* weight functions as defined in [MRVT].

3.1   Imaginary twists

Let ψ be the quadratic character of conductor 7. This is an odd character, and thus the weight function ω_7 associated to ψ is odd. To compute ω_7, we can use $b_0 = (1, 0, 2)$, of norm $44 \not\equiv 0 \pmod 7$. By computing $\langle (x, y, z), b_0 \rangle = 16x + 36z \equiv 2x + z \pmod 7$, we find that

$$\omega_7(x, y, z) := \left(\frac{2x + z}{7}\right).$$

We must now find a suitable spherical polynomial of degree 1; any homogeneous polynomial of degree 1 is indeed a spherical polynomial. Note that

$$Q(x + z, -y, -z) = Q(x, y, z),$$

so that $(x + z, -y, -z)$ is an automorphism of Q, and also

$$\omega_7(x + z, -y, -z) = \omega_7(x, y, z).$$

Thus, polynomials which are odd with respect to this involution, like y and z, will lead to null theta series; any other polynomial will give the same theta series as the unique even (i.e. invariant) polynomial $x + z/2$ (up to a constant). This is the natural candidate, although for the sake of simplicity we will use x instead.

$-d$	$c_-(d)$	$L(f,-d,2)$	$-d$	$c_-(d)$	$L(f,-d,2)$	$-d$	$c_-(d)$	$L(f,-d,2)$
-4	1	2.238791	-71	0	0.000000	-148	-12	1.432431
-8	1	0.791532	-79	-8	1.632461	-151	14	1.891883
-11	2	1.963697	-88	4	0.347136	-155	4	0.148499
-15	2	1.233180	-95	2	0.077371	-163	-10	0.860640
-23	-2	0.649489	-107	-2	0.064727	-179	-18	2.423088
-39	-6	2.647336	-116	12	2.064329	-183	-6	0.260453
-43	-6	2.286669	-120	18	4.414450	-184	2	0.028694
-51	8	3.147228	-123	4	0.210071	-191	24	3.908201
-67	2	0.130632	-127	-2	0.050056			

Table 3.1: Coefficients of g_- and central values for the imaginary twists of f.

We are now ready to compute

$$g_- := \frac{1}{2} \sum_{(x,y,z)\in\mathbb{Z}^3} x\,\omega_7(x,y,z)\,q^{Q(x,y,z)}$$

$$= \sum_{n=1}^{\infty} c_-(n)\,q^n$$

$$= q^4 + q^8 + 2q^{11} + 2q^{15} - q^{16} - 2q^{23} + O(q^{32}),$$

a weight $5/2$ modular form of level $4 \cdot 7^2$ and character ψ_1, in Shimura correspondence with $f \otimes \psi$. Here $\psi_1(n) := \left(\frac{-1}{n}\right)\psi(n)$.

Table 3.1 shows the values of $c_-(d)$ and $L(f,-d,2)$, where $-200 < -d < 0$ is a fundamental discriminant such that $\left(\frac{-d}{7}\right) = -1$. The formula

$$L(f,-d,2) = k_- \frac{c_-(d)^2}{d^{3/2}}$$

is satisfied, where

$$k_- := 8\,L(f,-4,2) = 17.910324143488857621563653980249050613 9323\ldots$$

Note that if $\left(\frac{-d}{7}\right) \neq -1$, i.e. $\left(\frac{d}{7}\right) \neq 1$, it is trivial that $c_-(d) = 0$, because the genus of Q only represents squares modulo 7 and $\omega_7 = 0$ for zeros modulo 7 of Q, and also that $L(f,-d,2) = 0$, since the sign of the functional equation for $L(f,-d,s)$ is negative for such d.

3.2   Real twists

For this we need to use an odd weight function ω_l, for a suitably chosen prime l. First of all we need $l \equiv 3 \pmod 4$ so that ω_l is odd, but we should also require that $L(f,-l,2) \neq 0$. From Table 3.1, the smallest such l is 11, for which $L(f,-11,2) \approx 2.238791$. In order to compute ω_{11}, we will use again $b_0 = (1,0,2)$, of norm $44 \equiv 0 \pmod{11}$. Now, $\langle (x,y,z), b_0 \rangle \equiv 5x + 3z \equiv$

d	$c_+(d)$	$L(f,d,2)$	d	$c_+(d)$	$L(f,d,2)$	d	$c_+(d)$	$L(f,d,2)$
1	1	0.599566	77	14	0.347846	141	84	2.526777
8	7	1.298369	85	-28	0.599825	149	0	0.000000
21	14	2.442273	88	42	1.281185	156	98	2.955305
28	-7	0.396576	92	0	0.000000	161	-42	1.035445
29	-28	3.009928	93	-28	0.524118	165	-84	1.996041
37	0	0.000000	105	42	1.965992	168	-42	0.971408
44	28	1.610550	109	28	0.413060	172	-28	0.208381
53	28	1.218258	113	14	0.097831	177	42	0.449134
56	-35	3.505267	120	-14	0.089397	184	-84	1.695001
57	-14	0.273074	133	-42	1.379076	193	42	0.394424
60	-42	2.275668	137	-28	0.293138	197	56	0.679989
65	14	0.224245	140	56	2.270132			

Table 3.2: Coefficients of g_+ and central values for the real twists of f.

$3(-2x + z) \pmod{11}$, with $\left(\frac{3}{11}\right) = +1$, so that

$$\omega_{11}(x,y,z) := \begin{cases} 0 & \text{if } 11 \nmid Q(x,y,z), \\ \left(\frac{-2x+z}{11}\right) & \text{if } 2x \not\equiv z \pmod{11}, \\ \left(\frac{x}{11}\right) & \text{otherwise.} \end{cases}$$

We claim that

$$\omega_{11}(x+z, -y, -z) = \omega_{11}(x,y,z).$$

Indeed, by the uniqueness of weight functions, the above equation is true up to a constant. It is thus enough to check the equality for a single nonzero value, such as $\omega_{11}(1+2, 0, -2) = \omega_{11}(1, 0, 2) = +1$. Therefore, the same considerations as above apply, and lead us to choose x as the spherical polynomial.

We can then define and compute

$$g_+ := \frac{1}{4} \sum_{(x,y,z)\in\mathbb{Z}^3} x\,\omega_{11}(x,y,z)\, q^{Q(x,y,z)/11}$$

$$= \sum_{n=1}^{\infty} c_+(n)\, q^n$$

$$= q - 3q^4 + 7q^8 - 5q^9 - 5q^{16} + 14q^{21} + O(q^{25}),$$

a weight $5/2$ modular form of level $4 \cdot 7$ and trivial character, in Shimura correspondence with f.

Table 3.2 shows the values of $c_+(d)$ and $L(f,d,2)$, where $0 < d < 200$ is a fundamental discriminant such that $\left(\frac{d}{7}\right) \neq -1$. The formula

$$L(f,d,2) = k_+ \frac{c_+(d)^2}{d^{3/2}} \cdot \begin{cases} 1 & \text{if } \left(\frac{d}{7}\right) = 1, \\ 2 & \text{if } \left(\frac{d}{7}\right) = 0, \\ 0 & \text{if } \left(\frac{d}{7}\right) = -1. \end{cases}$$

is satisfied, where

$$k_+ := L(f, 1, 2) = 0.5995661579686175665810611670752228207656156...$$

Note that if $\left(\frac{d}{7}\right) = -1$, it is trivial that $c_+(d) = 0$, because the genus of Q only represents squares modulo 7, and also that $L(f, d, 2) = 0$, since the sign of the functional equation for $L(f, d, s)$ is negative for such d.

References

[BSP1] S. Böcherer and R. Schulze-Pillot, *On a theorem of Waldspurger and on Eisenstein series of Klingen type*, Math. Ann. 288 (1990) 361–388

[BSP2] S. Böcherer and R. Schulze-Pillot, *Vector valued theta series and Waldspurger's theorem*, Abh. Math. Sem. Univ. Hamburg 64 (1994), 211–233

[E] M. Eichler, *The basis problem for modular forms and the traces of the Hecke operators*, In: Modular Functions of One Variable I, Lecture Notes Math. 320 (1973), 76–151.

[G] B. Gross, *Heights and the special values of L-series*, Canadian Math. Soc. Conf. Proceedings, volume 7, (1987) p. 115–187.

[MRVT] Z. Mao, F. Rodriguez-Villegas and G. Tornaría, *Computation of central value of quadratic twists of modular L-functions*, in this volume.

[PT] A. Pacetti and G. Tornaría, *Examples of Shimura correspondence for level p^2 and real quadratic twists*, in this volume.

[W] J-L. Waldspurger *Sur les coefficients de Fourier des formes modulaires de poids demi-entier*, J. Math. pures Appl. 60 (1981), p. 375–484.

Heuristics on class groups and on Tate-Shafarevich groups: The magic of the Cohen-Lenstra heuristics

Christophe Delaunay

When we have to study a number field or an elliptic curve defined over $\mathbb{Q}$, some groups may appear which make the explicit computations more complicated and which are, in a way, not very "welcome". These groups are the class groups of number fields and the Tate-Shafarevich groups of elliptic curves. A direct study of their general behavior is a very difficult problem. In [3], Cohen and Lenstra explained how to obtain precise conjectures for this purpose using a general fundamental heuristic principle. In [6], it is shown how to adapt the Cohen-Lenstra idea to Tate-Shafarevich groups using the analogy between number fields and elliptic curves. Understanding the behavior of Tate-Shafarevich groups is important in itself first but it may also be useful for studying the distribution of the special values of the L-functions $L(E, s)$ attached to elliptic curves. Indeed, the Birch and Swinnerton-Dyer conjecture relates the value $L(E, 1)$ to natural invariants of E including the order of the Tate-Shafarevich group. This paper sketches the Cohen-Lenstra philosophy in both cases of class groups and of Tate-Shafarevich groups. It is organized as follows:

In the first section, we describe the analogy between number fields and elliptic curves defined over $\mathbb{Q}$. In the second section, we recall the Cohen-Lenstra heuristic for class groups. Using the analogy of the first section, we adapt, in the third section, the heuristic for Tate-Shafarevich groups. Finally, we restrict the heuristic to the case of families of quadratic twists of an elliptic curve.

Acknowledgements. I am very pleased to be able to thank here the organizers of the Clay Mathematics Institute Special Week on Ranks of Elliptic Curves and Random Matrix Theory, held at the Isaac Newton Institute for Mathematical Sciences, for the invitation to participate and to give a lecture on the heuristics. I could benefit from many interesting discussions with participants whom I want to thank. I am very happy to thank H. Cohen very much for his availability and his many suggestions. I am also grateful to C. Liebendörfer for her remarks and advice she kindly gave me.

This text was prepared during a post-doctoral position at the École Polytechnique Fédérale de Lausanne in Switzerland. I am grateful to all the members of the Eva Bayer's "Chaire des structures algébriques et géométriques".

1 Analogy between number fields and elliptic curves defined over $\mathbb{Q}$

In [6], we made a study of Tate-Shafarevich groups of elliptic curves defined over $\mathbb{Q}$ similar to the one made in [3] about class groups of number fields. To do this, we used the deep analogy between number fields and elliptic curves defined over $\mathbb{Q}$ on the one hand and class groups and Tate-Shafarevich groups on the other hand. This analogy is summarized in this section. First, we give the following table which states the correspondences between the main invariants of number fields and of elliptic curves defined over $\mathbb{Q}$:

Elliptic curve $E/\mathbb{Q}$		Number field K
$E(\mathbb{Q})_{\text{tors}}$ rational torsion points	$\leftrightarrows$	$U(K)_{\text{tors}}$ roots of unity
$E(\mathbb{Q})$ Mordell-Weil group of E	$\leftrightarrows$	$U(K)$ unit group of K
$N(E)$ conductor of E	$\leftrightarrows$	$|D_K|$ absolute value of the discriminant of K
$\text{III}(E)$ Tate-Shafarevich group of E	$\leftrightarrows$	$Cl(K)$ class group of K
$R(E)$ regulator of E	$\leftrightarrows$	$R(K)$ regulator of K
$E(\mathbb{Z})$ integer points on E	$\leftrightarrows$	exceptional units of K

The torsion parts of the groups $E(\mathbb{Q})$ and $U(K)$ are both finite and easy to determine; furthermore they play the same role. For a number field K, the unit group $U(K)$ is a finitely generated abelian group and it is not difficult to compute its rank r since we have $r = r_1 + r_2 - 1$ where r_1 (resp. r_2) is the number of real (resp. complex) places of K. However, it may be difficult to find the units of K. The Mordell-Weil group of an elliptic curve $E(\mathbb{Q})$ is a finitely generated abelian group, its rank can be predicted by the Birch and Swinnerton-Dyer conjecture and it may also be difficult to compute rational points on $E(\mathbb{Q})$ if they have large denominators. The primes dividing the absolute value of the discriminant or the conductor are rather special in both cases. Another property is that there are only finitely many number fields (resp. elliptic curves/$\mathbb{Q}$) up to isomorphism with a bounded absolute value of the discriminant (resp. conductor). The class group of a number field is a finite abelian group and measures in a way the obstruction of the ideals to be principal. Whenever this group is non-trivial, the arithmetic in K is more complicated. Similarly, the Tate-Shafarevich group III of an elliptic curve is a finite abelian group (here the finite part is only conjectural but we assume this conjecture to be true) and it measures the obstruction of the "local-global" principle. When III is non-trivial, it can be more difficult to study the elliptic curve. The regulator of a number field (resp. an elliptic curve) is the absolute value of the determinant of a certain matrix which is defined with the help of a basis of the unit group (resp. a basis of the Mordell-Weil group). In the case of a real quadratic field (the rank of the unit group of such a field is 1), as well as in the case of a rank 1 elliptic curve, there exist analytic processes to find a unit of the number field and a non-torsion point of the elliptic curve (the processes are the Gauss construction for number fields and the Heegner point construction for elliptic curves). The integer points on an elliptic curve form a finite subset of the Mordell-Weil group and are exact analogues of the exceptional units of a number field; exceptional units are units u such that $1-u$ is also a unit and they form a finite subset of the unit group (this analogy was pointed out by H. Cohen). Finally, for a number field, we have the following exact sequence:

$$1 \to U(K)/U(K)^p \to S_p(K) \to Cl(K)[p] \to 1$$

where $S_p(K) = V_p(K)/K^{*p}$ with $V_p(K) = \{\gamma \in K^* | \gamma \mathbb{Z}_K = \mathcal{I}^p$ for some ideal $\mathcal{I} \subset K\}$ and where $\mathbb{Z}_K$ is the ring of integers of K. The set $V_p(K)$ is indeed a subgroup of the multiplicative group K^*: it is called the group of p-virtual units. The group $S_p(K)$ is called the p-Selmer group of the number field K (we refer to [2] for all this terminology and for some more information about those groups).

If $L(K, s)$ is the L-function associated to K (i.e. the Dedekind zeta function), we have:

$$L(K, s) \sim_{s=0} -s^r \frac{R(K)|Cl(K)|}{|U(K)_{\text{tors}}|}$$

where $r = r_1 + r_2 - 1$ is the the rank of $U(K)$.
For an elliptic curve E:

$$1 \to E(\mathbb{Q})/pE(\mathbb{Q}) \to S_p(E) \to \text{Ш}(E)[p] \to 1$$

where $S_p(E)$ is the p-Selmer group of E (cf. [11]), and if $L(E, s)$ is the L-function attached to E, then the Birch and Swinnerton-Dyer conjecture predicts that:

$$L(E, s) \sim_{s=1} (s-1)^r \frac{R(E)|\text{Ш}(E)|}{(|E(\mathbb{Q})_{\text{tors}}|)^2} \, c\,\Omega \tag{1.1}$$

where r is the rank of the Mordell-Weil group, c is the product of the Tamagawa numbers (it is a small integer) and Ω is the real period of E.
The exact sequences and the estimates of the L-functions are exact analogues. However, we should note that the main terms in the right-hand side of (1.1) are perfect squares:

- $R(E)$ is the determinant of a Gram matrix and so is naturally the square of a determinant.

- The order of the group $\text{Ш}(E)$ is a square (we assume it is finite).

- In the denominator, there is the square of the order of $E(\mathbb{Q})_{\text{tors}}$.

Cassels proved that there exists a bilinear alternating pairing:

$$\beta \,:\, \text{Ш} \times \text{Ш} \longrightarrow \mathbb{Q}/\mathbb{Z}$$

which is non-degenerate if the Tate-Shafarevich group is finite; we assume III to be a finite group. Then we will say that a couple (G, β) is a group of type S, if G is a finite abelian group and:

$$\beta : G \times G \longrightarrow \mathbb{Q}/\mathbb{Z}$$

is a non-degenerate alternating bilinear pairing.

Two groups (G_1, β_1) and (G_2, β_2) of type S are said to be isomorphic if there exists an isomorphism $\sigma : G_1 \to G_2$ such that:

$$\beta_2(\sigma(x), \sigma(y)) = \beta_1(x, y) \quad \text{for all } x, y \in G_1.$$

If (G, β) is a group of type S, then $G \simeq H \times H$ where H is a finite abelian group; in particular, this explains why the order of a Tate-Shafarevich group is a perfect square. Conversely, every group $G \simeq H \times H$, where H is a finite abelian group, can be endowed with a unique (up to isomorphism) structure of group of type S.

In the sequel, the letter p will always denote a prime number. For G a finite abelian group, we denote by G_p the p-part of G: that is G_p is the subgroup of G consisting of elements of order a power of p. Note that every finite abelian group can be written as the direct sum of its p-group. The subgroup G_p is thus a p-group (i.e. $|G_p| = p^n$ for some $n \in \mathbb{N}$) and then can be uniquely written as:

$$G_p \simeq \mathbb{Z}/p^{a_1}\mathbb{Z} \times \mathbb{Z}/p^{a_2}\mathbb{Z} \cdots \times \mathbb{Z}/p^{a_r}\mathbb{Z}$$

for some (unique) positive integers $a_1 \leqslant a_2 \leqslant \ldots \leqslant a_r \in \mathbb{N}$. The number r is called the p-rank of G. It is denoted by $r_p(G)$ and is also equal to the dimension over $\mathbb{Z}/p\mathbb{Z}$ of the $\mathbb{Z}/p\mathbb{Z}$-vector space G/pG.

The symbol $\sum_{G(n)}$ (resp. $\sum_{G^S(n)}$) means that the sum is over all isomorphism classes of finite abelian groups (resp. groups of type S) of order n. Note that $\sum_{G^S(n)} \equiv 0$ if n is not a perfect square. Finally, $\mathrm{Aut}(G)$ denotes the group of automorphisms of G and $\mathrm{Aut}^S(G)$ the group of automorphisms of (G, β) which preserve the pairing β.

2 Heuristics on class groups of quadratic number fields

The class group measures in a way how difficult it is to perform the explicit computations related to some underlying arithmetical problem. Then, we would like to understand how it behaves in general; are we lucky if, for example, the p-part of the class group of some number field is trivial or are they often trivial? In fact, for those questions, we have to restrict our study to natural families of number fields whose unit groups have the same rank. Unfortunately, and even for such natural families, a direct study of this problem is completely out of reach nowadays. In [3], however, Cohen and Lenstra proposed a wonderful heuristic principle that allows to give conjectural answers to many questions related to the general behavior of class groups in a natural family. We now sketch their philosophy in the first case, that is, in the case of class groups attached to quadratic imaginary number fields.

2.1   Imaginary quadratic number fields

Imaginary quadratic number fields have the form $K = \mathbb{Q}(\sqrt{D_K})$ where $D_K < 0$ is a fundamental discriminant. The unit group is a finite group (its rank is 0) and the discriminant of K is D_K.

For our purpose, we let F be a $\mathbb{C}$-valued function defined on isomorphism classes of finite abelian groups (because class groups are finite abelian groups).

Examples. We will look at the following ones:

$$F_{p\text{-}triv}(G) \;=\; \begin{cases} 1 & \text{if } G_p \simeq \{0\} \\ 0 & \text{else} \end{cases}$$

$$F_{cyclic}(G) \;=\; \begin{cases} 1 & \text{if } G \text{ is cyclic} \\ 0 & \text{else} \end{cases}$$

$$F_{p-rank}(G) \;=\; p^{r_p(G)}$$

Then we consider the following limit:

$$M_{Cl,0}(F) = \lim_{X \to \infty} \left(\frac{\displaystyle\sum_{|D_K| \leqslant X} F(Cl(K))}{\displaystyle\sum_{|D_K| \leqslant X} 1} \right) \tag{2.1}$$

where the sums are over all quadratic imaginary number fields K whose absolute value of the discriminant is bounded by X. Note that there are only finitely many such number fields, so that the term in the brackets of (2.1) is meaningful.

We have two problems: does the limit exist? If yes, what is its value? One moment's thought tells us that, in fact, this is exactly what we want to answer. For example, if we consider the function $F = F_{p\text{-}triv}$, then if the limit in (2.1) exists for F, this limit is precisely the frequency of class groups with trivial p-parts. But, as we mentioned above we cannot study this limit directly. The fundamental idea of Cohen and Lenstra is to say that class groups behave as random finite abelian groups G except that they have to be weighted by:

$$\frac{1}{|\operatorname{Aut}(G)|} \tag{2.2}$$

More precisely, we consider the following average:

Definition 1. *Let F be as above. The 0-average of F over finite abelian groups is defined by:*

$$M_{G,0}(F) = \lim_{X \to \infty} \left(\frac{\displaystyle\sum_{n \leqslant X} \sum_{G(n)} \frac{F(G)}{|\operatorname{Aut}(G)|}}{\displaystyle\sum_{n \leqslant X} \sum_{G(n)} \frac{1}{|\operatorname{Aut}(G)|}} \right).$$

If F is the characteristic function of a property $\mathcal{P}$, we will speak of 0-probability instead of 0-average.

From the theory of genera, the 2-part of the class group $Cl(K)$ behaves in a very special way and so we have to exclude it from our discussion and we denote by:

$$Cl_0(K) = \{x \in Cl(K) \text{ such that } x \text{ has odd order}\}$$

the odd part of $Cl(K)$.

If F is a function as above, we define the function $F \circ odd$ to be the function: $F \circ odd \; : \; G \mapsto F(G_0)$, where G_0 denotes the odd part of G. We can now formulate the Cohen-Lenstra heuristic:

Fundamental heuristic assumption for imaginary quadratic fields. *For all reasonable functions F, we have:*

$$M_{Cl,0}(F \circ odd) = M_{G,0}(F \circ odd)$$

The magic of the Cohen-Lenstra heuristic is that it works! Indeed, there are strong evidences to believe in this assumption. Furthermore, the value of $M_{G,0}(F \circ odd)$ can be computed for many interesting functions and we can be confident enough in the results it produces. In practice, in order to compute $M_{G,0}(F)$, we treat the numerator and the denominator of the definition of $M_{G,0}(F)$ separately. For this purpose we need the following Tauberian theorem:

Theorem 2. *Let $(c(n))_{n \geqslant 1}$ be a sequence of non-negative numbers and $D(z) = \sum_n c(n)/n^z$. If $D(z)$ converges for $\Re(z) > 0$ and if there exists $C \in \mathbb{C}$ such that $D(z) - C/z$ can be analytically continued to an open subset containing $\Re(z) \geqslant 0$, then, as X tends to infinity, we have:*

$$\sum_{n \leqslant X} c(n) \sim C \log(x)$$

In view of the definition 1 we would like to apply this theorem with $c(n) = \sum_{G(n)} F(G)/|\operatorname{Aut}(G)|$, leading to:

Definition 3. *Let F be a function as above. We define the two following Dirichlet series:*

$$\zeta_G(z) \;=\; \sum_{n \geqslant 1} \frac{1}{n^z} \sum_{G(n)} \frac{1}{|\operatorname{Aut}(G)|}$$

$$\zeta_{G,F}(z) \;=\; \sum_{n \geqslant 1} \frac{1}{n^z} \sum_{G(n)} \frac{F(G)}{|\operatorname{Aut}(G)|}$$

Cohen and Lenstra proved (cf. [3]):

Theorem 4. *We have:*

$$\zeta_G(z) = \prod_{j=1}^{\infty} \zeta(z + j)$$

where ζ is the Riemann zeta function.

From theorem 4 and theorem 2 we deduce a very good estimate for the denominator of $M_{G,0}(F)$:

Corollary 5. *We have:*

$$\sum_{n \leqslant X} \sum_{G(n)} \frac{1}{|\operatorname{Aut}(G)|} \sim \prod_{j=2}^{\infty} \zeta(j) \, \log(X).$$

For the numerator of $M_{G,0}(F)$ we do the same; in general for reasonable functions F, the Dirichlet series $\zeta_{G,F}$ satisfies the conditions of theorem 2, and we can deduce that:

$$\sum_{n \leqslant X} \sum_{G(n)} \frac{F(G)}{|\operatorname{Aut}(G)|} \sim C \log(X)$$

(for convenience $C = 0$ means that the left hand-side is $O(1)$). We then obtain:

$$M_{G,0}(F) = \frac{C}{\prod_{j=2}^{\infty} \zeta(j)}$$

By the same method we can compute $M_{G,0}(F \circ odd)$ and by the heuristic assumption this is the average of F over class groups.

2.2   Real quadratic fields

This case is a little bit more subtle since the rank of the unit group is 1. More generally, when the unit group is not a finite group, the Cohen-Lenstra heuristic is more technical ([3], [4]). In our case, the philosophy is to say that the odd part of a class group associated to a real quadratic field behaves as a random finite abelian group G of odd order divided by a random cyclic subgroup (we still have the weight $1/|\operatorname{Aut}(G)|$). With this idea we can also define a "1-average" over finite abelian group $M_{G,1}(F)$ and the heuristic predicts that:

Fundamental heuristic assumption for real quadratic fields: *For all reasonable functions F we have:*

$$M_{Cl,1}(F \circ odd) = M_{G,1}(F \circ odd)$$

where $M_{Cl,1}(F)$ is defined as in (2.1) except that the sums are over real quadratic number fields.

In fact, Cohen and Lenstra defined the u-average $M_{G,u}(F)$ of F over finite abelian groups for all $u \in \mathbb{N}$. In general, the u-average can be computed by a straightforward generalization of the method explained above ([3]):

Theorem 6. *Let F be a function as above, $u \in \mathbb{N}$ and suppose that $\zeta_{G,F}$ satisfies the conditions of theorem 2 then:*

$$if\ u = 0\ then\ M_{G,0}(F) \quad = \quad \lim_{z \to 0} \frac{\zeta_{G,F}(z)}{\zeta_G(z)},$$

$$if\ u > 0\ then\ M_{G,u}(F) \quad = \quad \frac{\zeta_{G,F}(u)}{\zeta_G(u)}.$$

2.3 Examples

Let us consider the function $F = F_{p\text{-}triv}$, where $p \neq 2$; then $F \circ odd = F$. The function $\zeta_{G,F}$ is exactly the function ζ_G without its p-part. From theorem 4 we have:

$$\begin{aligned}
\zeta_G(z) \quad &= \quad \prod_{j=1}^{\infty} \zeta(z+j) \\
&= \quad \prod_j \prod_{q\ \text{prime}} \left(1 - \frac{1}{q^{z+j}}\right)^{-1} \\
&= \quad \prod_{q\ \text{prime}} \prod_j \left(1 - \frac{1}{q^{z+j}}\right)^{-1}.
\end{aligned}$$

So the term $\prod_j \left(1 - \frac{1}{p^{z+j}}\right)^{-1}$ is exactly the p-Euler factor of $\zeta_G(z)$. Then we deduce:

$$\begin{aligned}
\zeta_{G,F}(z) \quad &= \quad \prod_{q \neq p} \prod_j \left(1 - \frac{1}{q^{z+j}}\right)^{-1} \\
&= \quad \zeta_G(z) \prod_j \left(1 - \frac{1}{p^{z+j}}\right).
\end{aligned}$$

Finally, we obtain:

$$M_{G,0}(F) \;=\; \prod_{j=1}^{\infty}\left(1 - \frac{1}{p^j}\right)$$

$$M_{G,1}(F) \;=\; \prod_{j=1}^{\infty}\left(1 - \frac{1}{p^{j+1}}\right).$$

By the Cohen-Lenstra heuristic, we deduce the conjecture:

Conjecture 7. *Let $p \neq 2$.*
The probability that p divides the order of the class group of an imaginary quadratic field is equal to:

$$1 - \prod_{j=1}^{\infty}\left(1 - \frac{1}{p^j}\right) = \frac{1}{p} + \frac{1}{p^2} - \frac{1}{p^5}\cdots$$

The probability that p divides the order of the class group of a real quadratic field is equal to:

$$1 - \prod_{j=1}^{\infty}\left(1 - \frac{1}{p^{j+1}}\right) = \frac{1}{p^2} + \frac{1}{p^3} + \frac{1}{p^4}\cdots$$

Let us consider some other examples that can be found in [3] (they are a little more technical because they involve the p-rank of finite abelian groups).
- The u-average of the function $F = F_{cyclic} \circ odd$ is equal to:

$$M_{G,u}(F) = \frac{\prod_{j \geqslant u+2}(1 - 1/2^j)}{1 + 1/2^{u+1}} \prod_{p}\left(\frac{1 - 1/p + 1/p^{u+2}}{1 - 1/p}\right) \prod_{j \geqslant 2}\frac{1}{\zeta(u+j)}$$

In particular, $M_{G,0}(F) \approx 0.98$.
- The u-average of the function $F(G) = p^{r_p(G)}$ is $1 + 1/p^u$ (note that $F \circ odd = F$ if $p \neq 2$). In particular, if $p = 3$, the 0-average of F is equal to 2 and the 1-average of F is $4/3$.

The Cohen-Lenstra heuristics have been checked by many numerical computations and they are very useful for understanding the behavior of class groups, even if the results are conjectural. For example, they explain why it is so difficult to find a non-cyclic $Cl_0(K)$.
The first theoretical result was obtained by Davenport and Heilbronn who proved (before the heuristics were formulated) that the average of the function $3^{r_3(Cl(K))}$ is equal to 2 (resp. $4/3$) in the case of imaginary (resp. real) quadratic fields.
The Cohen-Lenstra heuristics extend to many other families of number fields, and we refer to [3], [4], [15] for some generalizations.

3   Heuristics on Tate-Shafarevich groups of elliptic curves

As with class groups, we are annoyed by Tate-Shafarevich groups of elliptic curves. Thus, we use the analogy described in the first section and sketch the work in [6] which shows how the Cohen-Lenstra philosophy can be adapted to our case. To do this, we take into account the particular structure of Tate-Shafarevich groups, i.e., the structure of groups of type S.

3.1   Rank 0 case

By analogy, we consider a $\mathbb{C}$-valued function F defined on the isomorphism classes of groups of type S.

Examples. We will look at the following ones:

$$F_{p\text{-}triv}(G) \;=\; \begin{cases} 1 & \text{if } G_p \simeq \{0\} \\ 0 & \text{else} \end{cases}$$

$$F_{cyclic}(G) \;=\; \begin{cases} 1 & \text{if } G \text{ is the square of a cyclic group} \\ 0 & \text{else} \end{cases}$$

$$F_{p\text{-}rank=2r}(G) \;=\; \begin{cases} 1 & \text{if } r_p(G) = 2r \\ 0 & \text{else} \end{cases}$$

Note that we simply write G for a group of type S instead of (G, β), since there is only one group structure of type S for each group (up to isomorphism). We consider the following limit:

$$M_{\text{III},0}(F) = \lim_{X \to \infty} \left(\frac{\displaystyle\sum_{N(E) \leqslant X} F(\text{III}(E))}{\displaystyle\sum_{N(E) \leqslant X} 1} \right) \tag{3.1}$$

where the sums are over all isomorphism classes of rank 0 elliptic curves whose conductor is bounded by X (there are only finitely many such isomorphism classes). As for class groups, we have two questions: does the limit exist? If yes, what is its value? Furthermore, questions of this type are exactly what we would like to answer...

Definition 8. *We define:*

$$\zeta_{G^S}(z) \;=\; \sum_{n\geqslant 1}\frac{1}{n^z}\sum_{G^S(n)}\frac{1}{|\operatorname{Aut}^S(G)|}$$

$$\zeta_{G^S,F}(z) \;=\; \sum_{n\geqslant 1}\frac{1}{n^z}\sum_{G^S(n)}\frac{F(G)}{|\operatorname{Aut}^S(G)|}$$

We can prove:

Theorem 9. *We have:*

$$\zeta_{G^S}(z) = \prod_{j=1}^{\infty}\zeta(2z+2j+1).$$

The main difference with finite abelian groups is that the function $\zeta_{G^S}(z)$ converges for $z=0$ and that we have:

Corollary 10.

$$\zeta_{G^S}(0) = \sum_{n\geqslant 1}\sum_{G^S(n)}\frac{1}{|\operatorname{Aut}^S(G)|} = \prod_{j=1}^{\infty}\zeta(2j+1).$$

So, we have to adapt the definition for average over groups of type S:

Definition 11. *Let F be as above and $\alpha \geqslant 1$ (we will see later why we need α). Then, the 0-average of F over groups of type S is defined by:*

$$M_{G^S,0}(F,\alpha) = \lim_{X\to\infty}\left(\frac{\displaystyle\sum_{n\leqslant X}\sum_{G^S(n)}\frac{F(G)|G|^{\alpha}}{|\operatorname{Aut}^S(G)|}}{\displaystyle\sum_{n\leqslant X}\sum_{G^S(n)}\frac{|G|^{\alpha}}{|\operatorname{Aut}^S(G)|}}\right).$$

The exact analogue of definition 1 would have been to take $\alpha = 0$. But as we have shown before, the denominator converges for $\alpha = 0$ and this does not give a relevant average. For $\alpha \geqslant 1$ the denominator diverges. Another reason to insert α in definition 11 is that for reasonable functions F, the limit $M_{G,0}(F,\alpha)$ *does not depend on α if $\alpha \geqslant 1$* (this fact is an application of a generalization of theorem 2). In particular, it is not true that the limit does not depend on α if $\alpha < 1$. The *same phenomenon* already occurred for finite abelian groups; we could take the weight $|G|^{\alpha}/|\operatorname{Aut}(G)|$ in (2.2) and the results would not have depended on α for $\alpha \geqslant 0$. Once again, the situation is analogous to the one of class groups.

Since $M_{G,0}(F,\alpha)$ does not depend on α for $\alpha \geqslant 1$, we let:

$$M_{G^S,0}(F) = M_{G^S,0}(F,1).$$

Then we can compute $M_{G,0}(F)$ by using theorem 2 as for finite abelian groups. For instance, the function $\zeta_{G^S}(z-1) = \sum_n \frac{1}{n^z} \sum_{G^S(n)} \frac{|G|}{|\operatorname{Aut}^S(G)|}$ converges for $\Re(z) > 0$ and satisfies the conditions of theorem 2. Thus we deduce the following estimate for the denominator of $M_{G,0}(F)$:

Corollary 12. *As X tends to ∞ we have:*

$$\sum_{n \leqslant X} \sum_{G^S(n)} \frac{|G|}{|\operatorname{Aut}^S(G)|} \sim \frac{1}{2} \prod_{j=1}^{\infty} \zeta(2j+1) \, \log(X).$$

As regards the numerator, we expect that the function $\zeta_{G^S,F}(z-1)$ satisfies the conditions of theorem 2 so that:

$$\sum_{n \leqslant X} \sum_{G^S(n)} \frac{F(G)|G|}{|\operatorname{Aut}^S(G)|} \sim C \log(X).$$

And we would have:

$$M_{G^S,0}(F) = \frac{2C}{\prod_{j=1}^{\infty} \zeta(2j+1)}.$$

Now the heuristic idea is to assert that Tate-Shafarevich groups of rank 0 elliptic curves behave as random groups G of type S except that they have to be weighted by the weight $|G|/|\operatorname{Aut}^S(G)|$.

Fundamental heuristic assumption for rank 0 elliptic curves. *For all reasonable functions F we have:*

$$M_{\text{III},0}(F) = M_{G^S,0}(F).$$

3.2 Rank 1 case

As for class groups, the higher rank cases are a little bit more technical. In case of rank 1, we are interested in the following limit:

$$M_{\text{III},1}(F) = \lim_{X \to \infty} \left(\frac{\displaystyle\sum_{N(E) \leqslant X} F(\text{III}(E))}{\displaystyle\sum_{N(E) \leqslant X} 1} \right) \tag{3.2}$$

where now the sums are over all isomorphism classes of rank 1 elliptic curves with conductor bounded by X.

Definition 13. *Let F be as above and $u \geqslant 0$. We define $(c_u(F, n))_{n \geqslant 1}$ by:*

$$\sum_{n \geqslant 1} \frac{c_u(F, n)}{n^z} = \frac{\zeta_{G^S, F}(z + u)\zeta_{G^S}(z)}{\zeta_{G^S}(z + u)}.$$

The u-average of F over groups of type S is:

$$M_{G^S, u}(F) = \lim_{X \to \infty} \left(\frac{\displaystyle\sum_{n \leqslant X} nc_u(F, n)}{\displaystyle\sum_{n \leqslant X} \sum_{G^S(n)} \frac{|G|}{|\operatorname{Aut}^S(G)|}} \right).$$

Remarks. For reasonable functions F, the average $M_{G^S, u}(F)$ does not depend on $\alpha \geqslant 1$ if we replace in the definition $nc_u(F, n)$ by $n^\alpha c_u(F, n)$ and $|G|$ by $|G|^\alpha$. For $u = 0$, this is the same definition as in the section above.

Theorem 2 allows us to compute u-averages in many cases:

Proposition 14. *Let F be as above and suppose that $\zeta_{G^S, F}(z - 1)$ satisfies the conditions of theorem 2. We have:*

$$\text{if } u = 0 \text{ then } M_{G^S, 0} = \lim_{z \to 0} \frac{\zeta_{G^S, F}(z - 1)}{\zeta_{G^S}(z - 1)}$$

$$\text{if } u > 0 \text{ then } M_{G^S, u}(F) = \frac{\zeta_{G^S, F}(u - 1)}{\zeta_{G^S}(u - 1)}$$

Fundamental heuristic assumption for rank 1 elliptic curves. *For all reasonable functions F we have:*

$$M_{\text{III}, 1}(F) = M_{G^S, 1}(F).$$

Note that in [6], we formulated the heuristic for higher ranks by taking the $u/2$-average for the family of rank u elliptic curves. So the heuristic assumption here *is a correction of [6] in the rank 1 case*.

3.3 Examples

Let us consider the function $F = F_{p\text{-}triv}$. Then, as for finite abelian groups, we have:

$$\zeta_{G^S, F}(z) = \zeta_{G^S}(z) \prod_{j=1}^{\infty} \left(1 - \frac{1}{p^{2z+2j+1}} \right).$$

So we obtain the u-average of F:

$$M_{G^S, u}(F) = \prod_{j=1}^{\infty} \left(1 - \frac{1}{p^{2u+2j-1}} \right).$$

Then, we deduce:

-The 0-probability that p divides the order of a group of type S is equal to:

$$f_0(p) = 1 - \prod_{j=1}^{\infty}\left(1 - \frac{1}{p^{2j-1}}\right) = \frac{1}{p} + \frac{1}{p^3} - \frac{1}{p^4} + \frac{1}{p^5} - \frac{1}{p^6}\cdots \tag{3.3}$$

In particular $f_0(2) \approx 0.58$, $f_0(3) \approx 0.36$ and $f_0(5) \approx 0.21$.

-The 1-probability that p divides the order of a group of type S is equal to:

$$f_1(p) = 1 - \prod_{j=1}^{\infty}\left(1 - \frac{1}{p^{2j+1}}\right) = \frac{1}{p^3} + \frac{1}{p^5} + \frac{1}{p^7} - \frac{1}{p^8}\cdots \tag{3.4}$$

In particular $f_1(2) \approx 0.16$, $f_1(3) \approx 0.04$ and $f_1(5) \approx 0.01$.

We also consider some other examples that can be found in [6].
-The u-average of the function $F = F_{cyclic}$ is equal to:

$$M_{G^S,u}(F) = \prod_{p}\left(1 - \frac{1}{p^2} + \frac{1}{p^{2u+3}}\right) \frac{\zeta(2)}{\prod_{j\geqslant 1}\zeta(2u + 2j + 1)}.$$

In particular, $M_{G^S,0}(F) \approx 0.98$.
-The u-average of $F(G) = p^{r_p(G)}$ is equal to:

$$1 + p^{1-2u}. \tag{3.5}$$

-The u-average of the function $F = F_{p-rank=2r}$ is equal to:

$$M_{G^S,u}(F) = \frac{p^{-r(2u+2r-1)}}{\prod_{j\geqslant 1}(1 - 1/p^{2r})} \prod_{j\geqslant r+1}(1 - 1/p^{2u+2j-1}). \tag{3.6}$$

The heuristics as well as their consequences are out of reach. Furthermore it is difficult to check them numerically because there are too many elliptic curves and Tate-Shafarevich groups seem to appear for large conductors. There is no algorithm known to compute Tate-Shafarevich groups. The only thing we can do is to compute the (conjectural) order of Tate-Shafarevich groups using the Birch and Swinnerton-Dyer conjecture. Indeed all members in equation (1.1) are easily computable except $R(E)$ and $|\text{III}(E)|$. So if for some reason one can compute $R(E)$ (for rank 0 curves we simply have $R(E) = 1$), then we can deduce $|\text{III}(E)|$. If we have many data we can compare them with the heuristic predictions of type (3.3). We can also restrict the heuristics to some natural sub-families of elliptic curves (quadratic twists) for which the analogy with number fields seems to be even more deeper.

4   Quadratic twist families

Let E be an elliptic curve defined over $\mathbb{Q}$ with conductor N and let $L(E, s) = \sum_n a(n)n^{-s}$ be its L-function. From the work in [14], [12] and [1] E is known to be modular. This implies that its L-function can be analytically continued to the whole complex plane and satisfies a functional equation:

$$\Lambda(E, 2 - s) = \varepsilon\Lambda(E, s) \tag{4.1}$$

where $\varepsilon = \pm 1$ is the sign of the functional equation and:

$$\Lambda(E, s) = \left(\frac{\sqrt{N}}{2\pi}\right)^s \Gamma(s)L(E, s).$$

Note that the Birch and Swinnerton-Dyer conjecture implies $\varepsilon = (-1)^r$ where r is the rank of $E(\mathbb{Q})$. Let D be a fundamental discriminant (for simplicity we assume $(N, D) = 1$). Then the twisted L-function:

$$L(E, D, s) = \sum_n \left(\frac{D}{n}\right) a(n)n^{-s}$$

where $\left(\frac{D}{\cdot}\right)$ is the Kronecker symbol, corresponds to the quadratic twist E_D of E by D and has conductor $N_D = ND^2$. Then the function $L(E, D, s)$ satisfies a functional equation as (4.1) whose sign is $\varepsilon_D = \left(\frac{D}{-N}\right)$.

In this section, we consider the family of elliptic curves:

$$(E_D)_D \text{ where } D \text{ runs over all fundamental discriminant.}$$

In fact, there is another analogy between this family and the family of quadratic imaginary number fields. Indeed, from the work of Waldsurger ([13]), the values $L(E, D, 1)$ are related to the coefficients $c(|d|)$ of a 3/2-weight modular form; more precisely:

$$L(E, D, 1) = \kappa_E|D|^{-1/2}c(|D|)^2 \tag{4.2}$$

where κ_E is a constant depending only on E. Suppose that $c(|d|) \neq 0$ so that E_D has rank 0. Then replacing $L(E, D, 1)$ by its value predicted by the Birch and Swinnerton-Dyer conjecture, we deduce that the order $|\text{Ш}(E_D)|$ of the Tate-Shafarevich group of the rank 0 curve E_D is, up to some factors (namely the Tamagawa numbers), the square of the coefficients of a 3/2-weight modular form. We have exactly the same phenomenon for class groups (without the square). Indeed, the order of class groups of imaginary quadratic fields are, up to some normalization (namely, we have to consider the Hurwitz class numbers instead of the class numbers), the coefficients of a 3/2-weight modular form. Furthermore, using (4.2), Rubinstein ([10]) performed huge numerical experimentations and computed how often a given prime p divides the (conjectural) order of $\text{III}(E_D)$ for rank 0 quadratic twists of many elliptic curves E. His numerical results are in close agreement with the prediction (3.3) given by the heuristic except maybe for some "special" primes. These lead us to restrict the heuristics to the family (E_D).

If $\mathcal{T}$ is a set of prime numbers and F a function defined on isomorpism classes of groups of type S, then we define the function $F \circ \mathcal{T}$ by $F \circ \mathcal{T} \; : \; G \mapsto F(G_{\mathcal{T}})$, where $G_{\mathcal{T}}$ is the $\mathcal{T}$-part of G.

Heuristic assumption for rank 0 quadratic twists. *Let E be an elliptic curve defined over $\mathbb{Q}$. Then there exists a finite set $\mathcal{S}$ of prime numbers such that for all reasonable functions F we have:*

$$\lim_{X \to \infty} \left(\frac{\displaystyle\sum_{\substack{|D|<X \\ rk(E_D)=0}} F \circ \mathcal{T}(\text{III}(E_D))}{\displaystyle\sum_{\substack{|D|<X \\ rk(E_D)=0}} 1} \right) = M_{G^s,0}(F \circ \mathcal{T}) \qquad (4.3)$$

where the sum is over fundamental discriminants D such that the rank of E_D is 0 and where $\mathcal{T}$ is the set of prime numbers p with $p \notin \mathcal{S}$.

Heuristic assumption for rank 1 quadratic twists. *Let E be an elliptic curve defined over $\mathbb{Q}$. Then there exists a finite set $\mathcal{S}$ of prime numbers such that for all reasonable functions F we have:*

$$\lim_{X \to \infty} \left(\frac{\displaystyle\sum_{\substack{|D|<X \\ rk(E_D)=1}} F \circ \mathcal{T}(\text{III}(E_D))}{\displaystyle\sum_{\substack{|D|<X \\ rk(E_D)=1}} 1} \right) = M_{G^s,1}(F \circ \mathcal{T}) \qquad (4.4)$$

where the sum is over fundamental discriminants D such that the rank of E_D is 1 and where $\mathcal{T}$ is the set of prime numbers p with $p \notin \mathcal{S}$.

Remark: It is actually not clear which prime numbers have to be excluded form the discussion. Rubinstein's huge numerical data show that some primes behave in a rather special way. More precisely, those primes appear to be maybe the prime 2 and the odd primes ℓ dividing the order of the torsion sub-group of the curves belonging to the isogeny class of the curve E with the smallest conductor in the family in question (perhaps due to the fact that, in this case, the ℓ-part of the class group of $\mathbb{Q}(\sqrt{d})$ should have a weight on the ℓ-Selmer group of E_d. This is, indeed, what had been proved by Frey for some curves E [7]). However, the convergence for the prime 2 may be simply slower than for the others and so seems to be a special prime even if it is not.

In [8] and [9], Heath-Brown[1] studied the Selmer groups of the family of quadratic twists:

$$E_D \ : \ Dy^2 = x^3 - x.$$

When the rank of E_D is 0 or 1, it is not difficult to obtain information about the Tate-Shafarevich groups of E_D from its Selmer group. Furthermore, Heath-Brown considered all curves E_D and not only those that have rank 0 or 1. Nevertheless, there is a classical conjecture (the density conjecture) asserting that on average the curves E_D have either rank 0 or rank 1 (of course, this can be true only on average). The random matrix theory predicts very precise statements refining the density conjecture ([5]). Finally, the density conjecture and Heath-Brown's works imply the following rank 0 and rank 1 results:

Rank 0 case. Here we consider only D such that E_D has rank 0.

- The average of the function $2^{r_p(\text{Ш}(E_D))}$ over the curves E_D that have rank 0 is equal to 3.
- Let $r \in \mathbb{N}$. The probability that $r_2(\text{Ш}(E_D)) = 2r$ is equal to:

$$\prod_{j=1}^{\infty}(1 + 2^{-n})^{-1}\frac{2^r}{\prod_{1 \leqslant j \leqslant r}(2^j - 1)} \tag{4.5}$$

Rank 1 case. Here we consider only D such that E_D has rank 1.

- The average of the function $2^{r_p(\text{Ш}(E_D))}$ over the curves E_D that have rank 1 is equal to $3/2$.
- Let $r \in \mathbb{N}$. The probability that $r_2(\text{Ш}(E_D)) = 2r$ is equal to ([9]):

$$\prod_{j=1}^{\infty}(1 + 2^{-n})^{-1}\frac{2^{r-1}}{\prod_{1 \leqslant j \leqslant r-1}(2^j - 1)}. \tag{4.6}$$

These results should be compared with (3.5) and (3.6) with $p = 2$, $u = 0$ and $u = 1$. In fact, a little computation shows that they all agree! (Heath Brown's results and the link with the heuristics have been pointed out to me by E. Kowalski whom I thank here). In Heath-Brown's paper, it is suggested that the convergence should be extremely slow, so it would not be very surprising if the prime 2 behaved like a special prime in numerical computations although it is not. Heath-Brown's results and Rubinstein's data make the heuristics on Tate-Shafarevich groups even more believable in the case of quadratic twist families.

---

[1]Editors' comment: See also the article by D.R. Heath-Brown in this volume.

References

[1] C. Breuil, B. Conrad, F. Diamond and R. Taylor, *On the modularity of elliptic curves over $\mathbb{Q}$: wild 3-adic exercises*, J. Amer. Math. Soc. **14** (2001), no. 4, 843–939

[2] H. Cohen, *Advanced topics in computational algebraic number theory*, Graduate texts in Math. **193**, Springer-Verlag, New-York, (2000).

[3] H. Cohen and H. W. Lenstra, *Heuristics on class groups of number fields*, in Number theory (Noordwijkerhout, 1983), ed. H. Jager, Lecture Notes in Math. **1068**, Springer-Verlag (1984), pp. 33-62.

[4] H. Cohen and J. Martinet, *Etude heuristiques des groupes de classes des corps de nombres*, J. Reine Angew. Math. **404** (1990), pp. 39-76.

[5] J. Conrey, J. Keating, M. Rubinstein and N. Snaith, *On the frequency of vanishing of quadratic twists of modular L-functions*, Number theory for the millennium, I (Urbana, IL, 2000), A K Peters, Natick, MA, 2002, pp. 301–315.

[6] C. Delaunay, *Heuristics on Tate-Shafarevitch groups of elliptic curves defined over $\mathbb{Q}$*, Exp. Math. **10** (2001), no. 2, 191–196.

[7] G. Frey, *On the Selmer group of twists of elliptic curves with $\mathbb{Q}$-rational torsion points*, Canad. J. Math. ,**XL**, (1988), 649–665.

[8] D. R. Heath-Brown, *The size of the Selmer group for the congruent number problem, I*, Invent. Math. **111**, (1993), pp. 111-125.

[9] D. R. Heath-Brown, *The size of the Selmer group for the congruent number problem, II*, Invent. Math. **118**, (1994), pp. 331-370.

[10] M. Rubinstein, *Numerical data*, available at www.math.uwaterloo.ca/~mrubinst/L_function/VALUES/DEGREE_2/ ELLIPTIC/QUADRATIC_TWISTS/WEIGHT_THREE_HALVES/

[11] J. Silverman, *The arithmetic of elliptic curves*, Graduate texts in Math. **106**, Springer-Verlag, New-York, (1986).

[12] R. Taylor and A. Wiles, *Ring-theoretic properties of certain Hecke algebras*, Ann. of Math. (2) **141** (1995), no. 3, 553–572.

[13] J.-L. Waldspurger, *Sur les coefficients de Fourier des formes modulaires de poids demi-entier*, J. Math. Pures Appl. (9) **60** (1981), pp. 375-484.

[14] A. Wiles, *Modular elliptic curves and Fermat's last theorem*, Ann. of Math. (2) **141** (1995), no.3, 443–551.

[15] C. Wittmann, *p-class groups of certain extensions of degree p*, Math. Comp. **74** (2005), no. 250, 937–947.

Institut Camille Jordan
Université Claude Bernard Lyon 1
43, avenue du 11 novembre 1918
69622 Villeurbanne Cedex - France

A Note on the 2-Part of Ш for the Congruent Number Curves

D.R. Heath-Brown

The purpose of this note is to give a brief exposition of the results of the author's work [1,2]. The congruent number problem is described by the quadratic twists of the elliptic curve

$$E : y^2 = x^3 - x,$$

that is to say, by the curves

$$E_D : Dy^2 = x^3 - x$$

for positive square-free integers D. The L-functions for these twists have even functional equation for $D \equiv 1, 2, 3 \pmod 8$, and odd functional equation for $D \equiv 5, 6, 7 \pmod 8$. The group Ш is described in Silverman [3; page 297]. In order to describe the 2-part of $\text{Ш}(D)$ it will be convenient to make the following hypothesis, which we shall assume throughout this note.

Hypothesis. *There are $O(X(\log X)^{-2})$ positive square-free integers $D \leq X$ with $D \equiv 1, 2$ or $3 \pmod 8$ such that E_D has rank different from zero. Similarly there are $O(X(\log X)^{-2})$ positive square-free integers $D \leq X$ with $D \equiv 5, 6$ or $7 \pmod 8$ such that E_D has rank different from one.*

The order of the Selmer group $S^{(2)}$ is connected to the 2-part of Ш as follows. Let $\#S^{(2)} = 2^{2+s(D)}$, so that $s(D)$ is the upper bound for the rank of E_D which arises from the 2-descent process, see Silverman [3; page 281]. In the appendix to [2] Monsky showed that $s(D)$ is even for integers $D \equiv 1, 2, 3 \pmod 8$, and odd for values $D \equiv 5, 6, 7 \pmod 8$. Now define

$$\text{Ш}_2(D) = \{g \in \text{Ш}(D) : g \text{ has order } 1 \text{ or } 2\}.$$

Then if $D \equiv 1, 2, 3 \pmod 8$ and E_D has rank zero, we have

$$\#\text{Ш}_2(D) = 2^{s(D)},$$

while if $D \equiv 5, 6, 7 \pmod 8$ and E_D has rank one, then

$$\#\text{Ш}_2(D) = 2^{s(D)-1}.$$

The main result of [2] gives the frequency with which each value of $s(D)$ occurs. Let

$$\lambda = \prod_{n=0}^{\infty} (1 - 2^{-2n-1}) = 0.4194\ldots.$$

and set

$$d_k = \lambda \frac{2^k}{\prod_{1 \le j \le k}(2^k - 1)}, \quad (k = 0, 1, 2, \ldots).$$

Moreover we define

$$N(X; h) = \#\{D \le X : D \text{ square-free}, D \equiv h \pmod 8\},$$

$$N_s(X, k; h) = \#\{D \le X : D \text{ square-free}, s(D) = k, D \equiv h \pmod 8\}$$

and

$$N_{\text{III}}(X, k; h) = \#\{D \le X : D \text{ square-free}, \#\text{III}_2(D) = 2^{2k}, D \equiv h \pmod 8\}.$$

Theorem 1. *If* $h = 1$ *or 3 then*

$$\frac{N_s(X, 2k; h)}{N(X; h)} \to d_{2k}, \quad (X \to \infty),$$

while if $h = 5$ *or 7 then*

$$\frac{N_s(X, 2k + 1; h)}{N(X; h)} \to d_{2k+1}, \quad (X \to \infty).$$

Thus, under our hypothesis, it follows that if $h = 1$ *or 3 then*

$$\frac{N_{\text{III}}(X, k; h)}{N(X; h)} \to d_{2k}, \quad (X \to \infty),$$

while if $h = 5$ *or 7 then*

$$\frac{N_{\text{III}}(X, k; h)}{N(X; h)} \to d_{2k+1}, \quad (X \to \infty).$$

It is interesting to investigate the situation in which one restricts the number $\omega(D)$ of prime factors of D. When D is prime one finds that $s(D) = 2$ for every $D \equiv 1 \pmod 8$ and $s(D) = 0$ for every $D \equiv 3 \pmod 8$. When D is a product of two primes and $D \equiv 1 \pmod 8$ the cases $S(D) = 0, 2, 4$ occur with frequency $1/4, 5/8, 1/8$, while if $D \equiv 3 \pmod 8$ the cases $S(D) = 0, 2$ each occur with frequency $1/2$. As the number of prime factors grows these frequencies tend to the values $d_0, d_2, d_4, \ldots$, whether one restricts to values $D \equiv 1 \pmod 8$ or to $D \equiv 3 \pmod 8$. Thus one sees firstly that, for $\omega(D)$ small, the proportions depend heavily on the congruence value of D modulo 8, and secondly that the proportions differ from their limiting values quite significantly. The following table illustrates this. The figures aggregate the

	$k = 3$	$k = 5$	$k = 10$	$k = 20$	$k = \infty$
$s(D) = 0$	0.1875	0.1785	0.1905	0.2083	0.2097
$s(D) = 1$	0.3641	0.3650	0.4004	0.4163	0.4194
$s(D) = 2$	0.2607	0.2719	0.2883	0.2841	0.2796
$s(D) = 3$	0.1220	0.1239	0.0994	0.0784	0.0799
$s(D) = 4$	0.0531	0.0441	0.0174	0.0120	0.0107
$s(D) = 5$	0.0102	0.0135	0.0033	0.0009	0.0007
$s(D) = 6$	0.0024	0.0029	0.0006	0.0000	0.0000
$s(D) = 7$	0.0000	0.0002	0.0001	0.0000	0.0000

Table 1.1: Estimated Frequency of Selmer Ranks for $\omega(D) = k$

cases in which $D \equiv 1 \pmod 8$ and $D \equiv 3 \pmod 8$, and also the cases in which $D \equiv 5 \pmod 8$ and $D \equiv 7 \pmod 8$.

The table shows that when D has rather few prime factors the proportion of values $s(D) = 0, 1$ is less than in the limiting case. Even with $\omega(D) = 10$ the agreement is not very good. The reader should recall that for $D \leq 10^{10}$, say, one typically has $\omega(D)$ around 3. Thus one cannot expect the currently available numerical data to show good agreement with the theoretical limiting behaviour.

For the proof of the theorem, the starting point is the fact that the Selmer group $S^{(2)}$ has as elements those pairs

$$(a, b) \in \left(\frac{\mathbb{Q}^{\times}}{(\mathbb{Q}^{\times})^2} \right)^2$$

for which the simultaneous equations

$$abx^2 + Dy^2 = az^2, \quad abx^2 - Dy^2 = bw^2 \tag{1.1}$$

have non-trivial solutions in every completion of $\mathbb{Q}$. (There is a minor abuse of notation here, identifying $a \in \mathbb{Q}^{\times}/(\mathbb{Q}^{\times})^2$ with one of its coset representatives.) When D is odd one then finds, see [1], that $2^{s(D)}$ is given by the number of pairs for which a and b are positive divisors of D. Moreover the local conditions reduce to the requirement that the system (1.1) has solutions in $\mathbb{Q}_p$ for every prime divisor p of D. Finally, this last condition is satisfied if and only if the four equations

$$abx^2 + Dy^2 = az^2, \quad abx^2 - Dy^2 = bw^2, \quad 2abx^2 = az^2 + bw^2, \quad 2Dy^2 = az^2 - bw^2$$

are individually solvable in $\mathbb{Q}_p$.

This analysis immediately shows that $s(D) \leq 2\omega(D)$. Moreover one can easily read off the frequency of the different values of $s(D)$ when D has at most two prime factors, say. The theorem given above however requires a computation of all the integer moments of $2^{s(D)}$, and this turns out to require a rather lengthy combinatorial argument.

References

[1] D.R. Heath-Brown, The size of Selmer groups for the congruent number problem, *Invent. math.*, 111 (1993), 171-195.

[2] D.R. Heath-Brown, The size of Selmer groups for the congruent number problem, II, *Invent. math.*, 118 (1994), 331-370.

[3] J.H. Silverman, *The arithmetic of elliptic curves*, Graduate Texts in Mathematics, 106, (Springer-Verlag, New York, 1992).

Mathematical Institute
Oxford, UK
OX1 3LB

rhb@maths.ox.ac.uk

2-Descent Through the Ages

Sir Peter Swinnerton-Dyer

The main object of this note, which expands an expository lecture given at the conference, is to provide the reader with an account of the process of 2-descent on elliptic curves defined over $\mathbf{Q}$ which have the form

$$\Gamma : y^2 = (x - c_1)(x - c_2)(x - c_3)$$

— that is, elliptic curves all of whose 2-division points are rational. I have also included a description of the algorithm of Cassels [1] for 4-descents. My intention is to provide the tools needed for applications, in a way which requires minimal effort on the reader's part. I have therefore not included proofs, except in Appendix 2 which contains a proof/algorithm the full details of which may be needed for some applications. Instead, I have provided the necessary references.

This note describes the processes over $\mathbf{Q}$. But the statements of the theory over an arbitrary algebraic number field are not very different, except that the analogues of certain explicit results relating to the prime 2 are not known. On the other hand, some of the proofs are much harder.

We can clearly take the c_i to be integers. Let $\mathcal{B}$, the set of bad primes, be any finite set of primes containing 2, ∞ and all the odd primes dividing $(c_1 - c_2)(c_1 - c_3)(c_2 - c_3)$; thus $\mathcal{B}$ contains the primes of bad reduction for Γ. If $\mathcal{B}$ also contains some primes of good reduction, that is harmless.

The basic version of 2-descent, which goes back to Fermat, is as follows. (Good places to find proofs of the results that follow are Silverman [5] or Husemöller [3].) To any rational point (x, y) on Γ there correspond rational m_1, m_2, m_3 with $m_1 m_2 m_3 = m^2 \neq 0$ such that the three equations

$$m_i y_i^2 = x - c_i \quad \text{for} \quad i = 1, 2, 3 \tag{1.1}$$

are simultaneously soluble. We can multiply the m_i by non-zero squares, so that for example we can require them to be square-free integers; indeed one should really think of them as elements of $\mathbf{Q}^*/\mathbf{Q}^{*2}$, with a suitable interpretation of the equations which involve them. Denote by $\mathcal{C}(\mathbf{m})$ the curve given by the three equations (1.1), where $\mathbf{m} = (m_1, m_2, m_3)$. Looking for solutions of Γ is the same as looking for quadruples x, y_1, y_2, y_3 which satisfy (1.1) for some $\mathbf{m}$. For this purpose we need only consider the finitely many $\mathbf{m}$ for which the m_i are units at all primes outside $\mathcal{B}$; for if any m_i is divisible to an odd power by some prime p not in $\mathcal{B}$ then Γ is already insoluble in $\mathbf{Q}_p$.

One question of interest is the effect of *twisting* on the arithmetic properties of the curve Γ. If b is a nonzero rational, the twist of Γ by b is defined to be the curve

$$\Gamma_b : y^2 = (x - bc_1)(x - bc_2)(x - bc_3),$$

where we can regard b as an element of $\mathbf{Q}^*/\mathbf{Q}^{*2}$. The curve Γ_b is often written in the alternative form

$$v^2 = b(u - c_1)(u - c_2)(u - c_3).$$

The analogue of (1.1) for Γ_b is

$$m_i y_i^2 = x - bc_i \quad \text{for} \quad i = 1, 2, 3;$$

we shall call the curve given by these three equations $\mathcal{C}_b(\mathbf{m})$. It is often natural to compare $\mathcal{C}(\mathbf{m})$ and $\mathcal{C}_b(\mathbf{m})$ for the same $\mathbf{m}$.

Provided one treats the m_i as elements of $\mathbf{Q}^*/\mathbf{Q}^{*2}$, the triples $\mathbf{m}$ form an abelian group under componentwise multiplication:

$$\mathbf{m}' \times \mathbf{m}'' \mapsto \mathbf{m}'\mathbf{m}'' = (m_1'm_1'', m_2'm_2'', m_3'm_3'').$$

The $\mathbf{m}$ for which $\mathcal{C}(\mathbf{m})$ is everywhere locally soluble form a finite subgroup, called the 2-*Selmer group*. This is computable, and it contains the group of those $\mathbf{m}$ for which $\mathcal{C}(\mathbf{m})$ is actually soluble in $\mathbf{Q}$. This smaller group is $\Gamma(\mathbf{Q})/2\Gamma(\mathbf{Q})$, where $\Gamma(\mathbf{Q})$, the group of rational points on Γ, is the *Mordell-Weil* group of Γ. The quotient of the 2-Selmer group by this smaller group is ${}_2\text{III}$, the group of those elements of the *Tate-Shafarevich group* which are killed by 2. One of the key conjectures in the subject is that the order of ${}_2\text{III}$ is a square.

The process of going from the curve Γ to the set of curves $\mathcal{C}(\mathbf{m})$, or the finite subset which is the 2-Selmer group, is called a 2-*descent*, or sometimes a *first descent*, and the curves $\mathcal{C}(\mathbf{m})$ themselves are called 2-*coverings*. The reason for this terminology is that there is a commutative diagram

$$
\begin{array}{ccc}
\Gamma & \longrightarrow & \Gamma \\
\| & \nearrow & \\
\mathcal{C}(\mathbf{m}) & &
\end{array}
\tag{1.2}
$$

in which the left hand map is biregular (but defined over $\mathbf{C}$ rather than $\mathbf{Q}$), the top map is multiplication by 2 and the diagonal map is given by $y = m y_1 y_2 y_3$. A 2-covering which is everywhere locally soluble, and therefore in the 2-Selmer group, can also be written in the form

$$\eta^2 = f(\xi) \quad \text{where} \quad f(\xi) = a\xi^4 + b\xi^3 + c\xi^2 + d\xi + e,$$

and many 2-coverings do arise in this way; but a 2-covering which is not in the 2-Selmer group cannot always be put into this form.

We now put this process into more modern language. In what follows, italic capitals will always denote vector spaces over $\mathbf{F}_2$, the finite field of two elements, and each of p and q will be either a finite prime or ∞. Write

$$Y_p = \mathbf{Q}_p^* / \mathbf{Q}_p^{*2}, \quad Y_{\mathcal{B}} = \oplus_{p \in \mathcal{B}} Y_p.$$

Let V_p denote the vector space of all triples (μ_1, μ_2, μ_3) with each μ_i in Y_p and $\mu_1\mu_2\mu_3 = 1$; and write $V_{\mathcal{B}} = \oplus_{p \in \mathcal{B}} V_p$. This is the best way to introduce these spaces, because it preserves symmetry; but the reader should note that the prevailing custom in the literature is to define V_p as $Y_p \times Y_p$, which is isomorphic to the V_p defined above but not in a canonical way. Next, write $X_{\mathcal{B}} = \mathfrak{o}_{\mathcal{B}}^* / \mathfrak{o}_{\mathcal{B}}^{*2}$ where $\mathfrak{o}_{\mathcal{B}}^*$ is the group of nonzero rationals which are units outside $\mathcal{B}$; and let $U_{\mathcal{B}}$ be the image in $V_{\mathcal{B}}$ of the group of triples (m_1, m_2, m_3) such that the m_i are in $X_{\mathcal{B}}$ and $m_1 m_2 m_3 = 1$. It is known that the map $X_{\mathcal{B}} \to Y_{\mathcal{B}}$ is an embedding and $\dim U_{\mathcal{B}} = \frac{1}{2} \dim V_{\mathcal{B}}$; both these depend on the requirement that $\mathcal{B}$ contains 2 and ∞. Finally, if (x, y) is a point of Γ defined over $\mathbf{Q}_p$ other than a 2-division point then the product of the three components in the triple $(x - c_1, x - c_2, x - c_3)$ is y^2 which is in $\mathbf{Q}_p^{*2}$; so this triple has a natural image in V_p. We can supply the images of the 2-division points by continuity; for example the image of $(c_1, 0)$ is

$$((c_1 - c_2)(c_1 - c_3), c_1 - c_2, c_1 - c_3), \tag{1.3}$$

and the image of the point at infinity is the trivial triple $(1, 1, 1)$, which is also the product of the three triples like (1.3). Thus we obtain a map $\Gamma(\mathbf{Q}_p) \to V_p$. This map, which is called the *Kummer map*, is a homomorphism. We denote its image by W_p; clearly W_p is the set of those triples $\mathbf{m}$ for which (1.1) is soluble in $\mathbf{Q}_p$. It is sometimes useful to have explicit descriptions of the W_p, so these are given in Appendix 1. The 2-Selmer group of Γ can now be identified with $U_{\mathcal{B}} \cap W_{\mathcal{B}}$ where $W_{\mathcal{B}} = \oplus_{p \in \mathcal{B}} W_p$; for as was noted above, (1.1) is soluble at every prime outside $\mathcal{B}$ if and only if the elements of $\mathbf{m}$ are in $X_{\mathcal{B}}$.

Over the years, many people must have noticed that

$$\dim W_{\mathcal{B}} = \dim U_{\mathcal{B}} = \tfrac{1}{2} \dim V_{\mathcal{B}}. \tag{1.4}$$

The next major step, which explains and may well have been inspired by this relation, was taken by Tate. He introduced the bilinear form e_p on $V_p \times V_p$, defined by

$$e_p(\mathbf{m}', \mathbf{m}'') = (m_1', m_1'')_p (m_2', m_2'')_p (m_3', m_3'')_p.$$

Here $(u, v)_p$ is the multiplicative Hilbert symbol with values in $\{\pm 1\}$, defined by

$$(u, v)_p = \begin{cases} 1 & \text{if } ux^2 + vy^2 = 1 \text{ is soluble in } \mathbf{Q}_p, \\ -1 & \text{otherwise.} \end{cases}$$

The Hilbert symbol is symmetric and multiplicative in each argument:

$$(u, v)_p = (v, u)_p \quad \text{and} \quad (u_1 u_2, v)_p = (u_1, v)_p (u_2, v)_p.$$

Effectively it is a replacement for the quadratic residue symbol, with the advantage that it treats the primes 2 and ∞ in just the same way as any other prime. Its other key property is the Hilbert product formula

$$\prod_p (u,v)_p = 1,$$

where the product is taken over all p including ∞; the left hand side is meaningful because $(u,v)_p = 1$ whenever p is an odd prime at which u and v are units.

The bilinear form e_p is non-degenerate and alternating on $V_p \times V_p$; we use it to define $e_{\mathcal{B}} = \prod_{p\in\mathcal{B}} e_p$, which is a non-degenerate alternating bilinear form on $V_{\mathcal{B}} \times V_{\mathcal{B}}$. (For a bilinear form with values in $\{\pm 1\}$, "symmetric" and "skew-symmetric" are the same and they each mean that $e(\mathbf{m}',\mathbf{m}'') = e(\mathbf{m}'',\mathbf{m}')$; "alternating" means that also $e(\mathbf{m},\mathbf{m}) = 1$.) It is known from class field theory that $U_{\mathcal{B}}$ is a maximal isotropic subspace of $V_{\mathcal{B}}$. Tate showed that W_p is a maximal isotropic subspace of V_p, and therefore $W_{\mathcal{B}}$ is a maximal isotropic subspace of $V_{\mathcal{B}}$. (The proof of this, which is difficult, can be found in Milne [4].) This explains (1.4); and it also shows that the 2-Selmer group of Γ can be identified with both the left and the right kernel of the restriction of $e_{\mathcal{B}}$ to $U_{\mathcal{B}} \times W_{\mathcal{B}}$.

For both aesthetic and practical reasons, one would like to show that this restriction is symmetric or skew-symmetric — these two properties being the same. But to make such a statement meaningful we need an isomorphism between $U_{\mathcal{B}}$ and $W_{\mathcal{B}}$; and though they have the same structure as vector spaces it is not obvious that there is a natural isomorphism between them. The way round this obstacle was first shown in [2]. It requires the construction inside each V_p of a maximal isotropic subspace K_p such that $V_{\mathcal{B}} = U_{\mathcal{B}} \oplus K_{\mathcal{B}}$ where $K_{\mathcal{B}} = \oplus_{p\in\mathcal{B}} K_p$. Assuming that such spaces K_p can be constructed, let $t_{\mathcal{B}} : V_{\mathcal{B}} \to U_{\mathcal{B}}$ be the projection along $K_{\mathcal{B}}$ and write

$$U_{\mathcal{B}}' = U_{\mathcal{B}} \cap (W_{\mathcal{B}} + K_{\mathcal{B}}), \quad W_{\mathcal{B}}' = W_{\mathcal{B}}/(W_{\mathcal{B}} \cap K_{\mathcal{B}}) = \bigoplus_{p\in\mathcal{B}} W_p'$$

where $W_p' = W_p/(W_p \cap K_p)$. The map $t_{\mathcal{B}}$ induces an isomorphism

$$\tau_{\mathcal{B}} : W_{\mathcal{B}}' \to U_{\mathcal{B}}',$$

and the bilinear function $e_{\mathcal{B}}$ induces a bilinear function

$$e_{\mathcal{B}}' : U_{\mathcal{B}}' \times W_{\mathcal{B}}' \to \{\pm 1\}.$$

The bilinear functions $U_{\mathcal{B}}' \times U_{\mathcal{B}}' \to \{\pm 1\}$ and $W_{\mathcal{B}}' \times W_{\mathcal{B}}' \to \{\pm 1\}$ defined respectively by

$$\theta_{\mathcal{B}}^{\flat} : u_1' \times u_2' \mapsto e_{\mathcal{B}}'(u_1', \tau_{\mathcal{B}}^{-1}(u_2')) \quad \text{and} \quad \theta_{\mathcal{B}}^{\sharp} : w_1' \times w_2' \mapsto e_{\mathcal{B}}'(\tau_{\mathcal{B}} w_1', w_2') \qquad (1.5)$$

are symmetric. (For the proof, see [2] or [8].) Here the images of $w_1' \times w_2'$ under the second map and of $\tau_{\mathcal{B}} w_1' \times \tau_{\mathcal{B}} w_2'$ under the first map are the same.

The 2-Selmer group of Γ is isomorphic to both the left and the right kernel of $e'_{\mathcal{B}}$, and hence also to the kernels of the two maps (1.5).

There is considerable freedom in choosing the K_p, and this raises three obvious questions:

- Is there a canonical choice of the K_p?

- How small can we make U' and W'?

- Can we ensure that the functions (1.5) are not merely symmetric but alternating?

These questions were first raised and also to a large extent answered in [6]; proofs of the assertions which follow can be found there. The motive for ensuring that the functions (1.5) are alternating is that it implies that the ranks of these functions are even; this means that their coranks, which are equal to the dimension of the 2-Selmer group, are congruent mod 2 to $\dim U'_{\mathcal{B}}$ and $\dim W'_{\mathcal{B}}$.

The answer to the first question appears to be negative, though there is little freedom in the optimum choice of the K_p — particularly if one wishes to obtain not merely Lemma 1.1 but Theorem 1.2. Since $U'_{\mathcal{B}} \supset U_{\mathcal{B}} \cap W_{\mathcal{B}}$, the best possible answer to the second question would be that we can achieve $U'_{\mathcal{B}} = U_{\mathcal{B}} \cap W_{\mathcal{B}}$; we shall do this by satisfying the stronger requirement

$$W_{\mathcal{B}} = (U_{\mathcal{B}} \cap W_{\mathcal{B}}) \oplus (K_{\mathcal{B}} \cap W_{\mathcal{B}}). \tag{1.6}$$

For suppose that (1.6) holds; then $W_{\mathcal{B}} + K_{\mathcal{B}} = (U_{\mathcal{B}} \cap W_{\mathcal{B}}) + K_{\mathcal{B}}$ and it follows immediately that

$$U'_{\mathcal{B}} = U_{\mathcal{B}} \cap (W_{\mathcal{B}} + K_{\mathcal{B}}) = U_{\mathcal{B}} \cap W_{\mathcal{B}}. \tag{1.7}$$

The motivation for (1.6) is that we want to make $W_{\mathcal{B}} \cap K_{\mathcal{B}}$ as large as possible — that is, to choose $K_{\mathcal{B}}$ so that as much of it as possible is contained in $W_{\mathcal{B}}$. But because $K_{\mathcal{B}}$ must be complementary to $U_{\mathcal{B}}$, only the part of $W_{\mathcal{B}}$ which is complementary to $W_{\mathcal{B}} \cap U_{\mathcal{B}}$ is available for this purpose.

Since the 2-Selmer group $U_{\mathcal{B}} \cap W_{\mathcal{B}}$ is identified with the left and right kernels of each of the functions (1.5), if (1.7) holds then these functions are trivial and therefore alternating. The formal statement of all this is as follows.

Lemma 1.1. *We can choose maximal isotropic subspaces $K_p \subset V_p$ for each p in $\mathcal{B}$ so that $V_{\mathcal{B}} = U_{\mathcal{B}} \oplus K_{\mathcal{B}}$. We can further ensure that*

$$W_{\mathcal{B}} = (U_{\mathcal{B}} \cap W_{\mathcal{B}}) \oplus (K_{\mathcal{B}} \cap W_{\mathcal{B}}),$$

which implies $U'_{\mathcal{B}} = U_{\mathcal{B}} \cap W_{\mathcal{B}}$. If so, the functions $\theta^{\flat}_{\mathcal{B}}$ and $\theta^{\natural}_{\mathcal{B}}$ defined in (1.5) are trivial.

For some applications it is convenient to have an explicit description of the construction of the K_p; this is given in Appendix 2. But the other properties of the K_p chosen in this way are not at all obvious. Hence it is advantageous to consider other recipes for choosing the K_p, for which (1.6) does not hold but we can still prove that the functions (1.5) are alternating.

For this purpose we write $\mathcal{B}$ as the disjoint union of $\mathcal{B}'$ and $\mathcal{B}''$, where we shall always suppose that 2 and ∞ are both in $\mathcal{B}'$. For any odd prime p we denote by T_p the subset of V_p consisting of those triples (μ_1, μ_2, μ_3) with $\mu_1\mu_2\mu_3 = 1$ for which each μ_i is in $\mathfrak{o}_p^*/\mathfrak{o}_p^{*2}$ — that is, each μ_i is the image of a p-adic unit. The main point of the following theorem is that for p in $\mathcal{B}''$ it enables us to replace the complicated inductive definition of K_p used in the proof of Lemma 1.1 by the much simpler choice $K_p = T_p$. How one chooses $\mathcal{B}''$ depends on the particular application which one has in mind.

Theorem 1.2. *Let $\mathcal{B}$ be the disjoint union of $\mathcal{B}' \supset \{2, \infty\}$ and $\mathcal{B}''$. We can construct maximal isotropic subspaces $K_p \subset V_p$ such that $V_\mathcal{B} = U_\mathcal{B} \oplus K_\mathcal{B}$,*

$$W_{\mathcal{B}'} = (U_{\mathcal{B}'} \cap W_{\mathcal{B}'}) \oplus (K_{\mathcal{B}'} \cap W_{\mathcal{B}'}) \tag{1.8}$$

and $K_v = T_v$ for all v in $\mathcal{B}''$; and (1.8) implies that $U'_{\mathcal{B}'} = U_{\mathcal{B}'} \cap W_{\mathcal{B}'}$. Moreover

$$U'_\mathcal{B} = \jmath_* U'_{\mathcal{B}'} \oplus \tau_\mathcal{B} W'_{\mathcal{B}''} = \jmath_* U'_{\mathcal{B}'} \oplus \left(\oplus_{p \in \mathcal{B}''} \tau_\mathcal{B} W'_p\right), \tag{1.9}$$

and the restriction of $\theta_\mathcal{B}^\flat$ to $\jmath_* U'_{\mathcal{B}'} \times \jmath_* U'_{\mathcal{B}'}$ is trivial.

If $\mathcal{B}'$ also contains all the odd primes p such that the $v_p(c_i - c_j)$ are not all congruent $\mathrm{mod}\ 2$, then we can choose the K_p for p in $\mathcal{B}'$ so that also $\theta_\mathcal{B}^\flat$ is alternating on $U'_\mathcal{B}$.

The appearance of $\jmath_* U'_{\mathcal{B}'}$ in and just after (1.9) calls for some explanation. Let u be any element of $U_{\mathcal{B}'}$; then u is in $U_\mathcal{B}$. Moreover, for p in $\mathcal{B}''$ the image of u in V_p is in $T_p = K_p$ and therefore in $K_p + W_p$; hence u is in $U'_\mathcal{B}$. In this way we define a map $U'_{\mathcal{B}'} \to U'_\mathcal{B}$ which is clearly an injection and which we denote by $\jmath_*$.

Lemma 1.1 is the special case of Theorem 1.2 in which $\mathcal{B}' = \mathcal{B}$ and $\mathcal{B}''$ is empty. But the proof of Lemma 1.1 is a necessary step (and indeed the most substantial step) in the proof of Theorem 1.2. Indeed, to prove Theorem 1.2 we construct the K_p for p in $\mathcal{B}'$ according to the recipe in Appendix 2; for the final sentence of the theorem we need the particular version of the recipe which involves the functions ϕ_i.

The main application of Theorem 1.2 is to twisted curves Γ_b, where we can clearly take b to be an integer. Let $\mathcal{S}$ denote the set of bad primes for Γ itself — that is, $2, \infty$ and the odd primes dividing $(c_1 - c_2)(c_1 - c_3)(c_2 - c_3)$; and let $\mathcal{B} \supset \mathcal{S}$ be the set of bad primes for Γ_b. If we are to apply any part of Theorem 1.2, $\mathcal{B}$ must also contain all the odd primes dividing b; and such applications are much simpler when b is a unit at every prime of $\mathcal{S}$. (We can always arrange this by treating Γ_b as the twist of Γ_c by b/c, where c is the largest divisor of b which is a unit outside $\mathcal{S}$.) To describe the effect of twisting, we shall denote by d_b the dimension of the 2-Selmer group of Γ_b regarded as a vector space over $\mathbf{F}_2$; we write $d = d_1$ for the dimension of the 2-Selmer group of Γ itself. It is now possible to prove results about $d_b - d$, the change in the dimension of the 2-Selmer group as one goes from Γ to Γ_b. There is reason to expect that statements about the parities of d and d_b will be simpler and much easier to prove than statements about their actual values. The two major statements known about d_b are Lemma 1.3 and Theorem 1.5; Lemma 1.3 is an easy consequence of the last sentence of Theorem 1.2, and Theorem 1.5 is an easy consequence of Lemma 1.4 below.

Lemma 1.3. *If b is in $\mathfrak{o}_p^*$ for every $p \in \mathcal{S}$, then $d_b \equiv \dim(U_{\mathcal{S}} \cap W_{\mathcal{S}}) \bmod 2$ where $W_{\mathcal{S}} = \oplus_{p \in \mathcal{S}} W_p$ and the W_p must be defined with respect to Γ_b and not with respect to Γ. Thus $d_b \bmod 2$ only depends on the classes of b in the k_p^*/k_p^{*2} for p in $\mathcal{S}$.*

To prove Lemma 1.4 we need to take $\mathcal{B}' = \mathcal{S} \setminus \{p\}$; thus the last sentence of Theorem 1.2 is not applicable though the rest of that theorem is.

Lemma 1.4. *Let p be an odd prime in $\mathcal{S}$ such that*

$$v_p(c_1 - c_2) > 0, \quad v_p(c_1 - c_3) = v_p(c_2 - c_3) = 0.$$

Let b in k^* be such that b is in k_q^{*2} for all q in $\mathcal{S}$ other than p and b is a quadratic non-residue at p. Then d and d_b have opposite parities.

It is not hard to prove the analogue of Lemma 1.4 for the case $p = \infty$, though the proof falls outside the machinery described in this note. The combination of this result and Lemma 1.4 yields Theorem 1.5. (The analogue of Lemma 1.4 for $p = 2$ can be confidently asserted, on the basis of a large amount of numerical evidence, and the proof of it probably requires no new ideas. But even the statement involves so extensive a separation of cases that it is unlikely soon to appear in print.)

Theorem 1.5. *Let b', b'' in k^* be such that b'/b'' is a unit at all $p \in \mathcal{S}$ and $b'/b'' \equiv 1 \bmod 8$. Let $\mathcal{S}^*$ be the set of $p \in \mathcal{S}$ for which b'/b'' is not in k_p^{*2}. Let $\mathcal{S}^{**}$ consist of the finite odd p in $\mathcal{S}^*$ for which the $v_p(c_i - c_j)$ are not all equal and the smallest two of them are even, together with ∞ if $b'/b'' < 0$. Then*

$$d_{b'} - d_{b''} \equiv \#\mathcal{S}^{**} \bmod 2.$$

We can define a 4-*covering* and a 4-*descent* (sometimes called a *second descent*) by extension of the diagram (1.2). Let $\mathcal{C}$ be a 2-covering of Γ; then a 4-covering of Γ above this 2-covering is a curve $\mathcal{D}$ which fits into the commutative diagram

$$
\begin{array}{ccccc}
\Gamma & \longrightarrow & \Gamma & \longrightarrow & \Gamma \\
\| & & \| & \nearrow & \\
\mathcal{D} & \longrightarrow & \mathcal{C} & &
\end{array}
$$

in which the vertical maps are biregular (but defined over $\mathbf{C}$ rather than $\mathbf{Q}$) and each upper map is multiplication by 2. If $\mathcal{C}$ is everywhere locally soluble, we say that it *admits a second descent* if we can find such a $\mathcal{D}$ which is everywhere locally soluble. If $\mathcal{C}$ is actually soluble in $\mathbf{Q}$, then it certainly admits a second descent; thus carrying out a second descent is a way of replacing the 2-Selmer group by a hopefully smaller group which however still contains $\Gamma(\mathbf{Q})/2\Gamma(\mathbf{Q})$. A second descent may therefore refine the information about the Mordell-Weil group which is obtained from the 2-descent.

In its classical form, the process of 4-descent was constructive but it was arithmetically unattractive, largely because it involved a field extension. But Cassels [1] has shown how to determine which elements of the 2-Selmer group do admit a second descent, while working entirely in $\mathbf{Q}$. He constructs an alternating bilinear form g on the 2-Selmer group, whose kernel consists of exactly those elements which admit a second descent. Let $\mathcal{S}$ again be the set of bad primes for Γ, with $\mathcal{S} \supset \{2, \infty\}$, and let $\mathbf{m}'$ and $\mathbf{m}''$ be two triples in $U_{\mathcal{S}}$ which represent elements of the 2-Selmer group of Γ. If i, j, k is any permutation of $1, 2, 3$ we denote by $\mathcal{C}_i(\mathbf{m}')$ the conic

$$m'_j y_j^2 - m'_k y_k^2 = (c_k - c_j) y_0^2. \tag{1.10}$$

In view of (1.1) there is a map $\mathcal{C}(\mathbf{m}') \to \mathcal{C}_i(\mathbf{m}')$; so $\mathcal{C}_i(\mathbf{m}')$ is everywhere locally soluble. Because $\mathcal{C}_i(\mathbf{m}')$ is a conic, this implies that it is soluble in $\mathbf{Q}$; so choose a rational point P_i on $\mathcal{C}_i(\mathbf{m}')$ and let $\mathsf{L}_i(y_0, y_j, y_k) = 0$ be the equation of the tangent to $\mathcal{C}_i(\mathbf{m}')$ at P_i. By abuse of language, we can treat L_i as a homogeneous linear form in y_0, y_j, y_k; strictly speaking, it is only defined up to multiplication by an element of $\mathbf{Q}^*$, but it will not matter which multiple we choose. For each p in $\mathcal{S}$, choose a p-adic point Q_p on the affine curve $\mathcal{C}(\mathbf{m}')$. Then g is defined by

$$g(\mathbf{m}', \mathbf{m}'') = \prod_{p \in \mathcal{S}} \prod_i (\mathsf{L}_i(\mathsf{Q}_p), m''_i)_p$$

where the bracket on the right is as usual the Hilbert symbol.

APPENDIX 1 — Explicit description of the W_p

The main purpose of this Appendix is to give an explicit description of the W_p. The calculations are sometimes simplified by using the fact that W_p is isotropic and contains the three triples like (1.3); thus if $\mathbf{m}$ is in W_p then the three results like

$$(c_1 - c_2, m_3)_p = (c_1 - c_3, m_2)_p$$

all hold. The case $p = \infty$, which is trivial, is Lemma 1.7. The case when p is odd, the simplest proof of which can be found in [6], is Lemma 1.7. The results for the case $p = 2$ are much more complicated; they can be found in [7] but are not reproduced here.

Lemma 1.6. *After renumbering, suppose that $c_1 > c_2 > c_3$. Then W_∞ consists of the classes of $(1, 1, 1)$ and $(-1, -1, 1)$.*

In Lemma 1.7 and Theorem 1.8, $a_1 \sim a_2$ will mean that a_1/a_2 is in k_p^{*2}.

Lemma 1.7. *Let p be an odd prime.*

If p divides all the $c_i - c_j$ to the same even power, then $W_p = (\mathfrak{o}_p^*/\mathfrak{o}_p^{*2})^2$. If p divides all the $c_i - c_j$ to the same odd power, then W_p consists of the classes of $(1, 1, 1)$ and the three triples like (1.3).

Now suppose that p does not divide all the $c_i - c_j$ to the same power. After renumbering, let

$$v_p(c_1 - c_2) > v_p(c_1 - c_3) = v_p(c_2 - c_3).$$

Denote by η the class of $c_1 - c_2$, by ε the class of $c_1 - c_3$ and $c_2 - c_3$, and by ν the class of quadratic non-residues $\bmod p$.

If $v(\varepsilon)$ is odd then W_p consists of the classes of

$$(1, 1, 1), \ (\eta\varepsilon, \eta, \varepsilon), \ (-\eta, -\eta\varepsilon, \varepsilon), \ (-\varepsilon, -\varepsilon, 1).$$

If $v(\eta)$ is odd and $v(\varepsilon)$ even then W_p consists of the classes of

$$(1, 1, 1), \ (\eta\varepsilon, \eta, \varepsilon), \ (\nu, \nu, 1), \ (\nu\eta\varepsilon, \nu\eta, \varepsilon).$$

If $v(\eta)$ and $v(\varepsilon)$ are both even and $\varepsilon \sim \nu$ then W_p consists of the classes of

$$(1, 1, 1), \ (\nu, \nu, 1), \ (\nu, 1, \nu), \ (1, \nu, \nu).$$

If $v(\eta)$ and $v(\varepsilon)$ are both even and $\varepsilon \sim 1$ then W_p consists of the classes of

$$(1, 1, 1), \ (\nu, \nu, 1), \ (p, p, 1), \ (p\nu, p\nu, 1).$$

A number of people have proved results of the form: let p be in $\mathcal{S}$ and assume that $\mathcal{C}(\mathbf{m})$ is locally soluble at all primes other than perhaps p; then provided that certain local conditions on Γ hold, $\mathcal{C}(\mathbf{m})$ is also locally soluble at p. The best approach to this kind of result is as follows. For any permutation i, j, k of $1, 2, 3$ let $\mathcal{C}_k(\mathbf{m})$ denote the conic

$$m_i y_i^2 - m_j y_j^2 = (c_j - c_i) y_0^2,$$

this being essentially the same as the notation of (1.10). The existence of a map $\mathcal{C}(\mathbf{m}) \to \mathcal{C}_k(\mathbf{m})$ implies that $\mathcal{C}_k(\mathbf{m})$ is also locally soluble everywhere except possibly at p. Since $\mathcal{C}_k(\mathbf{m})$ is a conic, it follows that $\mathcal{C}_k(\mathbf{m})$ is also locally soluble at p — a condition which is equivalent to

$$(m_i(c_j - c_i), m_k)_p = 1. \tag{1.11}$$

Hence $\mathcal{C}(\mathbf{m})$ is locally soluble at p provided that this is implied by the local solubility of the three $\mathcal{C}_k(\mathbf{m}')$ at p — that is, by the three conditions like (1.11).

The question is under what local conditions on Γ at p this holds. Such results can be read off from the description of W_p; but in fact we can decide this question without knowing W_p. For we do know that the order of W_p is 2, 4 or 8 according as p is ∞, odd or 2. It is therefore enough to count the set of triples $\mathbf{m}$ which satisfy the three equations like (1.11); for this set contains W_p, so that it is equal to W_p if and only if it has the same order as W_p. Even when $p = 2$, this calculation is trivial to program.

The conclusions for $p = \infty$ and p odd are given in the following theorem. Those for $p = 2$ are too complicated to justify explicit statement.

Theorem 1.8. *Suppose that $\mathcal{C}(\mathbf{m})$ is everywhere locally soluble except possibly at one prime p which is in $\mathcal{S}$. If $p = \infty$ then $\mathcal{C}(\mathbf{m})$ is also locally soluble at p. If p is odd then $\mathcal{C}(\mathbf{m})$ is also locally soluble at p except perhaps when $c_i - c_k \sim c_j - c_k \sim 1$ for some permutation i, j, k of $1, 2, 3$.*

APPENDIX 2 — Construction of the K_p

In this Appendix we show how to construct the K_p. We do in fact prove a more general result, but this is only because otherwise we would be forced into a needlessly complicated notation. The reader will see that (subject to the introduction of the temporarily mysterious functions ϕ_i) the hypotheses of Lemma 1.9 mimic the structure described in the main body of the text. I give here only that part of the proof which is really an algorithm for the construction; a complete proof can be found in [6].

Lemma 1.9. *Let the V_i be n vector spaces over $\mathbf{F}_2$, each equipped with a non-degenerate additive alternating bilinear form ψ_i with values in $\mathbf{F}_2$. Denote by ψ the sum of the ψ_i, which is a non-degenerate bilinear form on $V = \oplus V_i$. For each i let W_i be maximal isotropic in V_i, and let U be maximal isotropic in V with respect to ψ. Then there exist maximal isotropic subspaces $K_i \subset V_i$ such that $V = U \oplus K$ and*

$$W = (U \cap W) \oplus (K \cap W) \tag{1.12}$$

where $W = \oplus W_i$ and $K = \oplus K_i$. Moreover $U \cap (W + K) = U \cap W$.

Suppose also that there are functions ϕ_i on V_i with values in $\mathbf{F}_2$ which satisfy

$$\phi_i(\xi + \eta) = \phi_i(\xi) + \phi_i(\eta) + \psi_i(\xi, \eta) \tag{1.13}$$

for any ξ, η in V_i, and let ϕ on V be the sum of the ϕ_i. Assume that ϕ is trivial on U and ϕ_i is trivial on W_i. Then we can further ensure that ϕ_i is trivial on K_i and therefore ϕ is trivial on K.

Proof If any V_i has dimension greater than 2, we can decompose it as a direct sum of mutually orthogonal subspaces of dimension 2, on each of which the restriction of the bilinear form ψ_i is non-degenerate and each of which meets

W_i in a subspace of dimension 1. This only reduces our freedom to choose the K_i, and the triviality of ϕ_i on the old K_i will follow from its triviality on the new and smaller K_i by (1.13). Thus we can assume that every V_i has dimension 2 and every W_i has dimension 1. We proceed by induction on n, the case $n = 0$ being trivial.

We shall assume that the ϕ_i exist, noting in the appropriate place how to modify the argument to prove the first part of the lemma without using the existence of the ϕ_i. If we regard W_n as a subspace of V, either $W_n \subset U$ or W_n is not contained in U and therefore meets it only in the origin. In each of these cases, we shall choose an α_i in V_i with $\phi_i(\alpha_i) = 0$ and use it to generate K_i. After reordering, we can assume that either W_n is not contained in U or every W_i is contained in U and therefore $W \subset U$.

Since U is isotropic it cannot contain V_i; so if $W_n \subset U$ and therefore $W_i \subset U$ for each i, then each V_i contains just two elements which do not lie in U. Denote them by α_i' and α_i'', and let β_i be the nontrivial element of W_i; thus $\alpha_i'' = \alpha_i' + \beta_i$. Since $\phi_i(\beta_i) = 0$ it follows from (1.13) and the non-degeneracy of ψ_i on V_i that

$$\phi_i(\alpha_i') + \phi_i(\alpha_i'') = \psi_i(\alpha_i', \beta_i) = 1;$$

choose α_i to be whichever of α_i' and α_i'' satisfies $\phi_i(\alpha_i) = 0$. (If we do not assume the existence of the ϕ_i then we can take α_i to be either of α_i' and α_i''.) Let K_i be the vector space generated by α_i; thus

$$W_i = U \cap W_i = (U \cap W_i) \oplus (K_i \cap W_i)$$

for each i, which implies (1.12). Moreover $U \supset W$ and therefore $U = W$ because U and W have the same dimension. So

$$V = \oplus V_i = \oplus(W_i \oplus K_i) = W \oplus K = U \oplus K.$$

If U does not contain W_n, then the non-trivial element of W_n is not in U. Denote this element by α_n, so that $\phi_n(\alpha_n) = 0$ by hypothesis. Let K_n be the vector space generated by α_n; thus $K_n = W_n$ and

$$W_n = (U \cap W_n) \oplus (K_n \cap W_n). \tag{1.14}$$

The construction now proceeds by induction on n. Write

$$V^- = V_1 \oplus \ldots \oplus V_{n-1}, \quad U^- = V^- \cap (U \oplus W_n). \tag{1.15}$$

It is straightforward to show that U^- is maximal isotropic in V^-. For the pair U^-, V^- we must replace the question whether $U \supset W$ by the question whether $U \oplus W_n$ contains $W^- = W_1 \oplus \ldots \oplus W_{n-1}$. By the induction hypothesis for the pair $U^- \subset V^-$, there exist K_i maximal isotropic in V_i for each $i < n$ such that if $K^- = (K_1 \oplus \ldots \oplus K_{n-1})$ then $V^- = U^- \oplus K^-$ and

$$W^- = (U^- \cap W^-) \oplus (K^- \cap W^-). \tag{1.16}$$

The need to check the remaining details of the argument can be circumvented by an appeal to Cassels' Axiom: all vector space theorems are trivial.

When we apply Lemma 1.9 to the construction of the K_p for p in $\mathcal{B}'$ and the proof of Theorem 1.2, we replace i by p and ψ_i by e_p; but note that we have chosen to write e_p multiplicatively and ψ_i additively. For $\mathbf{m}$ in V_p we take $\phi_p(\mathbf{m})$ to be any one of the expressions

$$(m_i(c_i - c_j)(c_i - c_k), m_j(c_j - c_i)(c_j - c_k))_p,$$

whose values are easily shown to be equal. The significance of ϕ_p is as follows. The antipodal involution $(x, y) \mapsto (x, -y)$ on Γ induces an involution on each 2-covering $\mathcal{C}(\mathbf{m})$; in the notation of (1.1) this involution reverses the signs of y_1, y_2, y_3. The quotient of $\mathcal{C}(\mathbf{m})$ by this involution is a smooth projective curve $\mathcal{D}(\mathbf{m})$ of genus 0, which is given by

$$(c_2 - c_3)m_1 y_1^2 + (c_3 - c_1)m_2 y_2^2 + (c_1 - c_2)m_3 y_3^2 = 0; \qquad (1.17)$$

and $\phi_p(\mathbf{m})$ is just the class $[\mathcal{D}(\mathbf{m})]$ as an element of Br k_p.

References

[1] J.W.S.Cassels, Second descent for elliptic curves, J. reine angew. Math. 494(1998), 101-127.

[2] J-L.Colliot-Thélène, A.N.Skorobogatov and Sir Peter Swinnerton-Dyer, Hasse principle for pencils of curves of genus one whose Jacobians have rational 2-division points, Invent. Math. 134(1998), 579-650.

[3] D.Husemöller, Elliptic Curves, Graduate Texts in Mathematics, 111 (Springer, 1987).

[4] J.S.Milne, Arithmetic Duality Theorems (Academic Press, Boston, 1986).

[5] J.H.Silverman, The Arithmetic of Elliptic Curves, Graduate Texts in Mathematics, 106 (Springer, 1986).

[6] A.N.Skorobogatov and Sir Peter Swinnerton-Dyer, 2-Descent on elliptic curves and rational points on certain Kummer surfaces, Advances in Mathematics 198(2) (2005), 448-483.

[7] Sir Peter Swinnerton-Dyer, Rational points on certain intersections of two quadrics, in *Abelian varieties* (ed. W.Barth, K.Hulek and H.Lange) (de Gruyter, Berlin, 1995), 273-292.

[8] Sir Peter Swinnerton-Dyer, Some applications of Schinzel's hypothesis to diophantine equations, in *Number theory in progress* (ed. K.Győry, H.Iwaniec and J.Urbanowicz) (Berlin, 1999), 503-530.

Index